Corrosion Engineering Handbook

Corrosion Engineering Handbook

Edited by **Andreas Byers**

NY RESEARCH PRESS
New York

Published by NY Research Press,
23 West, 55th Street, Suite 816,
New York, NY 10019, USA
www.nyresearchpress.com

Corrosion Engineering Handbook
Edited by Andreas Byers

International Standard Book Number: 978-1-63238-099-9 (Hardback)

Printed in China.

Contents

Preface

There is a huge amount of material in our world which is used in one way or the other to basically manufacture objects for someone or the other's benefit. As is the way of the universe, nothing is immortal, not even metal. And this so called 'death' of a metal is usually called "Corrosion". In simplified words, corrosion is the gradual and eventual destruction of a material through various means, but most often through chemical reactions to its environment. People might equate corrosion with rusting but the two are not exactly the same. Rusting is a form of corrosion, but there are many other types of corrosion as well. Corrosion is a dangerous chemical process that affects many objects around us that perform tasks essential to the smooth running of our lives, both public and private. There are many existent ways of preventing and protecting against corrosion while more and more are being discovered now. The phenomenon of corrosion affects many industries and not a matter to be taken lightly. There have been studies released that look at how the overall corrosion generated in an industry has affected the economy.

This book is a collection of such data which relates something as commonplace as erosion and to much larger economic concerns and the research that went behind such a collation of data. I would like to thank all those whose hard work resulted in these researches as well as my friends and family who have stayed by my side and supported me in this endeavor.

Editor

1

Corrosion Behavior of Zinc-Graphite Metal Matrix Composite in 1 M of HCl

M. A. Afifi

Modern University for Technology and Information (MTI University), Cairo, Egypt

Correspondence should be addressed to M. A. Afifi; mohamedafifi@eng.mti.edu.eg

Academic Editors: G. Bereket, C. Gervasi, C. Gu, and C.-H. Hsu

This paper is aimed at investigating the corrosion behavior of Pure and Zinc-graphite particles with percentage of 1, 3, and 5%, respectively. The composites were fabricated by powder metallurgy method. Corrosion tests were performed according to ASTM standard. Corrosion rate was calculated and it is found that in all cases the corrosion rate was decreasing with the increase in exposure time. Meanwhile, the microstructure of composites was imaged and analyzed using optical microscope and scanning electron microscope. It is observed that the best corrosion resistance was zinc with 1% Graphite while Zinc with 3% and 5% Graphite composites did not enhance the corrosion resistance comparing to pure Zinc.

1. Introduction

Increasing use of advanced composites materials is being made especially in aircraft and aerospace vehicles. Metal matrix composites (MMCs) are currently being developed as possible structural materials, offering improved elastic modulus, strength, elevated temperature properties, and control over the coefficient of thermal expansion. The resulting high specific properties can provide significant weight saving in many components compared with conventional alloys [1].

The corrosion behavior of the composites in various environments that the material is likely to encounter is one important consideration in choosing a suitable material for a particular application. It follows from the literature that limited work has been done to study the corrosion behavior of MMCs. Though a few studies concerning the corrosion behavior of MMCs state that they suffer severer attack than the respective nonreinforced alloys [2]. It has been established that the corrosion behavior of MMCs is influenced by various factors like composition of the alloy used, the reinforcing particles used, its size and distribution in the matrix, the technique adopted for fabrication, and the nature of interface between the matrix and reinforcement. Even a very small change in any one of these factors can seriously affect the corrosion behavior of the material [3].

Corrosion behavior of zinc composites with nano-CeO_2, which were prepared by powder metallurgy method, was reported and it was shown that the corrosion has been improved significantly using mass fraction of 1% of CeO_2 [4]. Seah et al. have investigated the corrosion characteristics of cast ZA-27 alloy reinforced with graphite particles (1, 3, and 5% wt) which results in increasing corrosion rate [5].

In the investigation carried out by Seah et al., the effect of the graphite particles content on corrosion behavior of ZA-27 alloy in 1 M HCl solution was studied. Figure 1 shows the effect of graphite content with %wt loss of ZA composite, which is a noble conductor that can induce galvanic corrosion (see Figure 1). Several hypotheses were made, such as the presence of the graphite network hindering the reaction between HCl and the matrix [5]. Another possibility is that zinc ions in solution, acting as a cathode inhibitor [6], may have eventually precipitated as zinc hydroxide on the cathodic graphite particles if the local pH resided in the correct range. This would also explain the decreasing corrosion rates with time that was observed. Similar behavior was noticed in graphite-aluminum couples in solutions containing zinc chloride [7].

This phenomenon is surprising because graphite is very noble compared to zinc and these two elements form an excellent galvanic cell in the HCl solution. One would expect that, as the graphite content is increased, the corrosion of the zinc should be aggravated instead of being reduced. Such a mechanism is commonly seen in grey cast iron in which the graphite present acts as a cathode, causing corrosion of the iron, even in relatively mild environments, as documented by Fontana [8].

Historically, galvanic corrosion has been reported to occur mostly in bimetallic couples. With the ever-increasing use of nonmetallic materials, galvanic corrosion is now being identified in many situations where a metal is in contact with a nonmetallic material (e.g., galvanic corrosion of metals occurs in metal-reinforced polymer matrix composites and graphite metal matrix composites [9, 10], in processing of semiconducting minerals [11], in contact with conducting polymers [12], with semiconducting metal oxides [13], and with conducting inorganic compounds [14]). It has been found that minerals, in general, exhibit potentials more noble than most metals and, therefore, may cause galvanic corrosion of metals used in processing equipment [11].

In view of this, we have produced Pure Zinc and Zinc based alloy with Graphite of mass fractions 1, 3, and 5 wt% and aimed to investigate the corrosion behavior of composites on 1 M HCL. The reinforcements have a size less than 4 μm. Zinc-Graphite MMCs were fabricated using conventional powder metallurgy (P/M) route. The effect of weight percent on the corrosion behavior of Zn-Graphite MMCs was extensively investigated.

2. Experimental

2.1. Material. Pure Zn was used as a matrix material; the Zn powder has purity of about 99.5% (wt.-%). The Zn powders have size range between 1 and 15 μm. The Graphite particulates have a size less than 4 μm.

2.2. Preparation of Composites by Powder Metallurgy Method. The Zn MMCs were fabricated according to the following procedures: first, the reinforcement particles and Zn powders were mixed together by a mixer, where Paraffin oil was added by 3% weight percent of the charge to lubricate the surface of powders and reduce the heat generated from friction during mixing. The green powders were put into a cylindrical die made from H13 tool steel.

The green powders were then compacted under 100 MPa for 15 minutes using the hydraulic press shown in Figure 1. After that, the powders were heated to about 310°C in the die during compression. Heating the mould was conducted using electrical resistance heaters around the die. After reaching the desired temperature, the powders were again hot-pressed under 320 MPa for 15 minutes and then release the pressure and remove the product from the die. After releasing the samples from the die, they were cut to a thickness of 5 mm, making the dimensions of samples identical to 15 mm diameter and 5 mm thickness.

2.3. Corrosion Test. The corrosion test was conducted at room temperature (27°C) using conventional corrosion rate measurement according to ASTM G1-03 and ASTM G31-72 [15, 16].

FIGURE 1: The hydraulic press.

The area of the specimen, subjected to corrosion, was calculated before performing corrosion test using the following equation (according to ASTM G31–72):

$$A = \frac{\pi}{2\left(D^2 - d^2\right)} + t\pi D + t\pi d, \quad (1)$$

where t is thickness, D is diameter of the specimen, and d is diameter of the mounting hole.

The corrodents used for the tests were 1 M hydrochloric acid. Small cylindrical discs of diameter 15 mm and thickness 5 mm were polished using emery paper in order to obtain a smooth and identical surface finish on all specimens, then washed in distilled water, followed by acetone, and then allowed to dry thoroughly [3]. They were finally weighted accurately to an accuracy of three decimal digits.

The initially weighted specimens were immersed in the corrosive environment and taken out at 24 h intervals for testing up to 96 hr. To avoid crevice corrosion, the specimens were suspended in the solution with a plastic string. The specimens were exposed to the test solution for several hours up to 96 hr. The corroded surface was removed by immersing the specimen on 100 mg ammonium per sulphate for 5 min. The cleaned specimens were dried and weighed to an accuracy of three decimal digits. The corrosion rate was calculated and compared with other reinforcements.

The corrosion rate (CR), from the mass loss, was calculated (according to ASTM Standard [6, 7] using the following equation:

$$\text{CR} = \frac{(K \times W)}{(A \times T \times D)}, \quad (2)$$

where CR is corrosion rate (mm/year), K is a constant equal to 8.76×10^4, T is time of exposure in hours to the nearest 0.01 hour, A is area in cm^2, W is mass loss in grams, and D is density in g/cm^3.

The microstructure was examined using CETI optical microscope and SEM Model Quanta FEG 250.

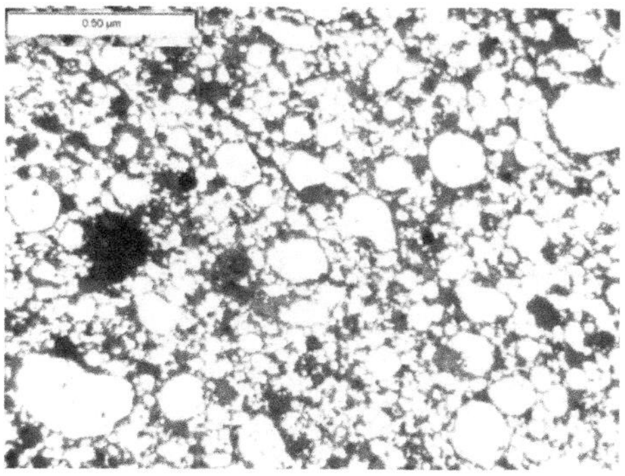

FIGURE 2: Microstructure of pure Zn matrix.

TABLE 1: The average corrosion rates of pure Zinc.

Exposure duration (hr)	Average corrosion rate (mm/year)
24	80
48	44
72	28
96	20

TABLE 2: The average corrosion rate of Zn/Gr MMCs.

Gr (wt.%)	Exposure duration (hr)	Average corrosion rate (mm/year)
1	24	53
	48	36
	72	20
	96	15
3	24	92
	48	42
	72	36
	96	27
5	24	130
	48	60
	72	40
	96	31

3. Results and Discussions

Figures 2 and 3 show micrographs of the microstructure of Pure Zn matrix as well as the Zn/Graphite composites before performing corrosion test.

As shown in figures, graphite ceramic particles were successfully dispersed into the Zn matrix. The micrographs indicated that the production of bulk MMCs using conventional P/M technique used in the current study is effective. The ceramic particulates distribution in the Zn matrix was fairly uniform. However, on increasing wt% of Graphite to 3 and 5% agglomerations start to appear in the microstructure.

The agglomeration of the particles in MMCs was reported by many workers [17–20]. For example, Xiandong et al. [17] have different observations; all pores are filled by graphite and the graphite flakes are uniformly distributed throughout the matrices and the interfaces are well integrated. The pores found on the microstructure may be result of inadequate mixing of graphite powder. Similar observations by Martinez-Flores et al. [18] have been reported; it was observed that the reason for reduction of grain growth with increasing Al_2O_3 is its accommodated around the big metal particles, producing some sort of thermal insulation for the metal grains on higher amounts of Al_2O_3; the effect of insulation is bigger, resulting in a retardation of grain growth inside the metal particles. Božić et al. [19] worked on Zn–Al–Cu alloy reinforced with Al_2O_3 particles produced by powder metallurgy method; it was observed that Al_2O_3 particles agglomerate around Zn grains, and the presence of the agglomeration has detrimental effect on the mechanical properties of composite. Seyed Reihani [20] states that, as the reinforcement of SiC is increased in volume, porosity andagglomeration of SiC particles become more evident. Gnjidic et al. [21] have similar observations, by using adequate mixing method; agglomeration of SiC is found and the bond between the particles within the agglomeration is weak and the rearrangement that occurred during hot pressing resulted in theformation of large pores. Wang et al. [22], who study characterization of a powder metallurgy SiC/Cu–Al composite, find that SiC grains disperse homogeneously in the boundary region or inside the grown Al grains and therelatively homogeneous microstructure leads to the highest hardness and as a result-the composite is inhomogeneous.

Tables 1 and 2 list the average corrosion rates of pure Zn as well as Zn/Graphite MMCs; the average is of 10 samples.

3.1. Effect of Exposure Time. It is clear from Figure 4 that corrosion rates of the Zn/Graphite MMCs as well as the pure Zn were decreased with increasing exposure duration. This may be attributed to the formation of passive layer on the surface of the samples and reduction of chlorides content in the acidic solution [23]. Sharma et al. [24] studied corrosion of ZA-27/Zircon particulate composites in 1 M HCl. They reported that corrosion rate of ZA-27/Zircon particulate composites was decreased with increasing the exposure time.

The same observation was observed by the same authors in another investigation when they studied the corrosion of ZA-27/Glass fiber composites [25] and Bobić et al. on working on Corrosion of Aluminium and Zinc-Aluminium Alloys Based Metal-Matrix Composites [26].

Different observations were found by Yildiz and Kaplan [27] on studying corrosion of Zn–Al alloy. They found that corrosion rate revealed an altering property by showing sometimes an increase and sometimes a decrease in time as a result of presence of Al content.

3.2. Effect of Reinforcement Weight Percent. In Zn/Graphite MMCs, it has been found that increasing the weight percent of Graphite particulates tends to increase the corrosion rates of the Zn/Graphite MMCs. It was found also that Zn/1% Graphite MMC improved the corrosion resistance to 1 M HCL in comparison to Zinc/3% Graphite and Zinc/5% Graphite and even better than Pure Zinc. Fontana [28]

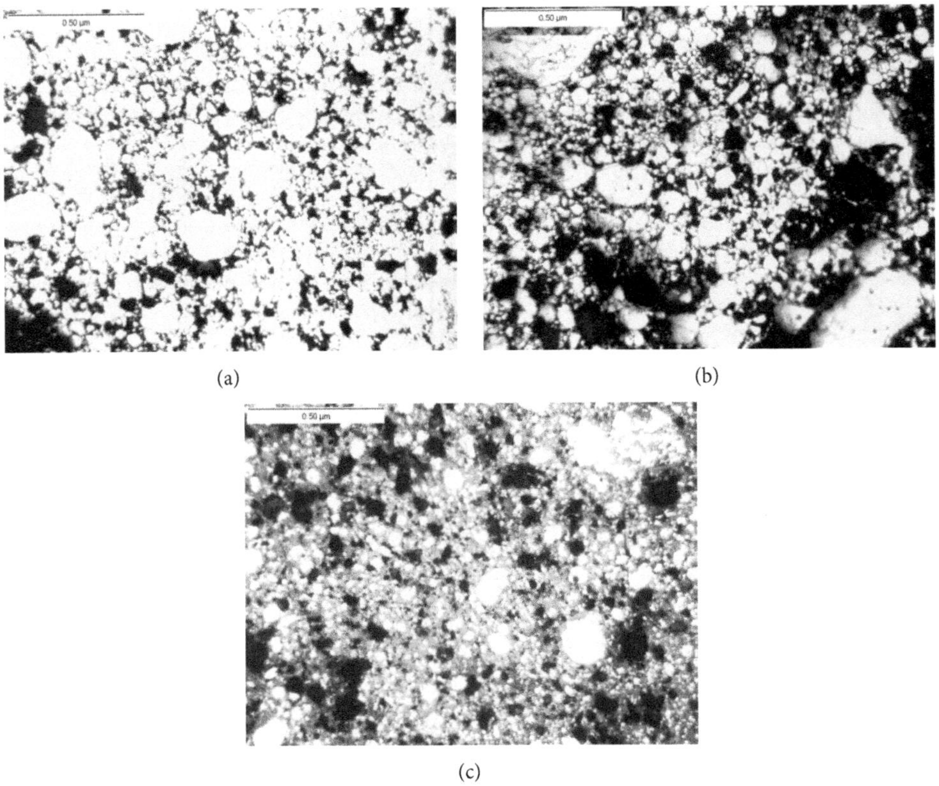

FIGURE 3: The microstructure of Zn/Gr composites containing (a) 1 wt.-%, (b) 3 wt.-%, and (b) 5 wt.-% of Graphite particulates.

reported that relatively large amount of Al_2O_3 present in the Zn alloy matrix confers a certain amount of corrosion resistance because the oxide layer is quite stable in neural and many acid solutions. The effect of Graphite reinforcement in ZA-27 magnesium alloy was reported by Seah et al. [5]; it was observed that, with increasing graphite content, corrosion rate decreases. This phenomenon was surprising because graphite is very noble compared to zinc and these two elements form an excellent galvanic cell in the HCl solution [28]. Figure 4 shows that, with increasing graphite content, the corrosion rate increases as a result of galvanic coupling.

In contrast, the observations noticed in the current investigation, in which increasing the amount of the ceramic particulates tends to increase the corrosion behavior of the MMCs, were reported by Hihara and Latanision [29]. It was observed when studying the galvanic corrosion of Al MMCs that, with increasing SiC content, a galvanic coupling between Al and SiC occurred which reduces the corrosion resistance of the MMCs.

It is believed that the contrary results reported by the investigators may be result of the difference in the fabrication methods, the nature, amount, shape, and size of the reinforcements and the matrix material. Several investigations showed that the aforementioned factors play an important role in determining the characteristics of MMCs [4, 24].

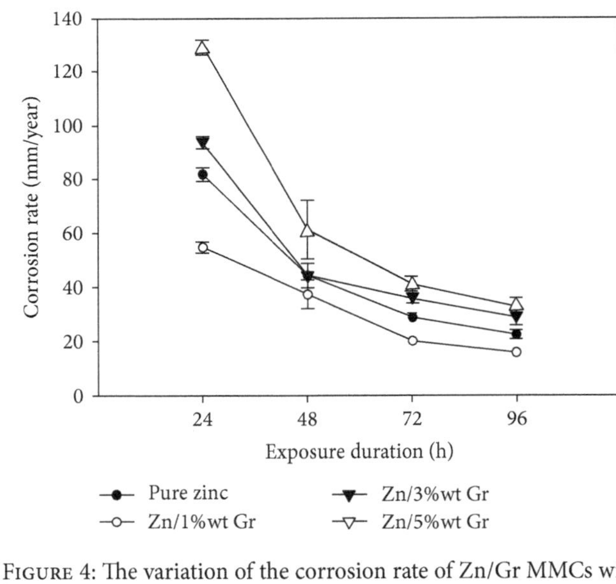

FIGURE 4: The variation of the corrosion rate of Zn/Gr MMCs with the exposure duration.

3.3. Examination of the Corroded Surfaces of MMCs. The surface of the pure Zn matrix underwent degradation, especially along the grain boundaries. These grain boundaries provide preferential corrosion initiation sites because of the discontinuity in the surface oxide film due to the changed substrate structure. Such a discontinuity would facilitate the passage of the hydrogen ions to the metal, which when contacted by ions would suffer localized corrosion (Figure 5).

Figures 6, 7, and 8 show optical micrographs of the corroded surfaces of Zn/Graphite MMCs containing 1, 3, and

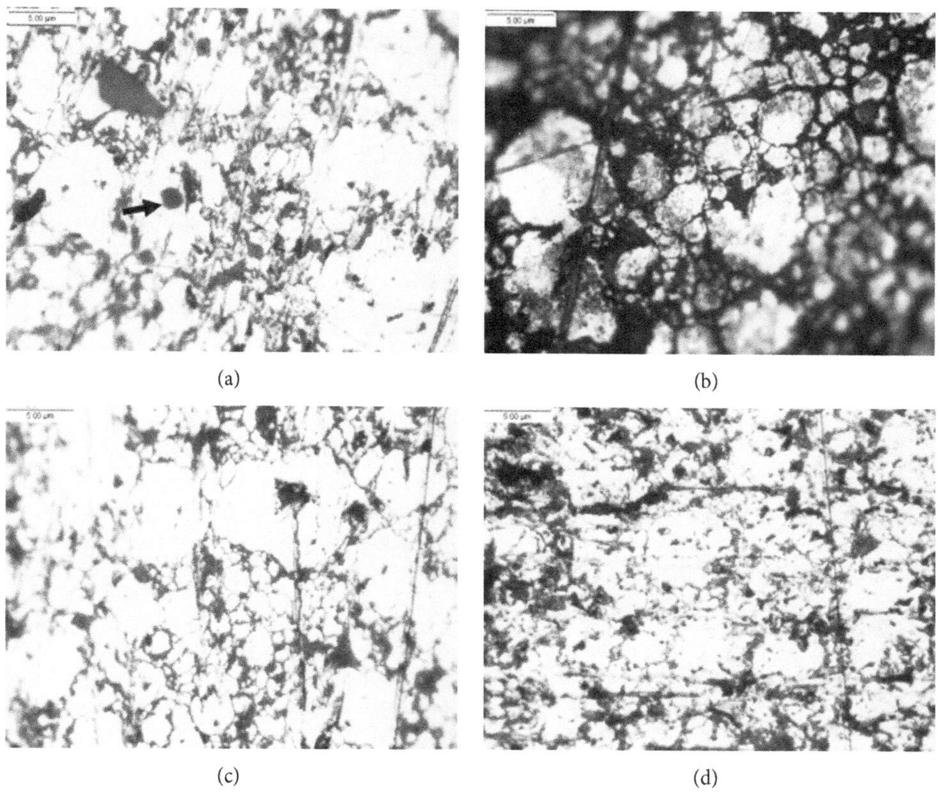

FIGURE 5: Optical micrographs of the corroded surfaces of unreinforced Zn matrix after immersion in 1 M HCl solution for (a) 24 hr, (b) 48 hr, (c) 72 hr, and (d) 96 hr.

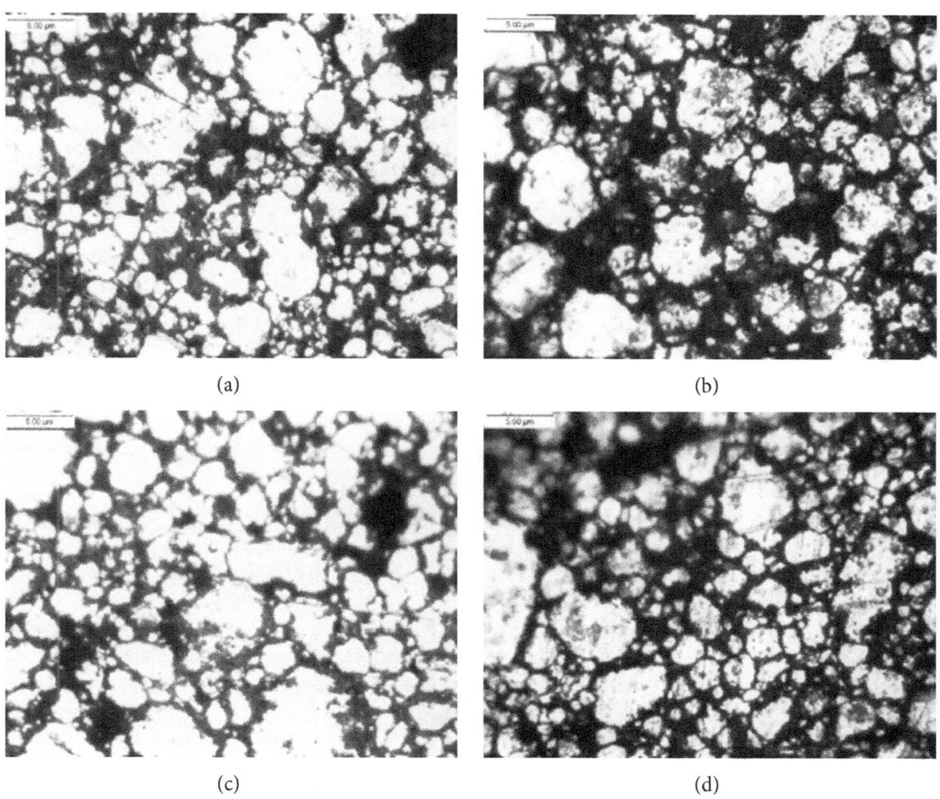

FIGURE 6: Optical micrographs of the corroded surfaces of Zn/Graphite (1 wt.-%) MMCs after immersion in 1 M HCl solution for (a) 24 hr, (b) 48 hr, (c) 72 hr, and (d) 96 hr.

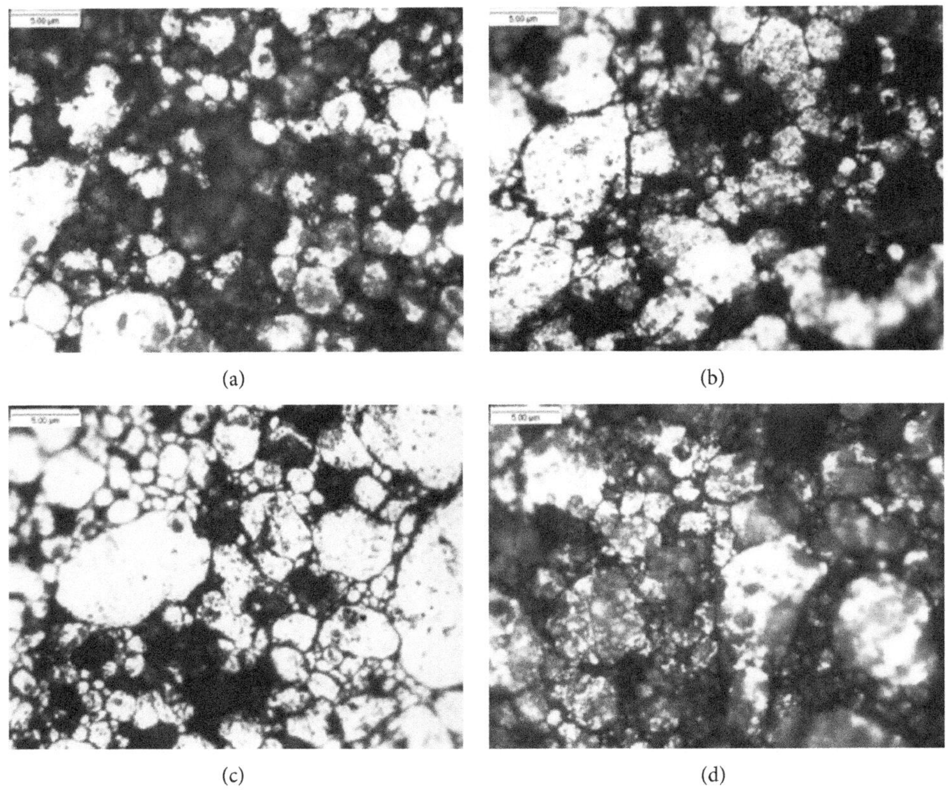

FIGURE 7: Optical micrographs of the corroded surfaces of Zn/Graphite (3 wt.-%) MMCs after immersion in 1 M HCl solution for (a) 24 hr, (b) 48 hr, (c) 72 hr, and (d) 96 hr.

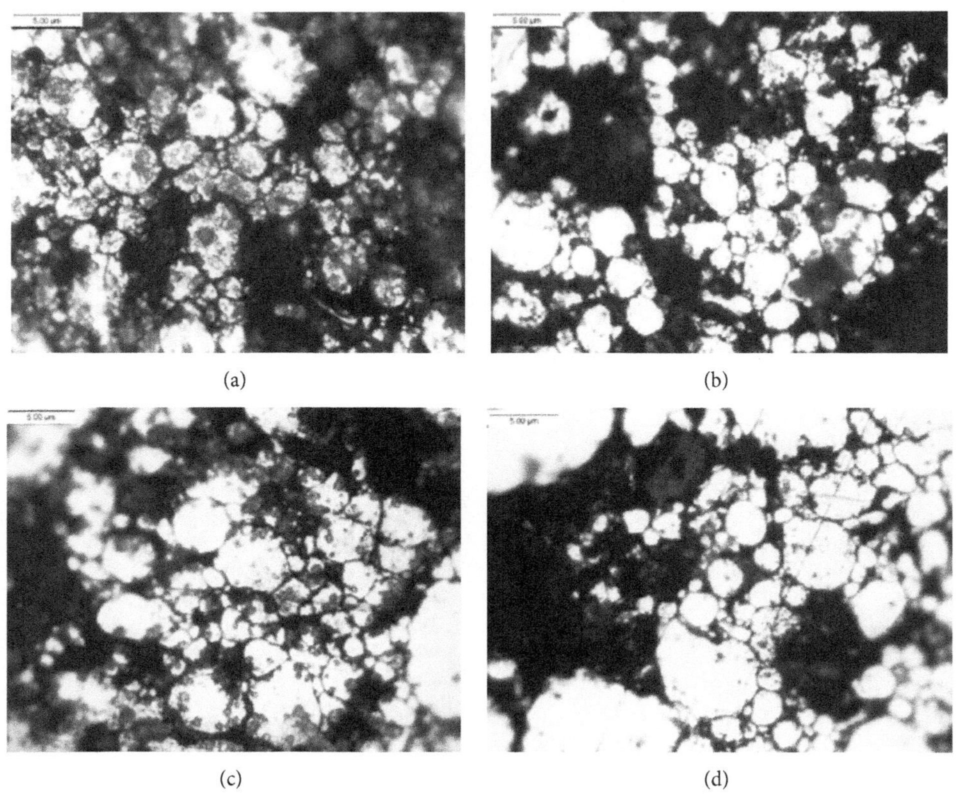

FIGURE 8: Optical micrographs of the corroded surfaces of Zn/Graphite (5 wt.-%) MMCs after immersion in 1 M HCl solution for (a) 24 hr, (b) 48 hr, (c) 72 hr, and (d) 96 hr.

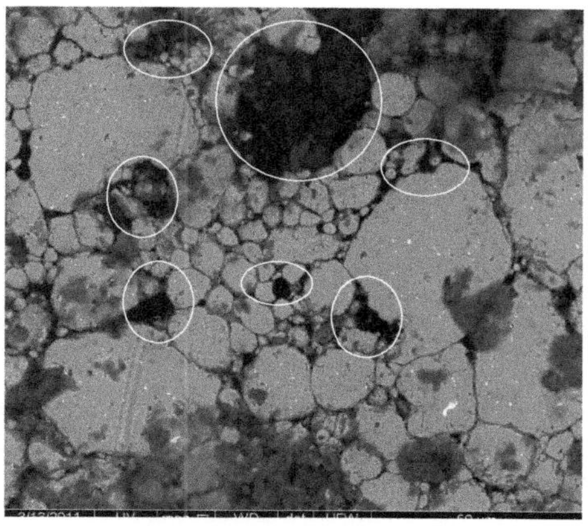

FIGURE 9: SEM of Zinc/3% Graphite showing galvanic corrosion performed between grains.

5 wt.-%, respectively, after exposure in 1 M HCl solution for 24, 48, 72, and 96 hours. It is clear from the micrographs that galvanic corrosion took place at the corroded surface of Zn/Graphite containing 3 and 5%. It has been reported that the graphite particles make graphite network interfering with the reaction between the acid and metallic matrix [5], and so, graphite protects zinc grains from corrosion and, on increasing graphite content, a galvanic coupling occurs.

It is clear from corrosion surface images and corrosion rate results that Zn/Graphite (1 wt.-%) MMC has the best corrosion resistance, when compared to Zn and Zn/Graphite (3 and 5 wt.-%) MMC. Figure 3(a) shows the microstructure of Zn/Gr (2 wt.-%) MMC, where graphite particles were successfully dispersed into the Zn matrix and as a result, when exposed to the corrosive medium, Graphite particles isolate Zn grains from the exposure to HCl solution protecting the composite from corrosion and enhancing the corrosion resistance. Fontana describes a possible mechanism by which graphite present in a metal can actually protect it from corrosion, as in the case of cast iron in strong acids. The better resistance is probably due to the graphite network interfering with the reaction between the acid and the metallic matrix [30].

Figures 7 and 8 show the corroded surface of Zn/Graphite (3 wt.-%) and Zn/Graphite (5 wt.-%) where deep pits and gaps are found causing an increase in the mass loss and corrosion rate. Moreover, Figure 9 shows that gaps were found as a result of the galvanic action between Zinc and Graphite by using SEM under ×2000; this is because, on increasing the weight percentage, the agglomeration was increased and as a result galvanic coupling occurs. Similar observations were found when working on higher weight percentage of graphite like Hihara and Latanision [29] when studying the galvanic corrosion of Aluminum matrix composites. A galvanic action between Al and Graphite was found and pitting was found on the corroded surface. Similar observations were obtained [31, 32]; Payan et al. [31] studied galvanic coupling between Al and Graphite fibers and found poor resistance to corrosion in aqueous environments (chloride solution) and found galvanic action between Al and Graphite; Buonanno et al. [32] obtained similar observations when studying corrosion of Graphite-Al MMCs.

In sum, it is found that corrosion resistance was improved on using Zinc/1% Graphite in comparison to Pure Zinc and Zinc/3% Graphite and Zinc/5% Graphite under static corrosion on 1 M HCl medium for all periods till 96 Hr.

4. Conclusions

(1) Zn/Gr MMCs were successfully fabricated using powder metallurgy (PM) method.

(2) Increasing the weight percent of the ceramic particulates in the fabricated MMCs increased the particles clusters volume fraction (agglomeration percent).

(3) The corrosion rates of Zn/Graphite MMCs as well as the pure Zn were decreased with increasing exposure duration. However, increasing the weight percent of the Graphite particles tends to increase the corrosion rate of the composite materials.

(4) Zn/Graphite MMCs containing 1 wt.-percent of Graphite particulates exhibited the best corrosion resistance comparing to Zinc, Zinc/3% Graphite, and Zinc/5% Graphite as a result of graphite network phenomena.

Conflict of Interests

There is no conflict of interests or any financial support.

References

[1] A. Brown, "How to deposit DLC successfully," *Materials World*, vol. 1, pp. 20–26, 1993.

[2] S. L. Pohlman, "Corrosion and electrochemical behavior of boron/aluminum composites," *Corrosion*, vol. 34, pp. 156–162, 1987.

[3] D. M. Aylor and P. J. Moran, "Effect of reinforcement on the pitting behavior of aluminum-base metal matrix composites," *Journal of the Electrochemical Society*, vol. 132, no. 6, pp. 1277–1281, 1985.

[4] Q. Wang, "Preparation and properties of nano-CeO2/Zn composites," *Transactions of Nonferrous Metals Society of China*, p. 623, 2007.

[5] K. H. W. Seah, F. S. C. Sharma, and B. M. Girish, "Corrosion characteristics of ZA-27-graphite particulate composites," *Corrosion Science*, vol. 39, no. 1, pp. 1–7, 1997.

[6] B. P. Bofardi, "Control of environmental variables in water reticulating systems," in *Metals Handbook*, vol. 13 of *Corrosion*, p. 487, ASM International, Materials Park, Ohio, USA, 9th edition, 1987.

[7] L. H. Hihara and R. M. Latanision, "Suppressing galvanic corrosion in graphite/aluminum metal-matrix composites," *Corrosion Science*, vol. 34, no. 4, pp. 655–665, 1993.

[8] M. G. Fontana, in *Corrosion Engineering*, p. 89, McGraw-Hill, New York, NY, USA, 3rd edition, 1987.

[9] F. Bellucci, "Galvanic corrosion between nonmetallic composites and metals I. Effect of metal and of temperature," *Corrosion*, vol. 47, no. 10, pp. 808–819, 1991.

[10] L. H. Hihara and R. M. Latanision, "Galvanic corrosion of aluminum-matrix composites," *Corrosion*, vol. 48, no. 7, pp. 546–552, 1992.

[11] D. A. Jones and A. J. P. Paul, "Galvanic interactions between alloys and minerals in sulfuric acid," *Corrosion*, vol. 50, no. 7, pp. 516–521, 1994.

[12] G. Schick, "Avoiding galvanic corrosion problems in the telephone cable plant," in *Galvanic Corrosion*, H. P. Hack, Ed., ASTM STP 978, pp. 283–290, American Society for Testing and Materials, Philadelphia, Pa, USA, 1988.

[13] S. M. Wilhelm, "Galvanic corrosion caused by corrosion products," in *Galvanic Corrosion*, H. P. Hack, Ed., ASTM STP 978, pp. 23–34, American Society for Testing and Materials, Philadelphia, Pa, USA, 1988.

[14] V. Brusic, M. Russak, R. Schad et al., "Corrosion of thin film magnetic disk: galvanic effects of the carbon overcoat," *Journal of the Electrochemical Society*, vol. 136, no. 1, pp. 42–46, 1989.

[15] ASTM Standard, "Standard practice for laboratory immersion corrosion testing of metals," American Society for Testing and Materials G31-72, 2004.

[16] ASTM Standard, "Standard practice for preparing, cleaning, and evaluating corrosion test specimens," American Society for Testing and Materials G1-03, 2004.

[17] S. Xiandong, L. Chengping, L. Zhuoxuan, and O. Liuzhang, "The fabrication and properties of particle reinforced cast metal matrix composites," *Journal of Materials Processing Technology*, vol. 63, no. 1–3, pp. 426–431, 1997.

[18] E. Martinez-Flores, J. Negrete, and G. Torres Villasenor, "Structure and properties of Zn-Al-Cu alloy reinforced with alumina particles," *Materials & Design*, vol. 24, no. 4, pp. 281–286, 2003.

[19] D. Božić, J. Stašić, and V. Rajković, "Microstructures and mechanical properties of ZA27-Al2O3 composites obtained by powder metallurgy process," *Science of Sintering*, vol. 43, no. 1, pp. 63–70, 2011.

[20] S. M. Seyed Reihani, "Processing of squeeze cast Al6061–30vol% SiC composites and their characterization," *Materials & Design*, vol. 27, no. 3, pp. 216–222, 2006.

[21] Z. Gnjidic, D. Bozic, and M. Mitkov, "The influence of SiC particles on the compressive properties of metal matrix composites," *Materials Characterization*, vol. 47, no. 2, pp. 129–138, 2001.

[22] H. Wang, R. Zhang, X. Hu, C.-A. Wang, and Y. Huang, "Characterization of a powder metallurgy SiC/Cu-Al composite," *Journal of Materials Processing Technology*, vol. 197, no. 1–3, pp. 43–48, 2008.

[23] W. Feitknecht, "Studies on the influence of chemical factors on the corrosion of metals," *Chemistry & Industry*, vol. 36, pp. 1102–1109, 1959.

[24] S. C. Sharma, D. R. Somashekar, and B. M. Satish, "A note on the corrosion characterisation of ZA-27/zircon particulate composites in acidic medium," *Journal of Materials Processing Technology*, vol. 118, no. 1–3, pp. 62–64, 2001.

[25] S. C. Sharma, K. H. W. Seah, B. M. Satish, and B. M. Girish, "corrosion characteristics of ZA-27/glass-fibre composites," *Corrosion Science*, vol. 39, no. 12, pp. 2143–2150, 1997.

[26] B. Bobić, S. Mitrovic, M. Babic, and I. Bobić, "Corrosion of aluminium and zinc-aluminium alloys based metal-matrix composites," *Tribology in Industry*, vol. 31, no. 3-4, pp. 44–53, 2009.

[27] A. K. Yildiz and M. Kaplan, "Corrosion behaviour, microstructure and phase transitions of Zn-based alloys," *Bulletin of Materials Science*, vol. 27, no. 4, pp. 341–345, 2004.

[28] M. G. Fontana, in *Corrosion Engineering*, p. 236, McGraw-Hill, New York, NY, USA, 3rd edition, 1987.

[29] L. H. Hihara and R. M. Latanision, "Galvanic corrosion of aluminum-matrix composites," Tech. Rep., Office of Naval Research, Arlington, Va, USA, 1991.

[30] M. G. Fontana, in *Corrosion Engineering*, p. 320, McGraw-Hill, New York, NY, USA, 3rd edition, 1987.

[31] S. Payan, Y. le Petitcorps, J.-M. Olive, and H. Saadaoui, "Experimental procedure to analyze the corrosion mechanisms at the carbon/aluminum interface in composite materials," *Composites A*, vol. 32, no. 3-4, pp. 585–589, 2001.

[32] M. A. Buonanno, R. M. Latanision, L. H. Hihara, and J. F. Chiang, "Corrosion of graphite aluminum metal matrix composites," Tech. Rep., Office of Naval Research, Arlington, Va, USA, 1991.

2

Surface Protection of Carbon Steel by Hexanesulphonic Acid-Zinc Ion System

C. Mary Anbarasi[1] and Susai Rajendran[2,3]

[1] *PG Department of Chemistry, Jayaraj Annapackiam College for Women, Periyakulam 625601, India*
[2] *Corrosion Research Centre, PG and Research Department of Chemistry, GTN Arts College, Dindigul 624005, India*
[3] *Department of Chemistry, RVS School of Engineering and Technology, Dindigul 624005, India*

Correspondence should be addressed to C. Mary Anbarasi; anbuc_m@yahoo.co.in

Academic Editors: K. N. Allahar, L. Bazzi, A. Hermann, and B. Normand

Inhibition of corrosion of carbon steel in dam water by hexanesulphonic acid as its sodium salt $C_6H_{13}SO_3Na$ (SHXS) in the absence and presence of a bivalent cation zinc ion ($Zn^{2þ}$) has been investigated using weight loss method. Results of weight loss method indicate that inhibition efficiency (IE) increased with increase of inhibitor concentration. Polarization study reveals that SHXS-Zn^{2+} system controls the cathodic reaction predominantly. AC impedance spectra reveal that a protective film is formed on the metal surface. The nature of the metal surface has been analysed by Fourier Transform Infrared Spectroscopy (FTIR) and Atomic Force Microscopy (AFM).

1. Introduction

Corrosion is a naturally occurring phenomenon which deteriorates a metallic material or its properties because of a reaction with its environment. Corrosion can cause dangerous and expensive damage to everything from pipelines, bridges, and public buildings to vehicles, water, and wastewater systems, and even home appliances. It is one of the most serious problems in the oil and gas industry. The use of organic inhibitors is one of the most widely practical methods for protection of metals and alloys against corrosion. The efficiency of an organic compound as a corrosion inhibitor is closely associated with the chemical adsorption [1–4]. Most well-known organic inhibitors contain nitrogen, sulfur, and oxygen atoms due to their ability to form an adsorbed protective film at the metal/media interface. Studies report that the adsorption of organic inhibitors mainly depends on some physicochemical properties of the molecule, related to its functional groups, to the possible steric effects and electronic density of donor atoms: adsorption is supposed also to depend on the possible interaction of P-orbitals of the inhibitor with d-orbitals of the surface atoms, which induce greater adsorption of the inhibitor molecules onto the surface of carbon steel, leading to the formation of a corrosion protective film [5].

A survey of the available literature reveals that the corrosion inhibition of 2-naphthalenesulfonic acid, 2, 7-naphthalenedisulfonic acid, and 2-naphthol-3, 6-disulfonic acid on Armco-iron electrode in sulfuric acid has been investigated by Vračar and Draži. The inhibition efficiency changes with the number of functional groups substituted on the benzene ring and increases with concentration [6]. The inhibition action of 2-mercaptobenzoxazol, 2-mercapto benzimidazole, N-cetyl pyridinium bromide, and propargyl benzene sulphonate on the corrosion of carbon steel in acid media has also been studied by Prakash et al. The corrosion inhibitors are used to reduce corrosion damage in subsurface equipment in oil well fields. The corrosion inhibition activity was studied by gravimetric and potentiostatic polarization methods in presence of 20% HCl [7]. Manickavasagam et al. have reported the corrosion inhibition of poly(Styrene sulphonic acid) - doped polyaniline on carbon steel in acid media. The polymer acts as an anodic inhibitor. The adsorption of the compound on the metal surface obeys Temkin's adsorption isotherm [8]. Aliev has described the influence of salts of alkyl phenol sulphonic acid on the corrosion of ST3 steel. The protective effect increases with temperature. The investigated compounds inhibit corrosion of ST3 steel as a result of chemical adsorption [9]. Shakkthivel and Vasudevan have studied the effect of acrylic acid-diphenylamine sulphonic acid copolymer as threshold inhibitor for sulphate and carbonate scales in cooling water systems. The results show that the polymer acts as a very good antiscaling inhibitor both in the carbonate and sulphate brines. Copolymer of acrylic acid-diphenylamine sulphonic acid can be used safely in cooling water systems [10]. Perusal of several literatures reveals that there is no information regarding the use of hexanesulphonic acid (SHXS) in combination with zinc ion (Zn^{2+}) as corrosion inhibitor. This paper focuses on the IE of hexanesulphonic acid as its sodium salt (SHXS) in controlling corrosion of carbon steel immersed in dam water in the absence and presence of Zn^{2+}. The investigation is performed using weight loss method, polarization technique, and AC impedance spectroscopy. The morphology of the protective film was examined by FTIR and AFM techniques.

The medium which is used in the present study is dam water collected from Sothuparai dam in the state of Tamil Nadu, India, constructed across the Vaigai River, the water from which is used in cooling systems by the industries located downstream.

2. Experimental Method

The chemicals used in this study, sodium hexanesulphonate (inhibitor) and $ZnSO_47H_2O$ (Zn^{2+} ions) coinhibitor, were AR grade.

2.1. Preparation of the Specimen. Carbon steel specimens of size 1.0 cm × 4.0 cm × 0.2 cm, area 10 cm^2, and chemical composition 0.026% sulphur, 0.06% phosphorous, 0.4% manganese, 0.1% carbon, and the rest iron (density 7.87 gm/cm^3) were polished to a mirror finish and degreased with trichloroethylene and used for the weight loss method and surface examination studies.

2.2. Weight Loss Method. Carbon steel specimens were immersed in 100 mL of the medium containing various concentrations of the inhibitor (sodium hexanesulphonate) in the absence and presence of Zn^{2+} for 3 days. The weights of the specimens before and after immersion were determined using a Digital Balance (Model AUY 220 SHIMADZU). The corrosion products were cleaned with Clarke's solution prepared by dissolving 20 gms of Sb_2O_3 and 50 gms of $SnCl_2$ in one litre of Conc. HCl of specific gravity (1.9) [11]. The corrosion IE was then calculated using the equation

$$\mathrm{IE} = 100\left[1 - \left(\frac{W_2}{W_1}\right)\right]\%, \quad (1)$$

where W_1 is the weight loss value in the absence of inhibitor and W_2 is the weight loss value in the presence of inhibitor.

Corrosion rate was calculated using formula [12] millimeter per year

$$(\mathrm{mm/year}) = \frac{87.6W}{DAT}. \quad (2)$$

W = weight loss in milligrams, D = density of specimen in g/cm^3, A = area of specimen in cm^2, and T = exposure time in hours.

2.3. Potentiodynamic Polarization. Polarization studies were carried out in a CHI—electrochemical work station with impedance model 660A. It was provided with iR compensation facility. A three-electrode cell assembly was used. The working electrode was carbon steel. Saturated calomel electrode was the reference electrode. Platinum was the counter electrode. From polarisation study, corrosion parameters such as corrosion potential (E_{corr}), corrosion current (I_{corr}), Tafel slopes anodic = b_a, and cathodic = b_c were calculated and linear polarization resistance (LPR) was also calculated. The scan rate (V/S) was 0.01. Hold time at Ef (s) was zero and quiet time was two seconds.

2.4. AC Impedance Spectra. The instrument and cell set-up used for polarization study was used to record AC impedance spectra also. The real part (Z') and imaginary part (Z'') of the cell impedance were measured in ohms at various frequencies. Values of charge transfer resistance (R_t) and the double layer capacitance (C_{dl}) were calculated:

$$R_t = (R_s + R_t) - R_s, \quad (3)$$

where R_s = solution resistance

$$C_{dl} = \frac{1}{2}\pi Rtf_{max}, \quad (4)$$

where f_{max} = frequency at maximum imaginary impedance.

AC impedance spectra were recorded with initial $E_{(v)}$ = 0, high frequency (Hz) = 1 × 10^5, low frequency (Hz) = 1, amplitude (V) = 0.005, and quiet time (s) = 2.

2.5. Surface Examination Study. The carbon steel specimens were immersed in various test solutions for a period of 1 day. After immersion, the specimens were taken out and dried. The nature of the film formed on the surface of the metal specimen was analyzed by various surface analysis techniques.

2.5.1. Fourier Transform Infrared Spectra (FTIR). These spectra were recorded in a Perkin-Elmer-1600 spectrophotometer using KBr pellet. The FTIR spectrum of the protective film was recorded by carefully removing the film, mixing it with KBr and making the pellet.

2.5.2. Atomic Force Microscopy Characterization (AFM). The carbon steel specimens immersed in blank and in the inhibitor solution for a period of one day were removed, rinsed with double distilled water, dried, and subjected to

TABLE 1: Water analysis.

Parameters	Result
Appearance	Brownish
Total dissolved solids	100 mol/L
Electrical conductivity	140 μS/cm
pH	8.25
Total hardness as $CaCO_3$	50 mol/L
Calcium	10 mol/L
Magnesium	06 mol/L
Iron	1.2 mol/L
Nitrate	10 mol/L
Chloride	10 mol/L
Sulphate	02 mol/L

TABLE 2: The corrosion inhibition efficiencies and the corresponding corrosion rates (mm/year) of SHXS-Zn^{2+} system.

Inhibitor SHXS (mol/L) $\times 10^{-3}$	Zn^{2+} (mol/L)					
	0		5.2×10^{-5}		1.04×10^{-4}	
	IE %	CR mm/year	IE %	CR mm/year	IE %	CR mm/year
0	—	0.1127	12	0.0992	21	0.0890
0.27	10	0.1014	26	0.0834	57	0.0642
0.54	13	0.0981	29	0.0801	64	0.0721
0.81	19	0.0913	32	0.0766	71	0.0327
1.08	26	0.0834	36	0.0721	74	0.0293
1.35	31	0.0778	40	0.0676	83	0.0192

the surface examination. Atomic Force Microscope (Veeco diInnova model) was used to observe the samples' surface in tapping mode, using cantilever with linear tips. The scanning area in the images was 5 μm × 5 μm and the scan rate was 0.6 Hz.

3. Results and Discussion

3.1. Analysis of Weight Loss Study. The physicochemical parameters of dam water are given in Table 1.

Inhibition efficiency (IE) and the corrosion rates in millimeter per year (mm/year) of carbon steel immersed in dam water in presence of SHXS-Zn^{2+} system for a period of 3 days are given in Table 2.

It is found that the IE increases as the concentration of SHXS increases. As the concentration of Zn^{2+} increases, IE also increases. A synergistic effect exists between SHXS and Zn^{2+}. For example, 1.35×10^{-3} mol/L of SHXS has 31% IE. 1.04×10^{-4} mol/L of Zn^{2+} has 21% IE. However, interestingly the formulation consisting of 1.35×10^{-3} mol/L of SHXS and 1.04×10^{-4} mol/L of Zn^{2+} has 83% IE. That is, the mixture of inhibitors shows better IE than the individual inhibitors. The improvement in the protection efficiency can be attributed to the formation of a complex between Zn^{2+} and SHXS which results from the synergistic effect between the two inhibitors,

TABLE 3: Synergism Parameter (S_I).

SHXS (mol/L) $\times 10^{-3}$	I_1	SHXS-Zn^{2+} (5.2×10^{-5})		SHXS-Zn^{2+} (1.04×10^{-4})	
		$I'_{(1+2)}$	S_I	$I'_{(1+2)}$	S_I
0.27	0.10	0.26	1.0703	0.57	1.6535
0.54	0.13	0.29	1.0783	0.64	1.9090
0.81	0.19	0.32	1.0482	0.71	2.2066
1.08	0.26	0.36	1.0175	0.74	2.2485
1.35	0.31	0.40	1.0120	0.83	3.2065

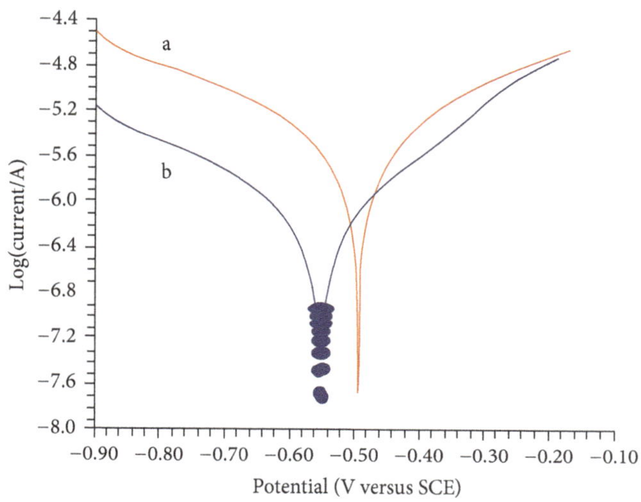

FIGURE 1: Polarisation curves of carbon steel immersed in various test solutions. (a) Dam water. (b) Dam water containing 1.35×10^{-3} mol/L of SHXS and 1.04×10^{-4} mol/L of Zn^{2+}.

Zn^{2+} and SHXS. As a result of this complex formation, the inhibitor molecules are readily transported from the bulk to the metal surface [13].

Synergism Parameter (S_I). Synergism parameters are indications of synergistic effect existing between inhibitors [14, 15]:

$$S_I = \frac{1 - I_{1+2}}{1 - I'_{1+2}}, \quad (5)$$

where $I_{1+2} = (I_1 + I_2) - (I_1 I_2)$, I_1 = surface coverage of inhibitor (SHXS), I_2 = surface coverage of inhibitor (Zn^{2+}), and I'_{1+2} = combined surface coverage of inhibitors (SHXS) and (Zn^{2+})

$$\text{Surface coverage} = \frac{\text{IE}}{100}. \quad (6)$$

I_2 for Zn^{2+} (5.2×10^{-5} mol/L) = 0.12 and I_2 for Zn^{2+} (1.04×10^{-4} mol/L) = 0.21.

The results are given in Table 3.

S_I value is found to be greater than one, indicating the synergistic effect existing between Zn^{2+} of concentrations 5.2 $\times 10^{-5}$ mol/L and 1.04×10^{-4} mol/L with various concentrations of SHXS.

TABLE 4: Corrosion parameters of carbon steel immersed in dam water in the presence and absence of inhibitor obtained by polarization method.

SHXS (mol/L)	Zn^{2+} (mol/L)	E_{corr} mV versus SCE	I_{corr} A/cm^2	b_a mV/dec	b_c mV/dec	LPR Ω cm^2
0	0	−494	2.66×10^{-6}	166	203	2.05×10^4
1.35×10^{-3}	1.04×10^{-4}	−572	5.31×10^{-7}	178	157	7.52×10^4

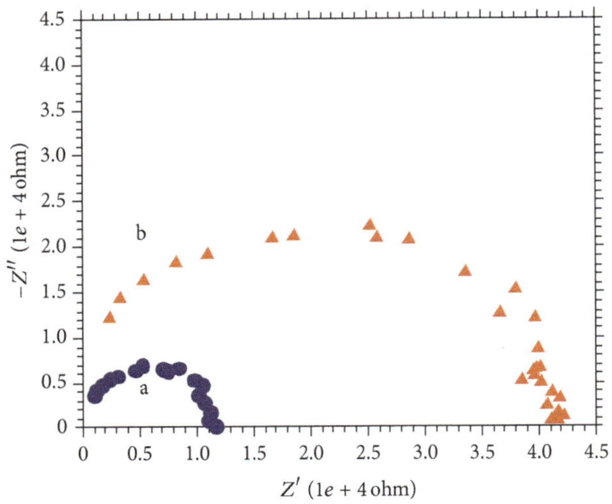

FIGURE 2: AC impedance spectra of carbon steel immersed in various test solutions. (a) Dam water. (b) Dam water containing 1.35×10^{-3} mol/L of SHXS and 1.04×10^{-4} mol/L of Zn^{2+}.

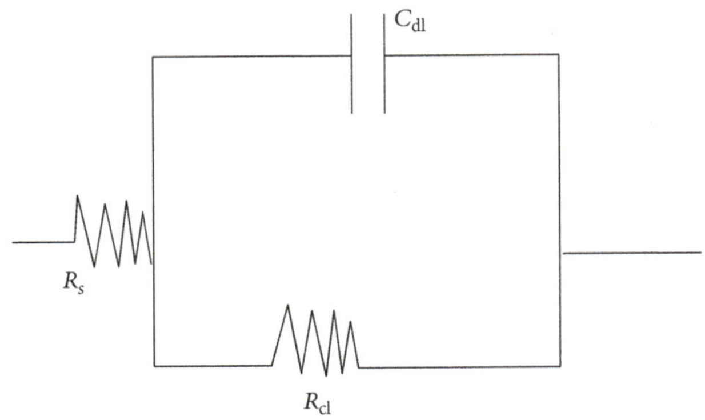

FIGURE 3: Equivalent electrical circuit diagram. R_s = solution resistance. R_{ct} = charge transfer resistance. C_{dl} = double layer capacitance.

3.2. Analysis of Polarization Curves. Figure 1 represents the potentiodynamic polarization curves of carbon steel in dam water in the absence and presence of the inhibitor system. The cathodic branch represents the oxygen reduction reaction, while the anodic branch represents the iron dissolution reaction. The electrochemical parameters such as corrosion potential (E_{corr}), corrosion current (I_{corr}), Tafel slopes (b_a and b_c), and linear polarization resistance (LPR) are given in Table 4.

When carbon steel is immersed in dam water, the corrosion potential is −494 mV versus SCE. The formulation consisting of SHXS (1.35×10^{-3} mol/L)-Zn^{2+} (1.04×10^{-4} mol/L) shifts the corrosion potential to −572 mV versus SCE. It is observed that the corrosion potential is shifted to the cathodic side. It is also observed that the shift in the cathodic slope (from 203 mV/dec to 157 mV/dec) is higher than the shift in the anodic slope (from 166 mV/dec to 178 mV/dec). Hence, it can be said that the inhibitor system predominantly controls the cathodic reaction [16]. The corrosion current value and LPR value for dam water are 2.66×10^{-6} A/cm^2 and 2.05×10^4 Ω cm^2.

For the formulation of SHXS (1.35×10^{-3} mol/L)-Zn^{2+} (1.04×10^{-4} mol/L), the corrosion current value has decreased to 5.31×10^{-7} A/cm^2, and the LPR value has increased to 7.52×10^4 Ω cm^2. The fact that the LPR value increases with decrease in corrosion current indicates adsorption of the inhibitor on the metal surface to block the active sites and inhibit corrosion and reduce the corrosion rate [17, 18].

3.3. Analysis of AC Impedance Spectra. AC impedance spectra have been used to detect the formation of film on the metal surface. If a protective film is formed, the charge transfer resistance (R_t) increases and double layer capacitance (C_{dl}) value decreases [19, 20]. Nyquist representations of carbon steel in dam water in the absence and presence of the inhibitor system are shown in Figure 2.

It is clear from the plots that the impedance response of carbon steel was significantly changed after addition of the inhibitors. The impedance diagrams obtained have an almost semicircular appearance. This indicates that the corrosion of carbon steel in aqueous solution is mainly controlled by a charge transfer process [21]. The deviation from the perfect semicircle shape (depression) is due to the frequency dispersion of interfacial impedance. This anomalous behavior is generally attributed to the inhomogeneity of the metal surface arising from surface roughness or interfacial phenomena [22]. The AC impedance parameters, namely, charge transfer resistance (R_t) and double layer capacitance (C_{dl}), are given in Table 5.

When carbon steel is immersed in dam water, R_t value is 1.08×10^4 Ω cm^2 and C_{dl} value is 8.36×10^{-10} F/cm^2. When SHXS (1.35×10^{-3} mol/L)-Zn^{2+} (1.04×10^{-4} mol/L) are added to dam water, R_t value increases from 1.08×10^4 Ω cm^2 to 3.93×10^4 Ω cm^2 and the C_{dl} decreases from 8.36×10^{-10} F/cm^2 to 2.30×10^{-10} F/cm^2. This suggests that a protective film is formed on the surface of the metal. Results obtained from impedance measurements can be interpreted in terms of the equivalent circuit of the electrical double-layer presented in Figure 3, which was used previously to model other interaction involving metal/solutions interface [23, 24].

TABLE 5: Impedance parameters of carbon steel in dam water in the presence and absence of inhibitor obtained by AC impedance method.

SHXS (mol/L)	Zn^{2+} (mol/L)	R_t $\Omega\,cm^2$	C_{dl} F/cm^2
0	0	1.08×10^4	8.36×10^{-10}
1.35×10^{-3}	1.04×10^{-4}	3.93×10^4	2.30×10^{-10}

This type of circuit is known as Randle circuit. Many authors also suggested Randle circuit for similar graphs [25, 26].

3.4. Analysis of FTIR Spectra. Earlier researchers have confirmed that FTIR spectrometer is a powerful instrument that can be used to determine the type of bonding for organic inhibitors adsorbed on the metal surface [27]. FTIR spectra have been used to analyze the protective film formed on metal surface. FTIR spectrum of pure SHXS is given in Figure 4(a). The FTIR spectrum of the film formed on the metal surface after immersion in the dam water for one day containing 250 ppm of SHXS and 30 ppm of Zn^{2+} is shown in Figure 4(b). The S=O stretching frequency has decreased from 1193 cm^{-1} to 1120 cm^{-1}. This indicates that the oxygen atom of S=O group has coordinated with Fe^{2+} formed on the metal surface resulting in the formation of Fe^{2+}-SHXS complex on the anodic sites of the metal surface. The peak at 1386 cm^{-1} is due to Zn–O stretching. The stretching frequency due to –OH appears at 3463 cm^{-1}. So, it is concluded that $Zn(OH)_2$ is formed on cathodic sites of the metal surface [28].

3.5. Atomic Force Microscopy Characterization. AFM is a powerful technique to investigate the surface morphology at nano- to microscale and has become a new choice to study the influence of inhibitor on the generation and the progress of the corrosion at the metal/solution interface [29–32]. The three-dimensional (3D) AFM morphology and the AFM cross-sectional profile for polished carbon steel surface (reference sample), carbon steel surface immersed in dam water (blank sample), and carbon steel surface immersed in dam water containing SHXS (1.35×10^{-3} mol/L)-Zn^{2+} (1.04×10^{-4} mol/L) are shown in Figures 5 and 6.

3.6. Root-Mean-Square Roughness, Average Roughness, and Peak-to-Valley Value. AFM image analysis was performed to obtain the average roughness, R_a, (the average deviation of all points roughness profile from a mean line over the evaluation length), root-mean-square roughness, R_q, (the average of the measured height deviations taken within the evaluation length and measured from the mean line), and the maximum peak-to-valley (P-V) height values (largest single peak-to-valley height in five adjoining sampling heights) [29]. Table 6 is a summary of (R_q), (R_a), and (P-V) value for carbon steel surface immersed in different environment.

In Figures 5(a) and 6(a) the surface topography of uncorroded metal surface is shown. The value of R_q, R_a, and

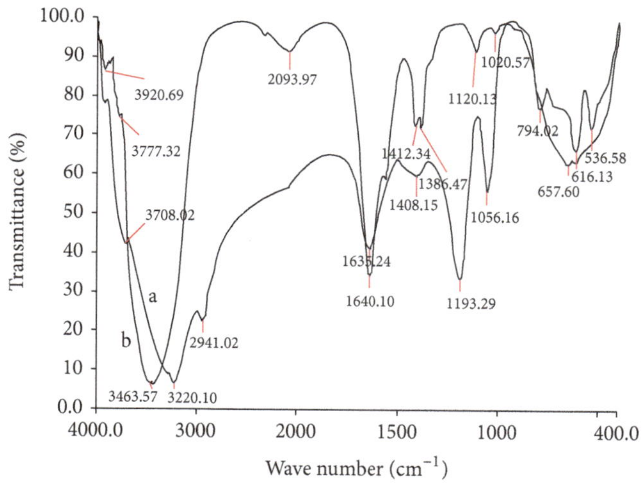

FIGURE 4: (a) FTIR spectrum of pure sodium hexanesulphonate. (b) FTIR spectrum of film formed on metal surface after immersion in dam water containing 1.35×10^{-3} mol/L of SHXS and 1.04×10^{-4} mol/L of Zn^{2+}.

P-V height for the polished carbon steel surface (reference sample) is 4.33 nm, 3.41 nm, and 35.28 nm, respectively. The slight roughness observed on the polished carbon steel surface is due to atmospheric corrosion.

Figures 5(b) and 6(b) show the pitted, corroded metal surface in the absence of the inhibitor immersed in dam water. The (R_q), (R_a), and (P-V) height values for the carbon steel surface are 31.9 nm, 24.9 nm, and 420.3 nm, respectively. These data suggest that carbon steel surface immersed in dam water has a greater surface roughness than the polished metal surface, which shows that the unprotected carbon steel surface is rougher and is due to the corrosion of the carbon steel in dam water environment.

Figures 5(c) and 6(c) show the steel surface after immersion in dam water containing SHXS (1.35×10^{-3} mol/L)-Zn^{2+} (1.04×10^{-4} mol/L). The (R_q), (R_a), and (P-V) height values for the carbon steel surface are 4.38 nm, 3.71 nm, and 27.21 nm, respectively. The (R_q), (R_a), and (P-V) height values are considerably less in the inhibited environment compared to the uninhibited environment. These parameters confirm that the surface is smoother. The smoothness of the surface is due to the formation of a compact protective film of Fe^{2+}-SHXS complex and $Zn(OH)_2$ on the metal surface thereby inhibiting the corrosion of carbon steel [29].

3.7. Mechanism of Corrosion Inhibition. With these discussions, a mechanism may be proposed for the corrosion inhibition of carbon steel immersed in dam water containing SHXS (1.35×10^{-3} mol/L)-Zn^{2+} (1.04×10^{-4} mol/L).

When the formulation consists of SHXS (1.35×10^{-3} mol/L)-Zn^{2+} (1.04×10^{-4} mol/L) in dam water, there is formation of SHXS-Zn^{2+} complex in solution.

TABLE 6: AFM data for carbon steel surface immersed in inhibited and uninhibited environment.

Samples	RMS (R_q) roughness (nm)	Average (R_a) roughness (nm)	Maximum peak-to-valley height (nm)
1. Polished carbon steel	4.33	3.41	35.28
2. Carbon steel immersed in dam water (blank)	31.9	24.9	420.3
3. Carbon steel immersed in dam water + SHXS (1.35×10^{-3} mol/L) + Zn^{2+} (1.04×10^{-4} mol/L)	4.38	3.71	27.21

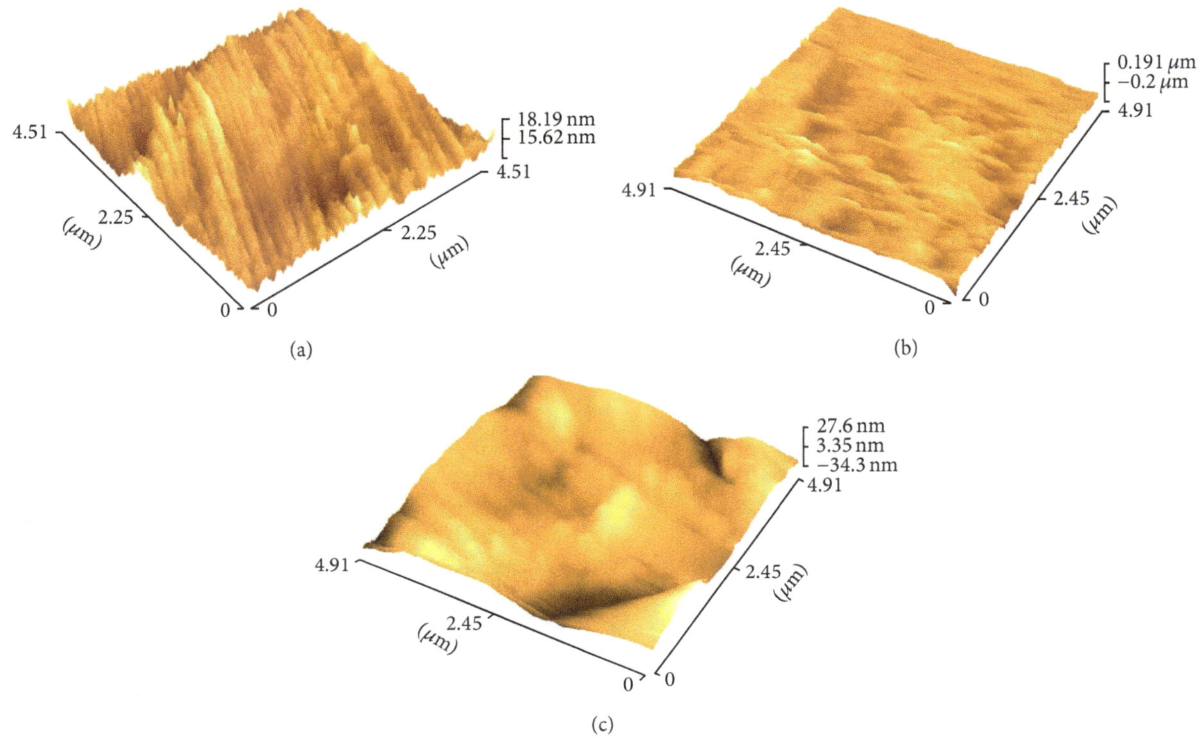

FIGURE 5: Three-dimensional AFM images of the surface of (a) polished carbon steel (control); (b) carbon steel immersed in dam water (blank); (c) carbon steel immersed in dam water containing 1.35×10^{-3} mol/L of SHXS and 1.04×10^{-4} mol/L of Zn^{2+}.

(1) When carbon steel is immersed in this solution SHXS-Zn^{2+} complex diffuses from the bulk of the solution towards the metal surface.

(2) SHXS-Zn^{2+} complex is converted into SHXS-Fe^{2+} complex on the anodic sites of the metal surface with the release of Zn^{2+} ion:

$$Zn^{2+}\text{-SHXS} + Fe^{2+} \longrightarrow Fe^{2+}\text{-SHXS} + Zn^{2+} \quad (7)$$

The released Zn^{2+} combines with OH^- to form $Zn(OH)_2$ on the cathodic sites of the metal surface.

$$Zn^{2+} + 2OH^- \longrightarrow Zn(OH)_2 \downarrow \quad (8)$$

(3) Thus the protective film consists of SHXS-Fe^{2+} complex and $Zn(OH)_2$.

(4) This accounts for the synergistic effect of SHXS-Zn^{2+} system.

4. Conclusions

The present study leads to the following conclusions.

The inhibition efficiency (IE) of SHXS in controlling corrosion of carbon steel immersed in dam water in the absence and presence of Zn^{2+} has been evaluated by weight loss method. The formulation consisting of 1.35×10^{-3} mol/L of SHXS and 1.04×10^{-4} mol/L of Zn^{2+} has 83% IE. Polarization study reveals that SHXS-Zn^{2+} system controls the cathodic reaction predominantly. AC impedance spectra and AFM study reveal that a compact protective film is formed on the metal surface.

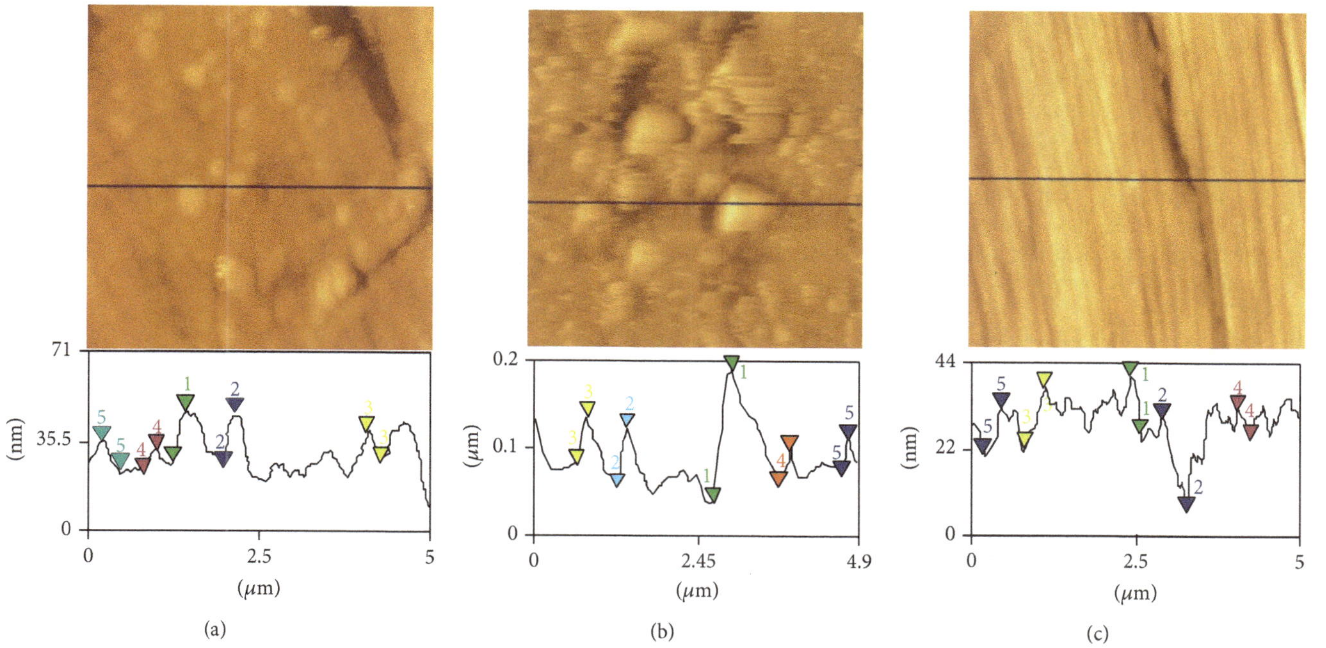

FIGURE 6: AFM cross-sectional images of the surface of (a) polished carbon steel (control); (b) carbon steel immersed in dam water (blank); (c) carbon steel immersed in dam water containing 1.35×10^{-3} mol/L of SHXS and 1.04×10^{-4} mol/L of Zn^{2+}.

Conflict of Interests

The authors declare that there is no conflict of interests regarding the publication of this paper.

Acknowledgment

The authors are thankful to their respective managements for their support and encouragement.

References

[1] Z. B. Vendrame and R. S. Gonçalves, "Electrochemical evidences of the inhibitory action of propargyl alcohol on the electrooxidation of nickel in sulfuric acid," *Journal of the Brazilian Chemical Society*, vol. 9, no. 5, pp. 441–448, 1998.

[2] R. S. Gonçalves and L. D. Mello, "Electrochemical investigation of ascorbic acid adsorption on low-carbon steel in 0.50 M Na_2SO_4 solutions," *Corrosion Science*, vol. 43, no. 3, pp. 457–470, 2001.

[3] R. S. Gonçalves, D. S. Azambuja, and A. M. Serpa Lucho, "Electrochemical studies of propargyl alcohol as corrosion inhibitor for nickel, copper, and copper/nickel 955/45 alloy," *Corrosion Science*, vol. 44, no. 3, pp. 467–479, 2002.

[4] W. X. Oliver and R. S. Gonclaves, "Electrochemical evidence of the protection efficiency of Furfural on the corrosion process of low carbon steel in ethanolic medium," *Journal of Brazil Chemical Society*, vol. 3, pp. 92–94, 1992.

[5] I. B. Obot, N. O. Obi-Egbedi, and S. A. Umoren, "Adsorption characteristics and corrosion inhibitive properties of clotrimazole for aluminium corrosion in hydrochloric acid," *International Journal of Electrochemical Science*, vol. 4, no. 6, pp. 863–877, 2009.

[6] L. M. Vračar and D. M. Draži, "Adsorption and corrosion inhibitive properties of some organic molecules on iron electrode in sulfuric acid," *Corrosion Science*, vol. 44, no. 8, pp. 1669–1680, 2002.

[7] D. Prakash, R. K. Singh, and R. Kumari, "Corrosion inhibition of mild steel in 20% HCl by some organic compounds," *Indian Journal of Chemical Technology*, vol. 13, no. 6, pp. 555–560, 2006.

[8] R. Manickavasagam, K. J. Karthik, M. Paramasivam, and S. V. Iyer, "Poly(styrene sulphonic acid)-doped polyaniline as an inhibitor for the corrosion of mild steel in hydrochloric acid," *Anti-Corrosion Methods and Materials*, vol. 49, no. 1, pp. 19–26, 2002.

[9] T. A. Aliev, "Influence of salts of alkylphenol sulfonic acid on the corrosion of St3 steel in HCl-kerosene systems," *Materials Science*, vol. 44, no. 5, pp. 665–671, 2008.

[10] P. Shakkthivel and T. Vasudevan, "Acrylic acid-diphenylamine sulphonic acid copolymer threshold inhibitor for sulphate and carbonate scales in cooling water systems," *Desalination*, vol. 197, no. 1-3, pp. 179–189, 2006.

[11] G. Wranglen, "Synergistic effect of 2-chloroethyl phosphonic acid and Zn2+," in *Introduction To Corrosion and Protection of Metals*, p. 236, Chapman and Hall, London, UK, 1985.

[12] M. G. Fontana, *Corrosion Engineering*, McGraw-Hill, New Delhi, India, 3rd edition, 2006.

[13] T. Umamathi, J. A. Selvi, S. A. Kanimozhi, S. Rajendran, and A. J. Amalraj, "Effect of Na_3PO_4 on the corrosion inhibition efficiency of EDTA: Zn^{2+} system for carbon steel in aqueous solution," *Indian Journal of Chemical Technology*, vol. 15, no. 6, pp. 560–565, 2008.

[14] S. Rajendran, S. Shanmugapriya, T. Rajalakshmi, and A. J. Amal Raj, "Corrosion inhibition by an aqueous extract of rhizome powder," *Corrosion*, vol. 61, no. 7, pp. 685–692, 2005.

[15] K. Anuradha, R. Vimala, B. Narayanasamy, J. A. Selvi, and S. Rajendran, "Corrosion inhibition of carbon steel in low chloride media by an aqueous extract of Hibiscus rosa-sinensis Linn,"

Chemical Engineering Communications, vol. 195, no. 3, pp. 352–366, 2008.

[16] C. M. Anbarasi and S. Rajendran, "Inhibition of corrosion of carbon steel by Heptane sulphonicacid: Zn^{2+} system," *Journal of Electrchemical Science and Engineering*, vol. 2, no. 1, pp. 1–18, 2012.

[17] C. M. Anbarasi, S. Rajendran, B. Narayanasamy, and A. Krishnaveni, "Surface analysis of inhibitor film formed by Butane sulphonic acid: Zn^{2+} system on carbon steel in aqueous medium," *Asian Journal of Chemistry*, vol. 24, pp. 5029–5034, 2012.

[18] J. Sathiyabama, S. Rajendran, and J. Arockia Selvi, "Eriochrome Black T as corrosion inhibitor for carbon steel in well water," *Bulletin of Electrochemistry*, vol. 22, pp. 363–370, 2006.

[19] J. Arockia Selvi, S. Rajendran, and J. Jeyasundari, "Analysis of nano film by atomic force microscopy," *Zastita Materijala*, vol. 50, pp. 91–98, 2009.

[20] V. Johnsirani, J. Sathiyabama, S. Rajendran, and A. Suriya Prabha, "Inhibitory mechanism of carbon steel corrosion in sea water by an aqueous extract of henna leaves," *ISRN Corrosion*, vol. 2012, Article ID 574321, 9 pages, 2012.

[21] S. Rajendrran, M. Agasta, R. Bama Devi, B. Shyamla Devi, K. Rajam, and J. Jeyasundari, "corrosion inhibition of aqueous extract of Henna leaves," *Zastita Materijala*, vol. 50, pp. 77–84, 2009.

[22] S. Rajendran, M. Manivannan, J. Wilson Sahayaraj et al., "Corrosion behavior of aluminium in methyl orange solution at pH 11," *Transactions of the SAEST*, vol. 41, no. 2, pp. 63–67, 2006.

[23] F. N. Grosser and R. S. Gonçalves, "Electrochemical evidence of caffeine adsorption on zinc surface in ethanol," *Corrosion Science*, vol. 50, no. 10, pp. 2934–2938, 2008.

[24] S. Martinez and M. Metikoš-Huković, "A nonlinear kinetic model introduced for the corrosion inhibitive properties of some organic inhibitors," *Journal of Applied Electrochemistry*, vol. 33, no. 12, pp. 1137–1142, 2003.

[25] M. Elayyachy, A. El Idrissi, and B. Hammouti, "New thio-compounds as corrosion inhibitor for steel in 1 M HCl," *Corrosion Science*, vol. 48, no. 9, pp. 2470–2479, 2006.

[26] S. K. Shukla and M. A. Quraishi, "4-Substituted anilinomethylpropionate: new and efficient corrosion inhibitors for mild steel in hydrochloric acid solution," *Corrosion Science*, vol. 51, no. 9, pp. 1990–1997, 2009.

[27] A. Lalitha, S. Ramesh, and S. Rajeswari, "Surface protection of copper in acid medium by azoles and surfactants," *Electrochimica Acta*, vol. 51, no. 1, pp. 47–55, 2005.

[28] R. M. Silverstein and X. Francis, *Webster Spectrometric Identification of Organic Compounds*, John Wiley and Sons, 6th edition, 2007.

[29] C. M. Anbarasi and S. Rajendran, "Surface protection of carbon steel by butanesulphonic acid-zinc ion system," *Research Journal of Chemical Sciences*, vol. 2, no. 12, pp. 21–26, 2012.

[30] B. Sherine, A. J. Abdul Nasser, and S. Rajendran, "Inhibitive action of hydroquinone: Zn^{2+} System in controlling the corrosion of carbon steel in well water," *International Journal of Engineering Science and Technology*, vol. 2, pp. 341–357, 2010.

[31] A. K. Singh and M. A. Quraishi, "Investigation of the effect of disulfiram on corrosion of mild steel in hydrochloric acid solution," *Corrosion Science*, vol. 53, no. 4, pp. 1288–1297, 2011.

[32] B. Wang, M. Du, J. Zhang, and C. J. Gao, "Electrochemical and surface analysis studies on corrosion inhibition of Q235 steel by imidazoline derivative against CO_2 corrosion," *Corrosion Science*, vol. 53, no. 1, pp. 353–361, 2011.

Adsorption and Corrosion Inhibition Characteristics of Some Nicotinamide Derivatives on Mild Steel in Hydrochloric Acid Solution

M. P. Chakravarthy and K. N. Mohana

Department of Studies in Chemistry, University of Mysore, Manasagangotri, Mysore 570 00, India

Correspondence should be addressed to K. N. Mohana; drknmohana@gmail.com

Academic Editors: I. Obot and M. Z. A. Rafiquee

The present investigation includes the study of corrosion inhibition effect of N-(1H-Indol-3-ylmethylene)-nicotinamide (IMN) and N-(3,4,5-trimethoxy-benzylidene)-nicotinamide (TMN) on mild steel in 0.5 M HCl by mass loss and electrochemical measurements. The obtained results showed that inhibition efficiency increased with the increasing concentration of inhibitors and decreased with increase in temperature and immersion time. Potentiodynamic polarization curves indicate that both IMN and TMN were acting as mixed type of inhibitors. Langmuir adsorption isotherm model was employed to determine the equilibrium of adsorption for inhibiting process in both inhibitors. Nyquist plots revealed that, as the concentration of the inhibitors increases, double layer capacitance (C_{dl}) and corrosion current (i_{corr}) decreases, while polarization resistance (R_p) increases. Various thermodynamic parameters for the adsorption of inhibitors on mild steel were computed and discussed. The passive film formed on the metal surface was characterized by FTIR, EDX, and SEM.

1. Introduction

Mild steel (MS) is cheap, strong, stiff, and widely used engineering material because of its good mechanical properties. It is the most important structural material exposed to outdoor conditions especially in acidic media [1] where corrosion is considerably more severe. Because of the general aggressiveness of acid solutions, inhibitors are commonly used to reduce the corrosive attack on metallic materials [2, 3]. Therefore, selection of an appropriate inhibitor ranging from rare earth elements [4] to organic compounds [5–8] for specific environment and metal is of great importance. In other words, the efficiency of these compounds is attributed to their steric factors, functional groups, lone pairs of electrons present on the hetero atoms, and pi-orbital character of donating electrons which determine the type of interaction between organic molecules and the metallic surfaces. The presence of corrosion inhibitors in a little amount blocks the corrosion sites and enhances the adsorption process, thereby increasing the life time of the metallic materials [9, 10].

A review of the literature reveals that the applicability of organic compounds as corrosion inhibitors for MS in acidic media has been recognized for a long time. A large number of organic compounds, particularly those containing nitrogen, oxygen, or sulphur in a conjugated system, are known to be applied as inhibitors to control acid corrosion of iron and steel. The inhibition process has been shown to occur via inhibitor adsorption, and the efficiency of an inhibitor strongly depends on the structure and chemical characteristics under particular experimental condition.

A large number of organic compounds containing nitrogen, oxygen, or sulphur have been used as inhibitors to control acid corrosion of iron and steel [11, 12]. The adsorption characteristics of organic molecules are also affected by sizes, electron density at the donor atoms, and orbital character of donating electrons [13–16]. Several organic compounds such as thiourea derivatives [17], thiophene derivatives [18], fatty acid oxadiazole derivatives [19], sulphonamide compounds [20], triazines [21], pyridine Schiff bases [22], fluoroquinolones [23], phenylhydrazone derivatives [24], fatty acid

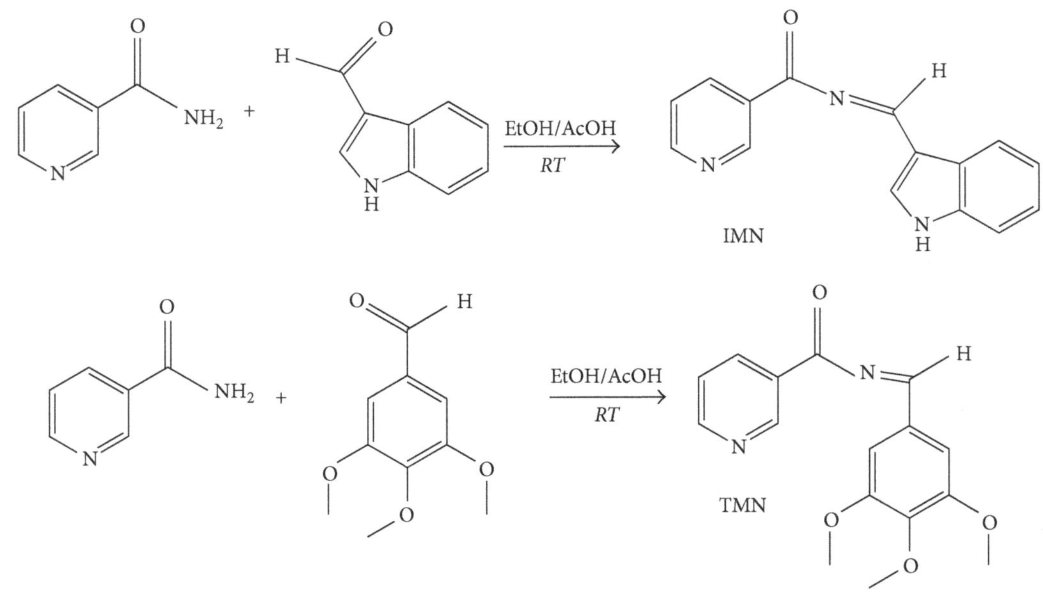

Figure 1: Synthetic schemes of IMN and TMN.

triazole derivatives [25],and lauric hydrazide derivatives [26] have been reported as anticorrosion substances.

In view of the literature cited above, the present work describes the study of inhibitive behaviour and adsorption mechanism of two nicotinamide derivatives such as N-(1H-Indol-3-yl-methylene)-nicotinamide (IMN) and N-(3,4,5-trimethoxy-benzylidene)-nicotinamide (TMN) on the corrosion of MS in 0.5 M HCl using mass loss and electrochemical techniques. The effect of temperature on the dissolution of MS in free and inhibited acid solutions has been studied. Various activation and adsorption thermodynamic parameters were computed. The passive film formed on the metal surface was characterized by FTIR, EDX, and SEM. Further, the inhibition performances of the two derivatives have been compared and discussed.

2. Experimental Procedure

2.1. Mild Steel Specimen Preparation. MS specimens used in the present study having the following chemical composition (in wt%): C 0.051; Mn 0.179; Si 0.006; P 0.005; S 0.023; Cr 0.051; Ni 0.05; Mo 0.013; Ti 0.004; Al 0.103; Cu 0.050; Sn 0.004; B 0.00105; Co 0.017; Nb 0.012; Pb 0.001; and the remainder iron. For all experiments, square type MS specimens of dimension 1 cm × 1 cm × 0.1 cm were used. The specimens were mechanically well abraded with different grades SiC (200–600) emery papers, degreased with benzene, washed with triply distilled water, and finally dried. All solvents and chemicals used were of AR grade and used as such. Triply distilled water was used in the preparation of the various concentrations of test solutions.

2.2. Synthesis of Inhibitors. IMN was synthesized by dissolving 0.61 g (5 mmol) of nicotinamide ($C_6H_6N_2O$, Mol. Wt. 122.12) in 15 mL of ethanol in a round bottom flask. To this 0.72 g (5 mmol) of indole-3-carboxaldehyde (C_9H_7NO, Mol. Wt. 145.16) in 15 mL ethanol was mixed and refluxed for 6 hr at room temperature in the presence of glacial acetic acid, and then the solution was concentrated using rotor vaporizer and kept for drying in vacuum. TMN was synthesized by dissolving 0.61 g (5 mmol) of nicotinamide in 15 mL of ethanol in a round bottom flask. To this a 0.98 g (5 mmol) of 3,4,5-trimethoxybenzaldehyde ($C_{10}H_{12}O_4$, Mol. Wt. 196.20) dissolved in 15 mL of ethanol was added and refluxed for 6 hrs with stirring at room temperature in the presence of glacial acetic acid. Then the solution was concentrated using rotor vaporizer and kept for drying in vacuum and the product obtained was collected. The synthetic scheme of IMN and TMN are shown in Figure 1.

The chemical structures of the compounds are characterized by FTIR and ^{1}H-NMR spectral studies. IMN ($C_{15}H_{11}N_3O$, Mol. Wt. 249.27): yield: 92%, melting range (°C): 148–150. FTIR (Nujol, cm^{-1}): 1462 (N–H), 1611 (N=C), 1677 (C=O). ^{1}H-NMR (400.15 MHz, DMSO-d$_6$) δ ppm: 7.21 (t, J = 7.20 Hz, 1H), 7.49 (t, J = 7.88 Hz, 2H), 7.51 (d, J = 3.28 Hz, 1H), 8.08 (d, J = 8.40 Hz, 2H), 8.18 (s, 1H), 8.20 (d, J = 4.00 Hz, 1H), 8.69 (d, J = 4.44 Hz, 1H), 9.87 (s, 1H), 12.10 (s, 1H). MS, m/z: 250 ($M + 1$). Elemental analyses found (calculated) for $C_{15}H_{11}N_3O$ (%): C, 72.21 (72.28): H, 4.39 (4.45): N, 16.80 (16.86).

TMN ($C_{16}H_{16}N_2O_4$, Mol. Wt. 300.31): yield: 90%, melting range (°C): 162–164. FTIR (Nujol, cm^{-1}): 1589 (N=C), 1701 (C=O). ^{1}H-NMR (400.15 MHz, DMSO-d$_6$) δ ppm: 3.83 (s, 9H), 6.73 (dd, J = 1.73, 7.88, 2H), 7.43 (t, J = 7.48 Hz, 1H), 8.03 (s, 1H), 8.28 (d, J = 7.80 Hz, 2H), 8.73 (s, 1H). MS, m/z: 301 ($M + 1$). Elemental analyses found (calculated) for $C_{16}H_{16}N_2O_4$ (%): C, 63.96 (63.99): H, 5.31 (5.37): N, 9.28 (9.33). Melting range was determined by Veego Melting Point VMP III apparatus. FTIR spectra were recorded using a Jasco FTIR 4100 double beam spectrophotometer. ^{1}H-NMR spectra were recorded on Bruker DRX-500 spectrometer at 400 MHz using DMSO-d_6 as solvent and TMS as an internal standard. Mass spectral data were obtained by LC/MSD Trap XCT. Elemental analyses were recorded on Vario-MICRO superuser V1.3.2 Elementar.

3. Methods

3.1. Mass Loss Measurements. Mass loss measurements were carried out by weighing the prepared cleaned and dried MS specimens before and after immersion in 0.5 M HCl solutions from one to five hours in the absence and presence of various concentrations of IMN and TMN at different temperatures (30–60°C). Triplicate experiments were performed in each case and the mean value of the mass loss was noted. From the mass loss data, corrosion rate (CR) in $\mathrm{mg\,cm^{-2}\,h^{-1}}$ and inhibition efficiency IE (%) were calculated using the following equations:

$$\mathrm{CR} = \frac{\Delta M}{St}, \quad (1)$$

where ΔM is the mass loss, "S" is the surface area of the specimen, and "t" is the immersion time.
Consider

$$\mathrm{IE}\,(\%) = \frac{(\mathrm{CR})_a - (\mathrm{CR})_p}{(\mathrm{CR})_a} \times 100, \quad (2)$$

where $(\mathrm{CR})_a$ and $(\mathrm{CR})_p$ are the corrosion rates in the absence and presence of inhibitor, respectively.

3.2. Potentiodynamic Polarization Measurements. The potentiodynamic polarization studies were carried out with MS specimen as working electrode in 0.5 M HCl solutions with different inhibitor's concentrations (200–500 ppm) with an exposed area of $1\,\mathrm{cm}^2$ and this working area remained precisely fixed throughout the experiment. All studies were made using three electrode cells consisting of MS sample as working electrode (WE), platinum counter electrode (CE), and saturated calomel electrode (SCE) as reference electrode. The temperature of the medium was maintained at 30°C using thermostatically controlled water bath. Potentiodynamic polarization studies were carried out using CH-instrument (model CHI660D). Before each experiment, the MS electrode was allowed to corrode freely and its open circuit potential (OCP) was recorded as a function of time upto 30 min. After this time, a steady state OCP corresponding to the corrosion potential (E_{corr}) of the working electrode was obtained. The polarization curves were recorded by changing the electrode potential automatically at a scan rate of 0.2 m V/s. The IE (%) was calculated from corrosion currents determined from the Tafel extrapolation plot method using the following relation:

$$\mathrm{IE}\,(\%) = \frac{(I_{\mathrm{corr}})_a - (I_{\mathrm{corr}})_p}{(I_{\mathrm{corr}})_a} \times 100, \quad (3)$$

where $(I_{\mathrm{corr}})_a$ and $(I_{\mathrm{corr}})_p$ are the corrosion current density ($\mu\mathrm{A\,cm^{-2}}$) in the absence and presence of the inhibitor, respectively.

3.3. Electrochemical Impedance Spectroscopy. Electrochemical impedance measurements were carried out in the form of Nyquist plots using the same CH-instrument. The electrochemical impedance spectroscopy (EIS) data were taken in the frequency range 10 kHz to 100 mHz. The double layer capacitance (C_{dl}) and the polarization resistance (R_P) were determined from Nyquist plots. The percentage inhibition efficiency, IE (%), was calculated from R_P values using the following expression:

$$\mathrm{IE}\,(\%) = \frac{1/(R_P)_a - 1/(R_P)_p}{1/(R_P)_a} \times 100, \quad (4)$$

where $(R_P)_a$ and $(R_P)_p$ are polarization resistances in the absence and presence of inhibitor, respectively.

3.4. FTIR, EDX, and SEM Studies. The surface characterization of uninhibited and inhibited MS specimens were carried out using FTIR, EDX, and SEM studies. The prepared MS specimens were immersed in 0.5 M HCl in the presence of inhibitors (500 ppm) for a period of 5 hr. Then the specimens were taken out and dried. The thin film that adhered on the MS surface was scrapped carefully and its FTIR spectra were recorded. Further, surface morphology was studied by energy dispersive X-ray spectroscopy (EDX) and scanning electron microscope (model JSM-5800) in the absence and the presence of inhibitors.

4. Results and Discussion

4.1. Mass Loss Studies. The CR and IE (%) in the absence and presence of various concentrations (200–500 ppm) of IMN and TMN in 0.5 M HCl solution and at different temperatures (30–60°C) were calculated by mass loss experiments, and the data were presented in Table 1. Investigation of mass loss data revealed the linear variation of mass loss with temperature, concentration, and time in inhibited and uninhibited solution. The mass loss was found to decrease and IE (%) observed to increase with increase in concentration of nicotinamide derivatives and the maximum inhibition efficiency was found at 500 ppm. This observation is also supported by electrochemical studies. After 6 hr of immersion time there is no any appreciable increase in the inhibition efficiency, this is due to desorption of the inhibitor molecules from the MS surface and instability of inhibitor film on the metal surface [27, 28].

The inhibition efficiency of the nicotinamide derivatives was found to depend on the concentration and the nature of the substituents. The increase in concentration of the inhibitors was accompanied by decrease in weight loss and increase in the inhibition efficiency. The inhibition efficiency obtained from mass loss measurements is lower than that for electrochemical experiments, because the mass loss experiment gives average corrosion rates whereas the electrochemical experiments gives instantaneous corrosion rates.

4.2. Effect of Temperature. The effect of temperature on CR and IE (%) was studied in 0.5 M HCl in the temperature range of 30–60°C in the absence and presence of different

TABLE 1: CR and IE (%) obtained from mass loss measurements of MS in 0.5 M HCl solution containing various concentrations of IMN and TMNat different temperatures.

T (°C)	C (ppm)	IMN CR (mg cm^{-2} h^{-1})	IMN IE (%)	TMN CR (mg cm^{-2} h^{-1})	TMN IE (%)
30	Blank	0.7200	—	0.7200	—
	200	0.2096	70.9	0.1262	82.5
	300	0.1555	78.4	0.0770	89.3
	400	0.1430	80.2	0.0644	91.1
	500	0.1123	84.4	0.0386	94.6
40	Blank	0.9490	—	0.9490	—
	200	0.2820	70.3	0.1783	81.2
	300	0.2412	74.6	0.1407	85.2
	400	0.2008	78.8	0.1005	89.4
	500	0.1570	83.5	0.0664	93.0
50	Blank	1.1520	—	1.1520	—
	200	0.3617	68.6	0.2357	79.5
	300	0.3100	73.1	0.1832	84.1
	400	0.2671	76.8	0.1436	87.5
	500	0.2214	80.8	0.0945	91.8
60	Blank	1.4350	—	1.4350	—
	200	0.4791	66.6	0.3292	77.1
	300	0.3916	72.7	0.2381	83.4
	400	0.3649	74.6	0.2105	85.3
	500	0.2913	79.7	0.1427	90.1

concentrations of inhibitors (Table 1). The inhibition efficiencies were found to decrease with increasing temperature from 30–60°C. Desorption of inhibitors is aided by increasing temperature. This proves that the inhibition occurs through the adsorption of the inhibitors on the metal surface. The corrosion current density increases with the rise of temperature and markedly pronounced in the absence of inhibitors. The activation parameters for the corrosion process were calculated from the Arrhenius type plot according to the following equation:

$$\text{CR} = k \exp^{-E_a/RT}, \quad (5)$$

where E_a is the apparent activation energy for corrosion process k is the Arrhenius preexponential factor, T is the absolute temperature, and R is the universal gas constant. The values of E_a for MS in 0.5 M HCl without and with various concentrations of inhibitors are obtained from the slopes of the plots of log CR versus $1/T$ (Figure 2) and are shown in Table 2. E_a values for inhibited systems are higher than those for the uninhibited systems suggest that dissolution of MS is slow [29]. Further, as the concentration of the inhibitors increases the values of E_a also increases. This means the presence of the inhibitors induces an energy barrier for corrosion reaction and the barrier increases with increasing concentration. At higher temperatures, there is an appreciable decrease in the adsorption of the inhibitors on the metal surface and a corresponding rise in the corrosion rate occurred.

Alternative Arrhenius plots of log CR/T versus $1/T$ (Figure 3) for MS dissolution in HCl medium in the absence and presence of different concentrations of IMN and TMN were used to calculate the values of activation thermodynamic parameters such as enthalpy of activation (ΔH_a) and entropy of activation (ΔS_a) using the following relation:

$$\text{CR} = \frac{RT}{Nh} \exp\left(\frac{\Delta S_a}{R}\right) \exp\left(\frac{-\Delta H_a}{\text{RT}}\right), \quad (6)$$

where R is the universal gas constant, T is the absolute temperature, N is the Avogadro's number, and h is Planks constant. The values of ΔH_a and ΔS_a were obtained from the slope and intercept of the above plot and presented in Table 2. The obtained ΔH_a values are in good agreement with the calculated values from the following equation:

$$\Delta H_a = E_a - RT. \quad (7)$$

The positive shift of enthalpy of activation (ΔH_a) with different inhibitors concentration reflects that the process of adsorption of the inhibitors on the MS surface is an endothermic process [30], and the negative values of entropy of activation (ΔS_a) represents association rather than dissociation of inhibitors indicating the decrease of system disorder due to the adsorption of inhibitor molecules on the MS surface [31–33]. The values of ΔS_a are higher for inhibited solutions than those for the uninhibited solution reflecting an increase in randomness on going from reactants to the activated complex. The increase in values of ΔS_a by the adsorption of inhibitor molecules on the MS surface could be regarded as quasisubstitution between the inhibitors molecules in the aqueous phase and water molecules on electrode surface [34]. In such condition, the adsorption of inhibitor molecules follow desorption of water molecules from the electrode surface and hence decrease the electrical capacity of MS.

4.3. Adsorption Isotherm. The efficiency of a corrosion inhibitor mainly depends on its adsorption ability on the metal surface. So, it is necessary to know the mechanism of adsorption and the adsorption isotherm that can give valuable information on the interaction of inhibitor and metal surface. The surface protection of MS depends upon how the inhibitor molecule will adsorbed on the metal surface and also ionization and polarization of molecules [35]. The degree of surface coverage (θ) as function of concentration (C) of the inhibitor was studied graphically by fitting it to various adsorption isotherms to find the best adsorption isotherm. Langmuir adsorption isotherm was found to be the best description for all the studied inhibitors on MS in 0.5 M HCl medium. According to this adsorption isotherm, θ is related to the inhibitor concentration, C, and adsorption equilibrium constant K_{ads} through the following expression:

$$\frac{C}{\theta} = \frac{1}{K_{\text{ads}}} + C. \quad (8)$$

TABLE 2: Values of activation parameters for MS in 0.5 M HCl medium in the absence and presence of various concentrations IMN and TMN.

Inhibitor	C (ppm)	E_a (kJ mol^{-1})	ΔH_a (kJ mol^{-1})	$\Delta H_a = E_a - RT$ (kJ mol^{-1})	ΔS_a (J mol^{-1} K^{-1})
Blank	0	19.0	16.4	16.5	−197.6
IMN	200	22.9	20.2	20.4	−191.3
	300	25.4	22.8	22.9	−184.9
	400	25.9	23.3	23.4	−184.3
	500	26.9	24.2	24.4	−183.2
TMN	200	26.4	23.8	23.92	−183.7
	300	30.8	28.1	28.23	−172.9
	400	32.8	30.2	30.29	−168.2
	500	35.9	33.3	33.38	−162.1

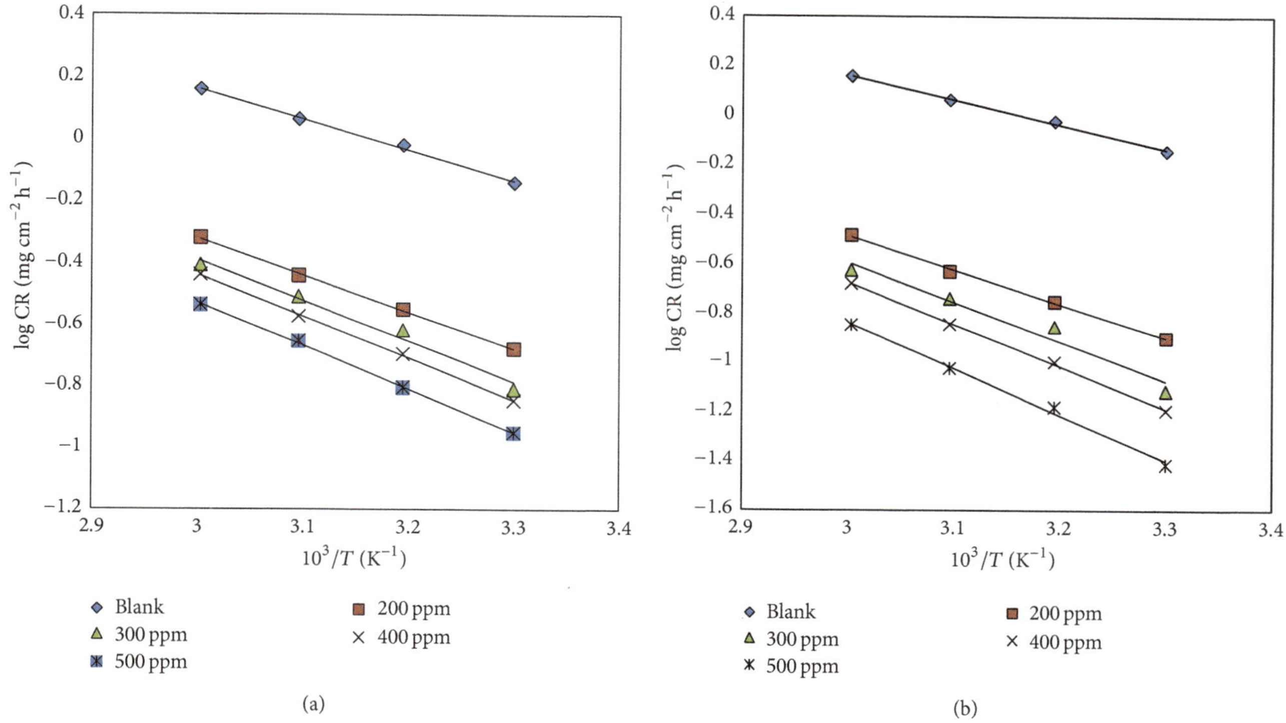

FIGURE 2: Plot of log CR versus $1/T$ for (a) IMN and (b) TMN.

The surface coverage was tested graphically by fitting a suitable adsorption isotherm. In the present case, the plots of C/θ versus C (Figure 4) yields straight lines with the linear correlation coefficient (R^2) values close to unity, which suggests that the adsorption of IMN and TMN in 0.5 M HCl medium on MS surface obeys the Langmuir adsorption isotherm. The slope ranges of these lines are 1.047–1.105 (IMN) and 0.960–0.996 (TMN) in the temperature range of 30–60°C, respectively, suggesting that the adsorbed molecules form monolayer on the MS surface and there is no interaction among the adsorbed inhibitor molecules. The free energy of adsorption was calculated using the following relation:

$$\Delta G^{\circ}_{\text{ads}} = -2.303RT \log 55.5K_{\text{ads}}, \quad (9)$$

where R is the universal gas constant, T is the absolute temperature, K_{ads} is the equilibrium constant for adsorption-desorption process, and 55.5 is the molar concentration of water in solution (mol L^{-1}). The other adsorption thermodynamic parameters such as enthalpy of adsorption ($\Delta H^{\circ}_{\text{ads}}$) and entropy of adsorption ($\Delta S^{\circ}_{\text{ads}}$) were obtained from the slope and intercept of the plot of lnK_{ads} versus $1/T$ (Figure 5) using

$$\log K_{\text{ads}} = \frac{1}{2.303}\left(-\frac{\Delta H^{\circ}_{\text{ads}}}{RT}\right) + \left(\frac{\Delta S^{\circ}_{\text{ads}}}{R}\right). \quad (10)$$

The calculated values of K_{ads}, $\Delta H^{\circ}_{\text{ads}}$, $\Delta G^{\circ}_{\text{ads}}$, and $\Delta S^{\circ}_{\text{ads}}$ over the temperature range 30–60°C are recorded in Table 3. Lagrenée et al. have reported that the higher the K_{ads} values, the stronger and more stable adsorbed layer is forming which results in the higher inhibition efficiency [36]. The

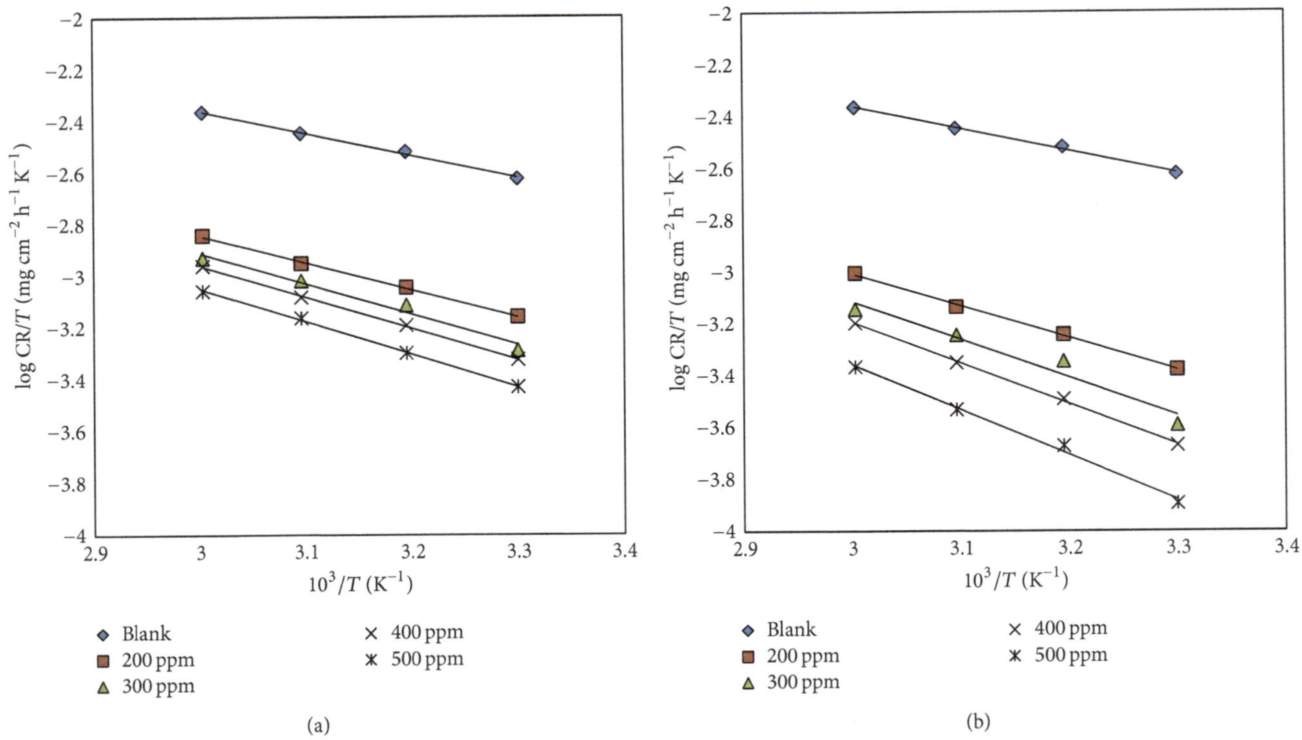

FIGURE 3: Alternative Arrhenius plots for MS dissolution in 0.5 M HCl medium in the absence and presence of (a) IMN and (b) TMN.

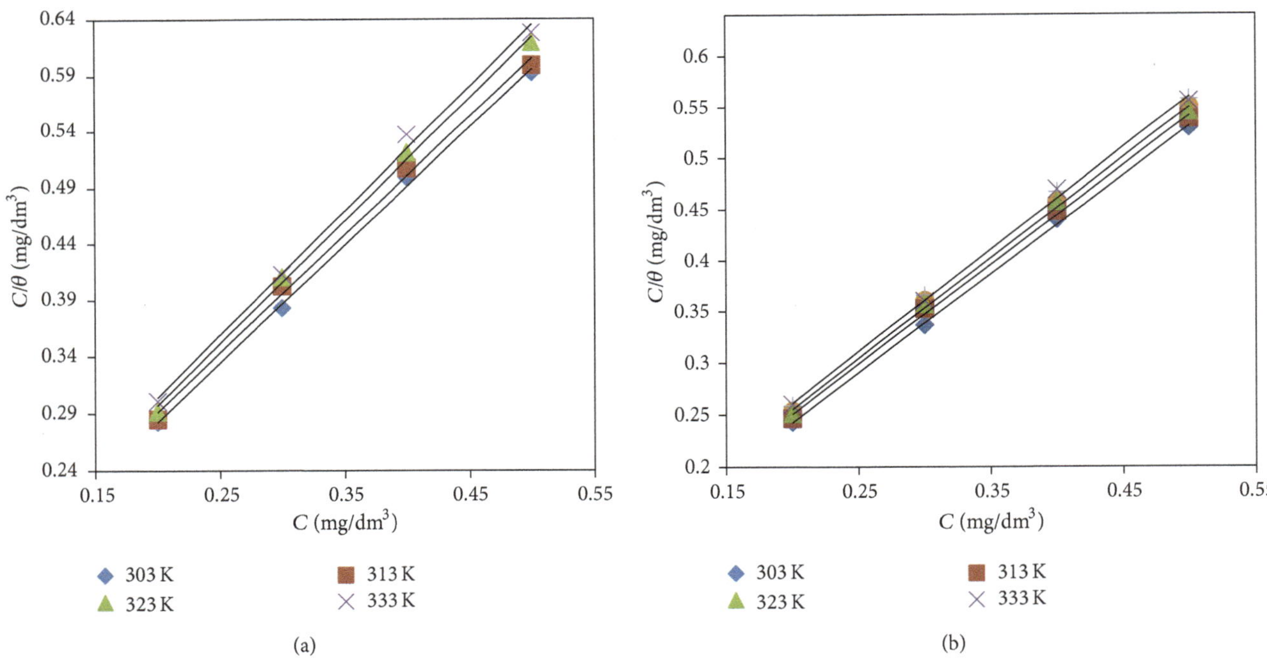

FIGURE 4: Langmuir's adsorption isotherm plots for the adsorption of (a) IMN and (b) TMN in 0.5 M HCl on MS surface at different temperatures.

negative values of ΔG°_{ads} indicate the spontaneous adsorption of inhibitor on the MS surface [37]. The values of ΔG°_{ads} are associated with water adsorption/desorption equilibrium which forms an important part in the overall free energy changes. In the present study, ΔG°_{ads} values for IMN and TMN are found to be in the range −34.15 to −37.17 and −35.07 to −37.94 kJ mol^{-1} in the temperature range of 30–60°C, respectively, which indicates charge transfer or electron sharing from the inhibitor molecules to the metal surface to form a coordinate type of bond [38, 39]. The calculated standard

TABLE 3: Thermodynamic parameters for adsorption of IMN and TMNon MS in 0.5 M HCl at different temperatures.

Inhibitor	T (°C)	R^2	Slope	K_{ads} ($L\,mol^{-1}$)	$-\Delta G^{\circ}_{ads}$ ($kJ\,mol^{-1}$)	ΔH°_{ads} ($kJ\,mol^{-1}$)	ΔS°_{ads} ($J\,mol^{-1}\,K^{-1}$)
IMN	30	0.998	1.047	13888	−34.2	0.405	8.18
	40	0.997	1.048	12820	−35.1		
	50	0.998	1.092	12345	−36.1		
	60	0.996	1.105	12195	−37.2		
TMN	30	0.999	0.960	20000	−35.1	0.653	7.72
	40	0.998	0.969	17857	−35.9		
	50	0.998	0.980	16949	−36.9		
	60	0.998	0.996	16129	−37.9		

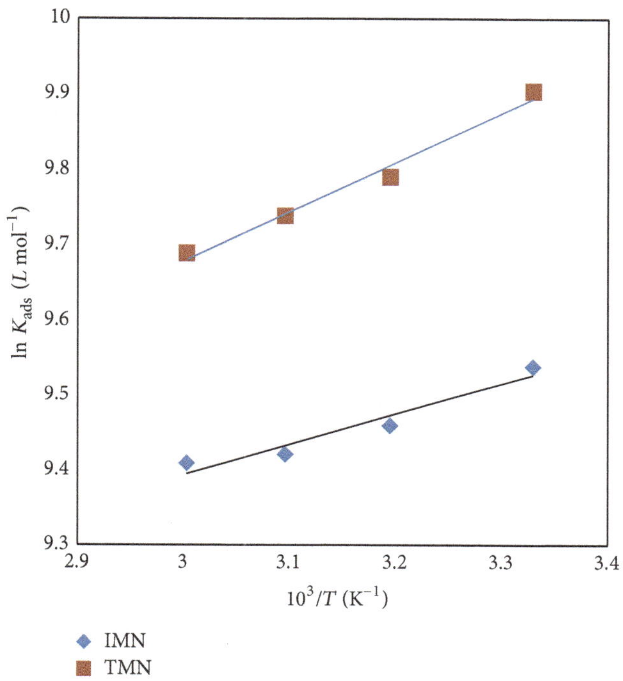

FIGURE 5: Plot of $\ln K_{ads}$ versus $1/T$ for (a) IMN and (b) TMN.

free energy of adsorption values was nearer to $-40\,kJ\,mol^{-1}$ indicating that the adsorption is more physisorption than chemisorption adsorption [40, 41].

4.4. IR Spectral Studies. FTIR spectra were recorded to understand the interaction of inhibitor molecules with the metal surface. Figures 6(a) and 7(a) show the IR spectra of pure IMN and TMN. Figures 6(b) and 7(b) represent the IR spectra of the scrapped samples obtained from the metal surfaces after corrosion experiments. It was found that peaks in the spectrum of pure compounds were changed in the spectrum of scrapped samples. The azomethine group stretching frequency for pure IMN and TMN were found to be at 1611 and 1589 cm^{-1}, and carbonyl stretching frequency were observed at 1677 and 1701 cm^{-1}, respectively. In the IR spectra of scrapped samples, the stretching frequencies of the azomethine group were found to disappear in both cases and carbonyl group was found to disappear in IMN and found to be shifted to 1663 cm^{-1} in case of TMN. These observations confirm that the azomethine group and carbonyl group of IMN and TMN are involved in the complex formation with the metal.

TABLE 4: Polarization parameters and corresponding inhibition efficiency (IE) for the corrosion of the MS in 0.5 M HCl without and with various concentrations of IMN and TMNat 30°C.

Inhibitor	C (ppm)	$-E_{corr}$ (mV)	i_{corr} ($\mu A\,cm^{-2}$)	IE (%)
Blank	0	613	1061	—
IMN	200	605	293.8	72.3
	300	592	230.1	78.3
	400	586	162.4	84.7
	500	511	160.5	84.9
TMN	200	531	187.1	82.4
	300	522	121.2	88.6
	400	515	86.49	91.8
	500	503	56.89	94.6

4.5. Potentiodynamic Polarization Studies. Polarization measurements have been carried out in order to gain knowledge regarding the kinetics of anodic and cathodic reactions. The obtained polarization curves for MS in 0.5 M HCl in the absence and presence of different concentrations of TMN and IMN at 30°C are shown in Figure 8. Tafel extrapolation plots showed that the addition of inhibitors hindered the acid attack on the MS electrode. The addition of inhibitors reduces both anodic and cathodic current densities indicating that these inhibitors exhibit cathodic and anodic inhibition effects; hence, they are relatively mixed type of inhibitors [42, 43].

The lower values of corrosion current densities facilitates slow electron transfer in the redox process; therefore, the rate of corrosion reaction becomes slower. Usually a low current density and long anodization time present a very good protection against corrosion due to the diminution of the porosity of the anodic films formed. Inspection of data in Table 4 clearly shows that as the concentration of the inhibitors increases there is a gradual decrease in the values of the corrosion potential and corrosion current. The values associated with electrochemical polarization parameters such

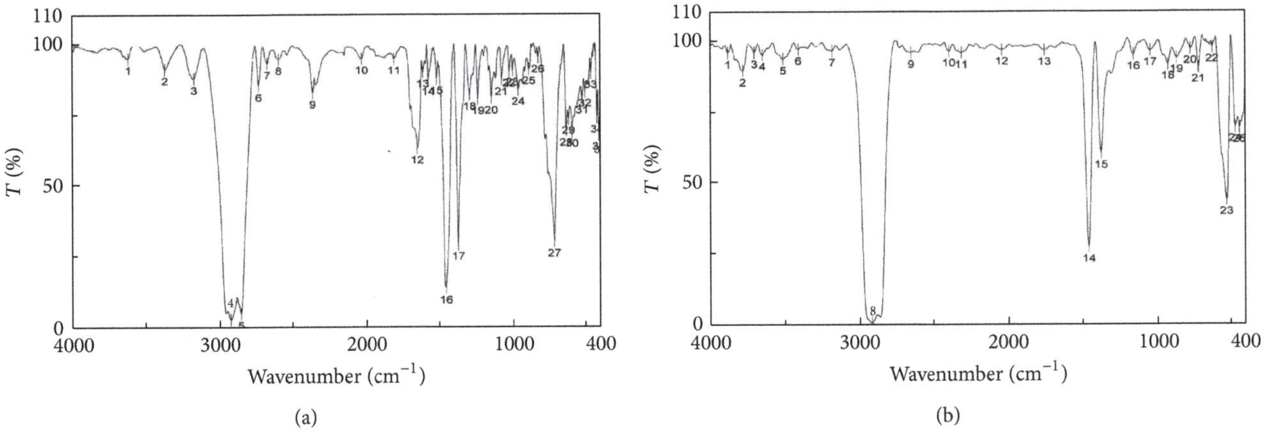

FIGURE 6: FTIR spectra of (a) IMN and (b) scratched MS surface adsorbed IMN film.

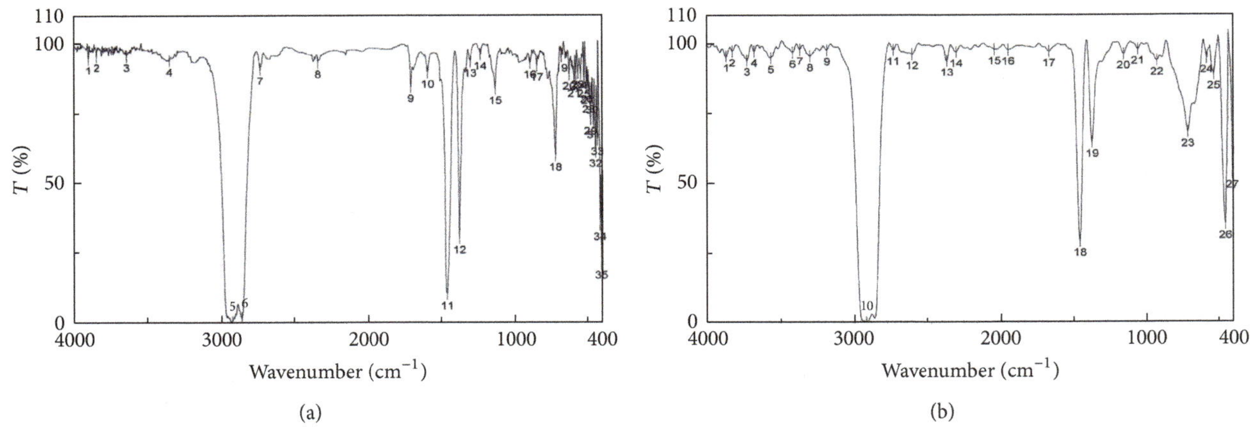

FIGURE 7: FTIR spectra of (a) TMN and (b) scratched MS surface adsorbed TMN film.

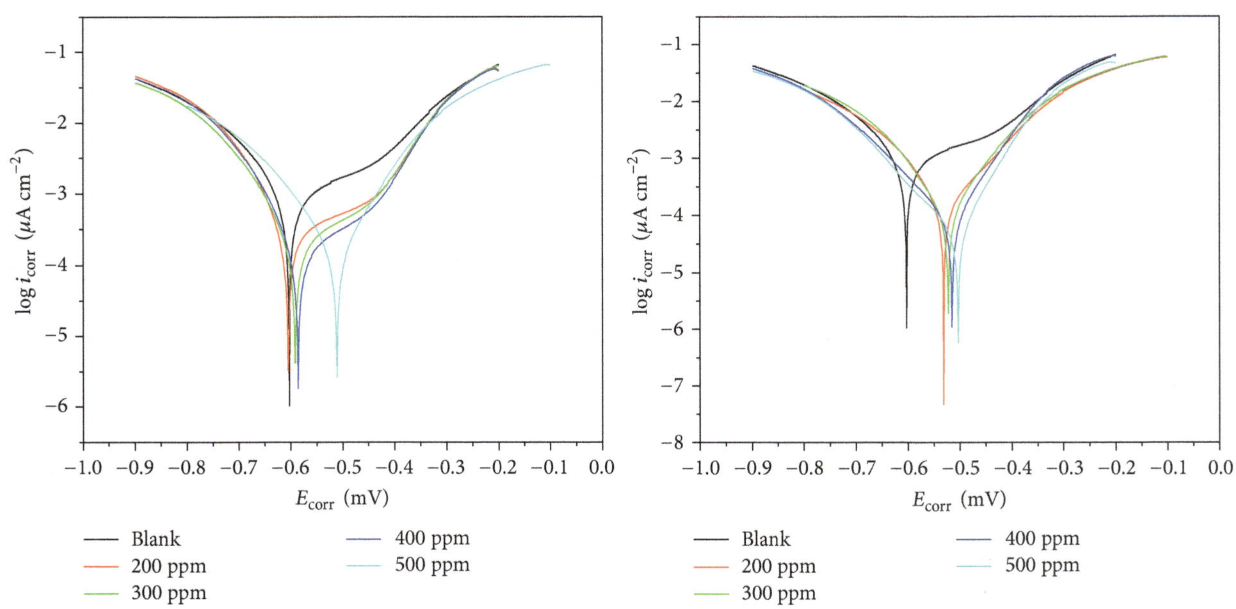

FIGURE 8: Polarization curves of MS in 0.5 M HCl in the presence of different concentrations of IMN and TMN.

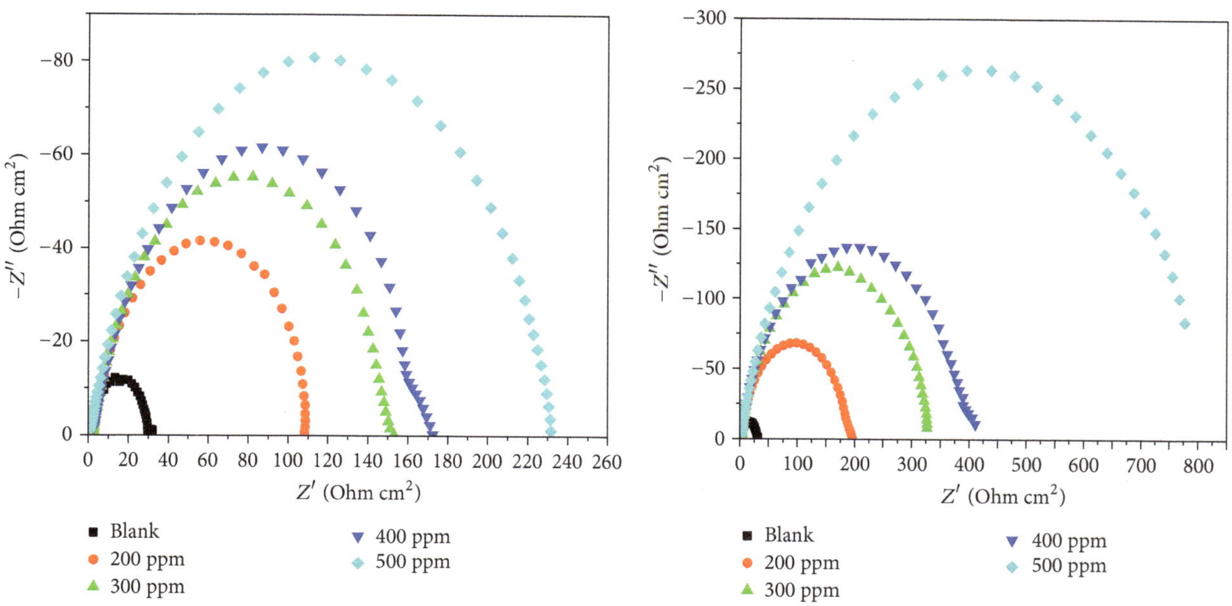

FIGURE 9: Nyquist plots for MS in 0.5 M HCl in the presence of different concentrations of IMN and TMN.

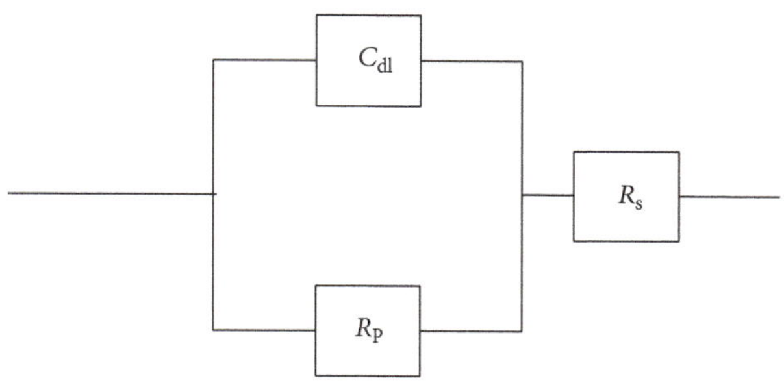

FIGURE 10: Equivalent circuit used to fit the impedance spectra.

as corrosion current density (i_{corr}), corrosion potential (E_{corr}), corrosion rate, and IE (%) were determined from the polarization plots as given in Table 4. The values of corrosion current densities were obtained by extrapolating the current-potential lines to the corresponding corrosion potentials. It is evident that IE (%) increases with inhibitors concentration, and protection action of IMN and TMN can be attributed to the electron density of the azomethine (–C=N–) group and this electron density varies with the substituents in the inhibitor molecules. The imine nitrogen can donate the lone pair of electrons to the metal surface more easily and hence reduce the corrosion rate. The IE (%) of the TMN is higher than that of IMN, which can probably be explained on the basis of the additional functional groups and also the nature of the hetero atoms in the TMN.

TABLE 5: Impedance parameters for the corrosion of MS in 0.5 M HCl in the absence and presence of different concentrations of IMN and TMN.

Inhibitor	C (ppm)	R_p (Ω cm^2)	C_{dl} (μF cm^{-2})	IE (%)
Blank	0	27.81	320.2	—
IMN	200	95.75	102.4	71.0
	300	128.4	91.07	78.3
	400	140.8	77.99	80.2
	500	181.5	61.45	84.7
TMN	200	160.0	81.34	82.6
	300	271.1	67.61	89.7
	400	316.1	48.73	91.2
	500	536.0	41.43	94.8

4.6. Electrochemical Impedance Spectroscopy. The Nyquist plots for MS in 0.5 M HCl solution with and without different concentrations of IMN and TMN are shown in Figure 9. The Nyquist plots were regarded as one part of a semicircle mostly referred to as frequency dispersion which could be attributed to different physical phenomenon such as roughness, heterogeneities, impurities, grain boundaries, and distribution of the surface active sites [44]. The electrochemical impedance parameters derived from the Nyquist plots and IE (%) are listed in Table 5. From the plots it is clear that the impedance response of MS in uninhibited acid solution has significantly changed after the addition of inhibitors to the

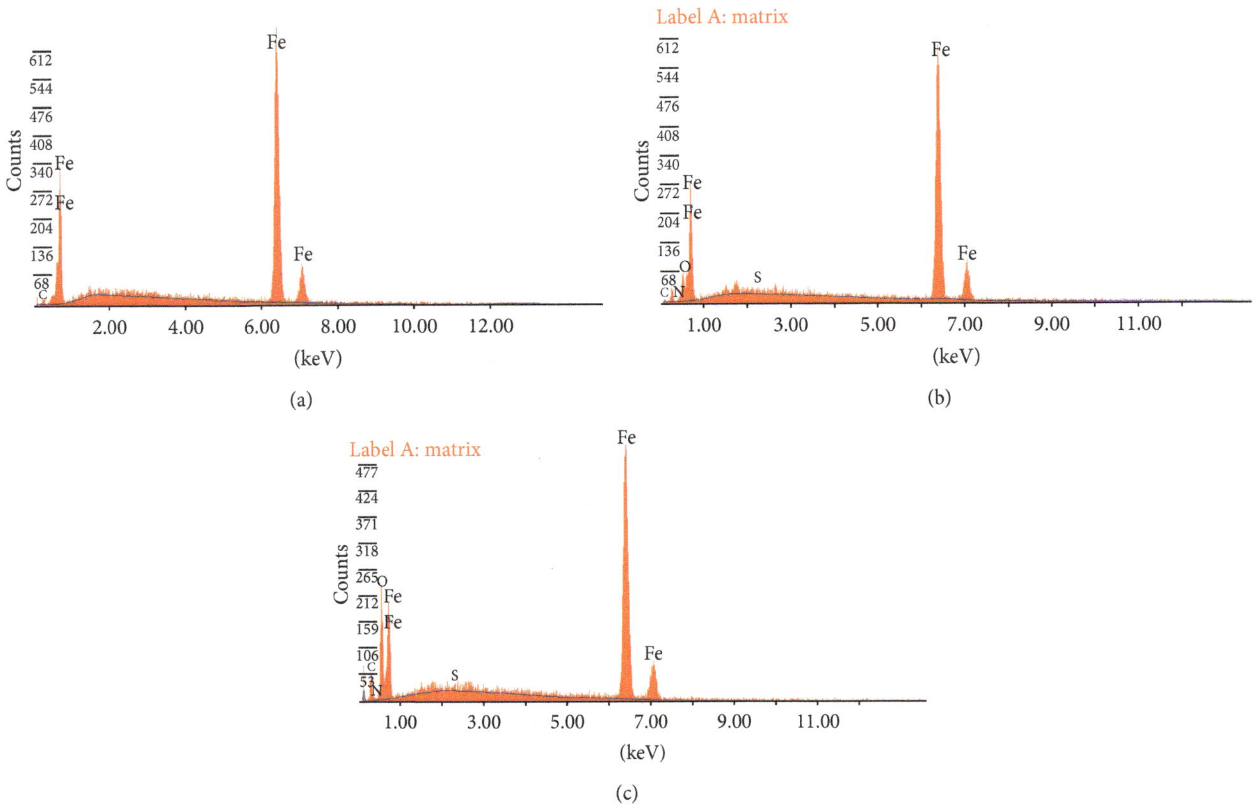

Figure 11: EDX images of (a) abraded MS surface, (b) MS in 500 ppm IMN, and (c) MS in 500 ppm TMN.

corrosive solution. This indicates that the impedance of the inhibited metal has increased with increasing concentration of inhibitors. The measured impedance data were based upon the equivalent circuit given in Figure 10, consists of constant double layer capacitance (C_{dl}) in parallel with charge transfer resistance (R_P) which is in series with solution resistance (R_s).

The value of R_P is a measure of electron transfer across the surface and inversely proportional to the corrosion rate. It was clear that R_P values in the absence of the inhibitors were always lower than those in the presence of the inhibitors. The increase in the R_P values in the presence of different concentrations of IMN and TMN indicate reduction in the MS corrosion rate with the formation of adsorbed protective film on the metal-solution interface [45, 46]. When the concentration was raised from 200 to 500 ppm, there was a gradual increase in the diameter of each semicircle of the Nyquist plot reflecting increasing R_P values from 27.81 $\Omega\,cm^2$ to 181.5 $\Omega\,cm^2$ and 536.0 $\Omega\,cm^2$ for IMN and TMN, respectively, which indicates the adsorption of inhibitor molecules on the metal surface.

As the adsorption of the inhibitors at the metal-solution interface increases, there is a gradual decrease in the double layer capacitance (C_{dl}) and corrosion current (i_{corr}) values indicating decrease in local dielectric constant of the electrical double layer [47–49]. The inhibitor decreases the C_{dl} values by replacing water molecules present at the electrode surface [50]. It was clear that as the immersion time increases the R_P values increases and C_{dl} values decreases which indicate the higher protection efficiency as a result of slow adsorption of these inhibitor molecules on to the surface of MS. However, when the immersion time is further enhanced, a sudden decrease in R_P values and increase in C_{dl} values were observed. This behaviour can be due to the instability of the passive film or desorption of the inhibitor molecules.

4.7. Mechanism of Inhibition. The inhibition effect of nicotinamide derivatives towards the corrosion of MS in 0.5 M HCl solution may be attributed to the adsorption of these compounds at the metal-solution interface. The adsorption of inhibitor is influenced by the nature of the metal, chemical structure of inhibitors, type of aggressive electrolyte, temperature, and the morphology of MS [51, 52]. The values of inhibition efficiency depend essentially on the electron density at the active centre of the inhibitor molecule. Chemisorption of these inhibitors arises from the donor-acceptor interactions between the free electron pairs of hetero atoms and π electrons of multiple bonds as well as phenyl group and vacant d-orbitals of iron [53, 54].

In the case of IMN, the inhibition effect is due to the interaction of π-electrons of phenyl and pyridine rings as well as the presence of electron donor groups (N, O and C=N) through which it forms bonds with MS. In the similar way the inhibition effect of TMN is due to π-electrons of phenyl and pyridine rings, presence of N, O, C=N, and $-OCH_3$, through which the inhibitors adsorb on the MS surface forming insoluble, stable, and uniform thin film. The highest inhibition

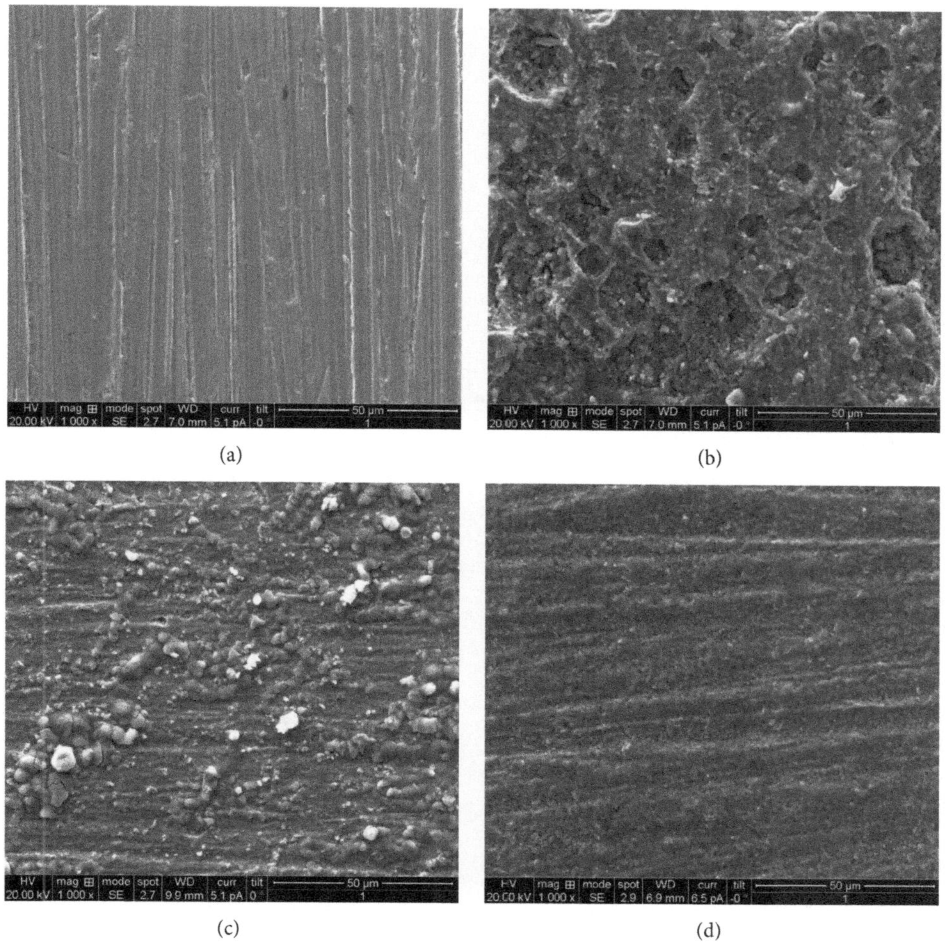

FIGURE 12: SEM images of (a) abraded MS surface, (b) MS in 0.5 M HCl, (c) MS in 500 ppm IMN, and (d) MS in 500 ppm TMN.

efficiency of TMN is due to the presence of electron donating three methoxy groups adjacent to the phenyl ring, which provides a high electron density and it was found that IE% increases with electron density.

4.8. EDX Analysis. EDX spectra were used to determine the elements present on MS surface before and after exposure to the inhibitor solution. The results are displayed in Figure 11. Figure 11(a) is the EDX spectrum of abraded MS sample and it is clear that the peak of oxygen is absent which confirm the absence of air formed oxide film. However, for inhibited solutions (Figures 11(b) and 11(c)) the presence of additional lines is a characteristics for the existence of N and O in the EDX spectra. These data showed that the N and O atom of inhibitors are involved in formation of bonds with the MS electrode surface. These results confirm those obtained from IR and SEM measurements.

4.9. SEM Analysis. The surface morphology of MS due to corrosion process was confirmed by the SEM images of the abraded and corroded MS surface in the absence and presence of inhibitors (Figures 12(a)–12(d)). Figure 12(a) represents the SEM image for abraded MS surface. Figure 12(b) is the SEM image of MS surface in 0.5 M HCl without inhibitor. It was found that the corroded MS surface contains large number of pits. However, SEM images of MS surface in the presence of inhibitors were observed to be smoother than that of MS surface in 0.5 M HCl alone (Figures 12(c) and 12(d)). These observations reveal that, the inhibitors form protective layer on the MS surface, which prevents the attack of acid as well as the dissolution of MS by forming surface adsorbed layer and thereby reducing the corrosion rate.

5. Conclusions

(i) The synthesized nicotinamide derivatives showed good inhibition efficiency for the corrosion of MS in 0.5 M HCl solutions and the inhibition efficiency was found to increase with increasing concentration of the inhibitors.

(ii) Mass loss and electrochemical measurements are in good agreement with each other and the inhibition efficiency of TMN is higher than IMN.

(iii) Langmuir adsorption isotherm model was employed to determine the equilibrium of adsorption for inhibiting process in both inhibitors.

(iv) Differences in inhibition efficiency between IMN and TMN are correlated with their chemical structures.

(v) The passive film formed on the metal surface was characterized by FTIR, EDX, and SEM.

Conflict of Interests

The authors declare that there is no conflict of interests regarding the publication of this paper.

Acknowledgment

One of the authors M. P. Chakravarthy is grateful to the University of Mysore, Mysore, for awarding SRF to carry out the research work.

References

[1] A. M. Fekry and R. R. Mohamed, "Acetyl thiourea chitosan as an eco-friendly inhibitor for mild steel in sulphuric acid medium," *Electrochimica Acta*, vol. 55, no. 6, pp. 1933–1939, 2010.

[2] H.-L. Wang, R.-B. Liu, and J. Xin, "Inhibiting effects of some mercapto-triazole derivatives on the corrosion of mild steel in 1.0 M HCl medium," *Corrosion Science*, vol. 46, no. 10, pp. 2455–2466, 2004.

[3] K. C. Emregül, R. Kurtaran, and O. Atakol, "An investigation of chloride-substituted Schiff bases as corrosion inhibitors for steel," *Corrosion Science*, vol. 45, no. 12, pp. 2803–2817, 2003.

[4] M. A. Arenas, A. Conde, and J. J. De Damborenea, "Cerium: a suitable green corrosion inhibitor for tinplate," *Corrosion Science*, vol. 44, no. 3, pp. 511–520, 2002.

[5] E. Cano, P. Pinilla, J. L. Polo, and J. M. Bastidas, "Copper corrosion inhibition by fast green, fuchsin acid and basic compounds in citric acid solution," *Materials and Corrosion*, vol. 54, no. 4, pp. 222–228, 2003.

[6] D.-J. Choi, Y.-W. Kim, and J.-G. Kim, "Development of a blended corrosion, scale, and microorganism inhibitor for open recirculating cooling systems," *Werkstoffe und Korrosion*, vol. 52, no. 9, pp. 697–704, 2001.

[7] G. Moretti, F. Guidi, and G. Grion, "Tryptamine as a green iron corrosion inhibitor in 0.5 M deaerated sulphuric acid," *Corrosion Science*, vol. 46, no. 2, pp. 387–403, 2004.

[8] S. A. Abd El-Maksoud, "The effect of organic compounds on the electrochemical behaviour of steel in acidic media-a review," *International Journal of Electrochemical Science*, vol. 3, pp. 528–555, 2008.

[9] M. A. Quraishi and R. Sardar, "Corrosion inhibition of mild steel in acid solutions by some aromatic oxadiazoles," *Materials Chemistry and Physics*, vol. 78, no. 2, pp. 425–431, 2003.

[10] M. Şahin, S. Bilgiç, and H. Yilmaz, "The inhibition effects of some cyclic nitrogen compounds on the corrosion of the steel in NaCl mediums," *Applied Surface Science*, vol. 195, no. 1–4, pp. 1–7, 2002.

[11] M. Abdallah, M. Al-Agez, and A. S. Fouda, "Phenylhydrazone derivatives as corrosion inhibitors for -α-brass in hydrochloric acid solutions," *International Journal of Electrochemical Science*, vol. 4, no. 3, pp. 336–352, 2009.

[12] M. Bouklah, N. Benchat, A. Aouniti et al., "Effect of the substitution of an oxygen atom by sulphur in a pyridazinic molecule towards inhibition of corrosion of steel in 0.5 M H_2SO_4 medium," *Progress in Organic Coatings*, vol. 51, no. 2, pp. 118–124, 2004.

[13] H.-L. Wang, R.-B. Liu, and J. Xin, "Inhibiting effects of some mercapto-triazole derivatives on the corrosion of mild steel in 1.0 M HCl medium," *Corrosion Science*, vol. 46, no. 10, pp. 2455–2466, 2004.

[14] M. Karakuş, M. Şahin, and S. Bilgiç, "An investigation on the inhibition effects of some new dithiophosphonic acid monoesthers on the corrosion of the steel in 1 M HCl medium," *Materials Chemistry and Physics*, vol. 92, no. 2-3, pp. 565–571, 2005.

[15] A. A. Rahim, E. Rocca, J. Steinmetz, M. J. Kassim, R. Adnan, and M. S. Ibrahim, "Mangrove tannins and their flavanoid monomers as alternative steel corrosion inhibitors in acidic medium," *Corrosion Science*, vol. 49, no. 2, pp. 402–417, 2007.

[16] W. Li, Q. He, C. Pei, and B. Hou, "Experimental and theoretical investigation of the adsorption behaviour of new triazole derivatives as inhibitors for mild steel corrosion in acid media," *Electrochimica Acta*, vol. 52, no. 22, pp. 6386–6394, 2007.

[17] S. D. Shetty, P. Shetty, and H. V. S. Nayak, "The inhibition action of N-furfuryl-N′-phenyl thiourea on the corrosion of mild steel in acid media," *Journal of the Serbian Chemical Society*, vol. 71, no. 10, pp. 1073–1082, 2006.

[18] K. F. Khaled and N. A. Al-Mobarak, "A predictive model for corrosion inhibition of mild steel by thiophene and its derivatives using artificial neural network," *International Journal of Electrochemical Science*, vol. 7, no. 2, pp. 1045–1059, 2012.

[19] M. Z. A. Rafiquee, N. Saxena, S. Khan, and M. A. Quraishi, "Some fatty acid oxadiazoles for corrosion inhibition of mild steel in HCl," *Indian Journal of Chemical Technology*, vol. 14, no. 6, pp. 576–583, 2007.

[20] E. E. Ebenso, T. Arslan, F. Kandemirli et al., "Quantum chemical studies of some Rhodanineazosulpha drugs as corrosion inhibitors for mild steel in acidic medium," *International Journal of Quantum Chemistry*, vol. 110, no. 14, pp. 2614–2636, 2010.

[21] O. I. Sizaya and A. P. Andrushko, "On the use of triazines as inhibitors of steel corrosion," *Protection of Metals*, vol. 40, no. 2, pp. 164–168, 2004.

[22] H. Ashassi-Sorkhabi, B. Shaabani, and D. Seifzadeh, "Corrosion inhibition of mild steel by some schiff base compounds in hydrochloric acid," *Applied Surface Science*, vol. 239, no. 2, pp. 154–164, 2005.

[23] S. Acharya and S. N. Upadhyay, "The inhibition of corrosion of mild steel by some fluoroquinolones in sodium chloride solution," *Transactions of the Indian Institute of Metals*, vol. 57, no. 3, pp. 297–306, 2004.

[24] M. Abdallah, M. Al-Agez, and A. S. Fouda, "Phenylhydrazone derivatives as corrosion inhibitors for -α-brass in hydrochloric acid solutions," *International Journal of Electrochemical Science*, vol. 4, no. 3, pp. 336–352, 2009.

[25] M. A. Quraishi and D. Jamal, "Fatty acid triazoles: novel corrosion inhibitors for oil well steel (N-80) and mild steel," *Journal of the American Oil Chemists' Society*, vol. 77, no. 10, pp. 1107–1111, 2000.

[26] M. A. Quraishi, V. Bhardwaj, and J. Rawat, "Prevention of metallic corrosion by lauric hydrazide and its salts under vapor phase conditions," *Journal of the American Oil Chemists' Society*, vol. 79, no. 6, pp. 603–609, 2002.

[27] A. Singh, A. K. Singh, and M. A. Quraishi, "Dapsone: a novel corrosion inhibitor for mild steel in acid media," *The Open Electrochemical Journal*, vol. 2, pp. 43–51, 2010.

[28] M. A. Quraishi and H. K. Sharma, "4-amino-3-butyl-5-mercapto-1,2,4-triazole: a new corrosion inhibitor for mild steel in sulphuric acid," *Materials Chemistry and Physics*, vol. 78, no. 1, pp. 18–21, 2003.

[29] T. Poornima, J. Nayak, and A. N. Shetty, "Corrosion inhibition of the annealed 18 Ni 250 grade maraging steel in 0. 67 M phosphoric acid by 3, 4-dimethoxybenzaldehydethiosemicarbazone," *Chemical Sciences Journal*, vol. 69, pp. 1–14, 2012.

[30] M. Abdallah, E. A. Helal, and A. S. Fouda, "Aminopyrimidine derivatives as inhibitors for corrosion of 1018 carbon steel in nitric acid solution," *Corrosion Science*, vol. 48, no. 7, pp. 1639–1654, 2006.

[31] S. S. Abd El-Rehim, S. A. M. Refaey, F. Taha, M. B. Saleh, and R. A. Ahmed, "Corrosion inhibition of mild steel in acidic medium using 2-amino thiophenol and 2-cyanomethyl benzothiazole," *Journal of Applied Electrochemistry*, vol. 31, no. 4, pp. 429–435, 2001.

[32] A. Yurt, A. Balaban, S. U. Kandemir, G. Bereket, and B. Erk, "Investigation on some Schiff bases as HCl corrosion inhibitors for carbon steel," *Materials Chemistry and Physics*, vol. 85, no. 2-3, pp. 420–426, 2004.

[33] L. Tang, X. Li, Y. Si, G. Mu, and G. Liu, "The synergistic inhibition between 8-hydroxyquinoline and chloride ion for the corrosion of cold rolled steel in 0.5 M sulfuric acid," *Materials Chemistry and Physics*, vol. 95, no. 1, pp. 29–38, 2006.

[34] A. K. Bhattacharya, T. K. Naiya, S. N. Mandal, and S. K. Das, "Adsorption, kinetics and equilibrium studies on removal of Cr(VI) from aqueous solutions using different low-cost adsorbents," *Chemical Engineering Journal*, vol. 137, no. 3, pp. 529–541, 2008.

[35] S. A. M. Refaey, F. Taha, and A. M. Abd El-Malak, "Corrosion and inhibition of 316 L stainless steel in neutral medium by 2-mercaptobenzimidazole," *International Journal of Electrochemical Science*, vol. 1, pp. 80–91, 2006.

[36] M. Lagrenée, B. Mernari, M. Bouanis, M. Traisnel, and F. Bentiss, "Study of the mechanism and inhibiting efficiency of 3,5-bis(4-methylthiophenyl)-4H-1,2,4-triazole on mild steel corrosion in acidic media," *Corrosion Science*, vol. 44, no. 3, pp. 573–588, 2002.

[37] S. A. Umoren and E. E. Ebenso, "The synergistic effect of polyacrylamide and iodide ions on the corrosion inhibition of mild steel in H_2SO_4," *Materials Chemistry and Physics*, vol. 106, no. 2-3, pp. 387–393, 2007.

[38] M. Elayyachy, A. El Idrissi, and B. Hammouti, "New thio-compounds as corrosion inhibitor for steel in 1 M HCl," *Corrosion Science*, vol. 48, no. 9, pp. 2470–2479, 2006.

[39] F. M. Bayoumi and W. A. Ghanem, "Corrosion inhibition of mild steel using naphthalene disulfonic acid," *Materials Letters*, vol. 59, no. 29-30, pp. 3806–3809, 2005.

[40] A. Yurt, A. Balaban, S. U. Kandemir, G. Bereket, and B. Erk, "Investigation on some Schiff bases as HCl corrosion inhibitors for carbon steel," *Materials Chemistry and Physics*, vol. 85, no. 2-3, pp. 420–426, 2004.

[41] X. Li, S. Deng, H. Fu, and T. Li, "Adsorption and inhibition effect of 6-benzylaminopurine on cold rolled steel in 1.0 M HCl," *Electrochimica Acta*, vol. 54, no. 16, pp. 4089–4098, 2009.

[42] M. Lebrini, F. Bentiss, H. Vezin, and M. Lagrenée, "Inhibiting effects of some oxadiazole derivatives on the corrosion of mild steel in perchloric acid solution," *Applied Surface Science*, vol. 252, no. 4, pp. 950–958, 2005.

[43] A. Y. Musa, A. A. H. Kadhum, A. B. Mohamad, M. S. Takriff, A. R. Daud, and S. K. Kamarudin, "On the inhibition of mild steel corrosion by 4-amino-5-phenyl-4H-1, 2, 4-trizole-3-thiol," *Corrosion Science*, vol. 52, no. 2, pp. 526–533, 2010.

[44] K. Jüttner, "Electrochemical Impedance Spectroscopy (EIS) of corrosion processes on inhomogeneous surfaces," *Electrochimica Acta*, vol. 35, no. 10, pp. 1501–1508, 1990.

[45] F. Bentiss, M. Lebrini, M. Lagrenée, M. Traisnel, A. Elfarouk, and H. Vezin, "The influence of some new 2,5-disubstituted 1,3,4-thiadiazoles on the corrosion behaviour of mild steel in 1 M HCl solution: AC impedance study and theoretical approach," *Electrochimica Acta*, vol. 52, no. 24, pp. 6865–6872, 2007.

[46] F. El-Taib Heakal, A. A. Ghoneim, and A. M. Fekry, "Stability of spontaneous passive films on high strength Mo-containing stainless steels in aqueous solutions," *Journal of Applied Electrochemistry*, vol. 37, no. 3, pp. 405–413, 2007.

[47] M. Lebrini, F. Bentiss, H. Vezin, and M. Lagrenée, "The inhibition of mild steel corrosion in acidic solutions by 2,5-bis(4-pyridyl)-1,3,4-thiadiazole: structure-activity correlation," *Corrosion Science*, vol. 48, no. 5, pp. 1279–1291, 2006.

[48] S. K. Shukla and M. A. Quraishi, "The effects of pharmaceutically active compound doxycycline on the corrosion of mild steel in hydrochloric acid solution," *Corrosion Science*, vol. 52, no. 2, pp. 314–321, 2010.

[49] F. Bentiss, M. Traisnel, and M. Lagrenee, "The substituted 1,3,4-oxadiazoles: a new class of corrosion inhibitors of mild steel in acidic media," *Corrosion Science*, vol. 42, no. 1, pp. 127–146, 2000.

[50] K. Parameswari, S. Chitra, C. N. Unnisa, and A. Selvaraj, "Effect of azlactones on corrosion inhibition of mild steel in acid medium," *Journal of Applied Sciences Research*, vol. 6, no. 8, pp. 1100–1110, 2010.

[51] S. Aloui, I. Forsal, M. Sfaira et al., "New mechanism synthesis of 1,4-benzothiazine and its inhibition performance on mild steel in hydrochloric acid," *Portugaliae Electrochimica Acta*, vol. 27, no. 5, pp. 599–613, 2009.

[52] I. Zaafarany, "Phenyl phthalimide as corrosion inhibitor for corrosion of C-Steel in sulphuric acid solution," *Portugaliae Electrochimica Acta*, vol. 27, no. 5, pp. 631–643, 2009.

[53] M. Behpour, S. M. Ghoreishi, M. Salavati-Niasari, and B. Ebrahimi, "Evaluating two new synthesized S-N Schiff bases on the corrosion of copper in 15% hydrochloric acid," *Materials Chemistry and Physics*, vol. 107, no. 1, pp. 153–157, 2008.

[54] I. Ahamad and M. A. Quraishi, "Bis (benzimidazol-2-yl) disulphide: an efficient water soluble inhibitor for corrosion of mild steel in acid media," *Corrosion Science*, vol. 51, no. 9, pp. 2006–2013, 2009.

4

Characterization of Corrosive Bacterial Consortia Isolated from Water in a Cooling Tower

Rajasekar Aruliah[1,2] and Yen-Peng Ting[1]

[1] *Department of Chemical & Biomolecular Engineering, National University of Singapore, Singapore 117576*
[2] *Environmental Molecular Microbiology Lab, Department of Biotechnology, Thiruvalluvar University, Serkkadu, Vellore 632 115, India*

Correspondence should be addressed to Rajasekar Aruliah; rajasekar.aruliah@gmail.com and Yen-Peng Ting; chetyp@nus.edu.sg

Academic Editors: M. Criado, W. Fürbeth, S. J. Lee, and Q. Qu

An analysis of a culturable corrosive bacterial community in water samples from a cooling tower was performed using traditional cultivation techniques and its identification based on 16S rRNA gene sequence. Seven aerobic bacterial species were identified: *Pseudomonas putida* ARTYP1, *Pseudomonas aeruginosa* ARTYP2, *Massilia timonae* ARTYP3, *Massilia albidiflava* ARTYP4, *Pseudomonas mosselii* ARTYP5, *Massilia* sp. ARTYP6, and *Pseudomonas* sp. ARTYP7. Although some of these species have commonly been observed and reported in biocorrosion studies, the genus *Massilia* is identified for the first time in water from a cooling tower. The biocorrosion behaviour of copper metal by the new species *Massilia timonae* ARTYP3 was selected for further investigation using a weight loss method, as well as electrochemical and surface analysis techniques (SEM, AFM, and FTIR). In contrast with an uninoculated system, thin bacterial biofilms and pitting corrosion were observed on the copper metal surface in the presence of *M. timonae*. The use of a biocide, bronopol, inhibited the formation of biofilm and pitting corrosion on the copper metal surface.

1. Introduction

In order to implement efficient monitoring and control strategies for the inhibition of biocorrosion, it is important to have knowledge of the microbial population responsible for this phenomenon, as well as interactions of different microorganisms with metallic surfaces [1–8]. In many industries, cooling towers are commonly used for heat transfer from recirculated water to the atmosphere, typically by means of trickling or spraying the water over a material with high surface area [9]. These towers generally have sizable water reservoirs, with temperature typically maintained between 25°C and 35°C. These conditions provide an ideal environment for microbial growth and propagation [10–13]. Both microbes and the substrates for microbial growth can either be present in the incoming water or be introduced from the atmosphere. Copper and copper alloys, which are used in many cooling tower systems, are known to be susceptible to microbiologically influenced corrosion (MIC) [10, 14]. Corrosion and its products have a negative impact on heat transfer and can cause a decrease in cooling efficiency of the cooling tower. Organisms responsible for MIC, including bacteria, microalgae and fungi readily attach themselves to the copper surface by excreting extracellular polymeric substances (EPS) to form a slime layer [15–18] and thereby initiate corrosion. A multilayer structure of microorganisms and their EPS have been reported to be entrapped between layers of different inorganic corrosion products on copper-based surfaces after exposure to natural seawater environment [19–24].

Interestingly, many traditional chemicals used water treatment, for example, antiscalants and zinc-based corrosion inhibitors, which are a source of nutrients that accelerate the growth of microbes in cooling towers [25]. Nonetheless, the control of corrosive bacterial fouling can be achieved through the application of effective biocides [26], or nitrate, or nitrite [27–30]. Environmental regulations and the development of water reservoirs in environmentally sensitive areas have spurred the development of easily degradable "green"

Table 1: Physicochemical characteristics of water from cooling tower.

Sl. no.	Physicochemical characteristics	Quantity
1	Temperature (°C)	25
2	pH	8.1
3	Total solids (mg/L)	341
4	Total hardness (mg/L)	150
5	Total alkalinity (mg/L)	165
6	Chloride (mg/L)	296

biocides that are less toxic to higher nontarget organisms, like fish [31]. Nitrite, a specific metabolic inhibitor of corrosive bacteria, especially sulphate reducing bacteria [32, 33], is also relatively nontoxic and inexpensive and has been successfully used to inhibit corrosive bacteria [34]. 2-Bromo-2-nitropropane-1,3-diol (bronopol) has a broad spectrum of antibacterial activity and is widely used as a preservative in pharmaceutical and cosmetic products and many industrial environments such as paper mills, oil exploration, and production facilities, as well as a disinfectant in cooling water [35, 36]. The present investigation describes 16S rRNA gene sequencing and the identification of bacterial communities (at the species level) present in water from a cooling tower. Microbiologically induced corrosion and the use of the biocide bronopol are also examined. As the literature on the identification of bacteria in such a system in tropical countries appears lacking, this work would be useful in the understanding of bacterial groups and biocides in cooling tower.

2. Materials and Methods

2.1. Sample Collection and Isolation of Bacteria. A sample of water was collected using sterilized conical flasks, from a cooling tower site located at the Faculty of Engineering at the National University of Singapore, and was transported in an ice box to the laboratory. The physiochemical parameters of the cooling water are summarized in Table 1. The water sample was serially diluted (10-fold) with sterile saline water (0.85% w/v sodium chloride) and plated using pour plate technique. Winogradsky nutrient medium [37], consisting of (grams per liter): 1.0 K_2HPO_4, 1.27 $NaNO_3$, 0.1 $CaCl_2$, 0.2 $MgSO_4{\cdot}7H_2O$, 0.05 $FeSO_4{\cdot}7H_2O$, 0.54 NH_4Cl, 1.27 $NaNO_3$, and 10 D-glucose anhydrous, was supplemented with 0.5 yeast extract and used as the medium. Viable bacteria count was enumerated on agar plates containing the supplemented Winogradsky nutrient medium (WNM) with 1.5% agar (pH 7). Experimental cultures were grown in conical flasks at 28 ± 2°C for two days. Bacterial growth was detected by a colour change in the medium (from green to red). Total viable count of bacteria was enumerated by pour plate method using sterile WNM plates and incubated at 28 ± 2°C for two days.

2.1.1. Partial Biochemical Characterization of the Isolates and Molecular Identification of Bacteria. Dissimilar aerobic bacteria isolated from the agar plates were identified according to Bergey's Manual of Determinative Bacteriology [38]. The isolated bacterial cultures were identified up to the genus level by their morphological and partial biochemical characterizations using the following: (i) gram staining, (ii) motility test, (iii) indole production, (iv) methyl red test, (v) Voges-Proskauer test, (vi) citrate utilization test, (vii) H_2S production test, (viii) carbohydrate fermentation test, (ix) catalase test, (x) oxidase test, (xi) starch, (xii) gelatin, and (xiii) lipid hydrolysis. In addition, citrate agar was used to detect any iron oxidizing activity of the isolates. Genomic DNA of the bacterial isolates was extracted according to Ausubel et al. [39]. Amplification of gene encoding small subunit ribosomal RNA was carried out using eubacterial 16S rRNA primers (forward primer 5′-AGAGTTTGATCCTGGCTCAG-3′ (*E. coli* positions 8 to 27) and reverse primer 5′-ACGGCTACCTTGTTACGACTT-3′ (*E. coli* positions 1494 to 1513)) [40]. Polymerase chain reaction (PCR) was performed with a 50 μL reaction mixture containing 2 μL (10 ng) of DNA as the template, with each primer at a concentration of 0.5 μM and 1.5 mM $MgCl_2$ and each dNTP at a concentration of 50 μM as well as 1 μL of *Taq* DNA polymerase and buffer as recommended by the manufacturer (MBI Fermentas). PCR was carried out with a Mastercycler Personal (Eppendorf) with the following program: initial denaturation at 95°C for 1 min, 40 cycles of denaturation (3 min at 95°C), annealing (1 min at 55°C), and extension (2 min at 72°C), followed by a final extension (at 72°C for 5 min). The amplified product was purified using GFX PCR DNA and Gel Band Purification kit (Amersham Biosciences) and cloned in pTZ57R/T vector according to the manufacturer's instructions (InsT/Aclone PCR Product Cloning Kit, MBI Fermentas) and transformants were selected on LB medium containing ampicillin (100 μg/mL) and X-gal (80 μg/mL). DNA sequencing was carried out using ABI PRISM 310 Genetic Analyzer (PE Applied Biosystems). For sequencing reaction, Big Dye Ready Reaction DyeDeoxy Terminator Cycle Sequencing kit (Perkin-Elmer) was used.

2.1.2. Phylogenetic Analysis of the Bacterial Isolates. The sequences obtained were analysed with BLAST search version 2.2.20 [41] and tools of Ribosomal Database Project II Release 10 (http://rdp.cme.msu.edu) for taxonomic hierarchy of the sequences. Multiple sequence alignments were performed using CLUSTAL X2 [42] with a collection of taxonomically related sequences obtained from National Center for Biotechnology Information (NCBI) Taxonomy Homepage (http://www.ncbi.nlm.nih.gov/Taxonomy/taxonomyhome.html/) and Ribosomal Database Project-II Release 10 (http://rdp.cme.msu.edu). Phylogenetic and similitude analyses were done with the common 16S rRNA gene regions and all alignment gaps were treated as missing data. The paired similitude and pairwise distance calculations using the transversion/transition weighting ($R = s/v$) and the Kimura-2-parameter model [43] were performed with the MEGA version 4.1 program [44]. The phylogenetic trees were constructed (neighbor-joining method) and 1000 bootstrap replications were carried out to validate internal branches [45]. MatGAT v. 2.01 software [46] was used to calculate the similitude percentages among sequences.

2.1.3. Nucleotide Sequence Accession Numbers. The 16S rRNA sequence data reported in this work has been deposited in the GenBank database, under the accession numbers from FJ755909 to FJ755915.

2.2. Biocorrosion Studies and Surface Analysis. Copper metal (≥99% Cu) was used in the study of the MIC behaviour of the bacteria *M. timonae* ARTYP3. Different size coupons (10 mm diameter disc of 2 mm thickness and 50 × 10 mm rectangular coupons of 2 mm thickness) were sequentially ground with a series of grit silicon carbide papers (grades 180, 500, 800, 1200, and 1500) to a smooth surface and finally polished to a mirror finish surface using 0.3 micron alumina powder. The polished coupons were rinsed with deionized water and surface was sterilized by immersion in an aqueous solution of 70% ethanol for one minute and finally dried in a desiccator. The prepared coupons were used for biocorrosion studies and surface analysis (using weight loss method, EIS, SEM-EDX, and FTIR). The polished coupons of both shapes and sizes were introduced into a 500 mL conical flask containing 250 mL sterilized cooling tower water with 1% of WNM broth as the control system (System 1), while a sterilized cooling tower water with 1% of WNM broth inoculated with 1 mL of *M. timonae* bacterial culture (about 10^5 CFU/mL) was used as the experimental system (System 2). Another experimental system (System 3) consisted of System 2 with the addition of 5 ppm biocide (2-bromo-2-nitropropane-1, 3-diol) (bronopol). Bronopol is commonly used as a microbiocide or microbiostat for the control of slime-forming bacteria, fungi, and algae [47]. Corrosion experiments were initiated by hanging polished copper coupons on a nylon string in all the systems. After 10 days of incubation, the coupons were removed from each system for electrochemical, FTIR, and SEM-EDX analyses. Three coupons were exposed in each conical flask with duplicate flasks for each system. In total, six coupons were used for the weight loss for each system; the average weight loss and standard deviations (SD) were calculated. Impedance and potentiodynamic polarization data were obtained using an Autolab PGSTAT potentiostat. All experiments were performed in a three-electrode electrochemical cell, with a platinum electrode as the counter electrode and an Ag/AgCl electrode as the reference electrode. The EIS measurements (using duplicate coupons) were performed ex-situ; the coupons that were removed from the appropriate systems served as the working electrode by embedding them in a sample holder of the corrosion cell (purchased from Metrohm Pte, Ltd.). The working electrode had an exposed surface area of 0.785 cm^2. 500 mL of the media was transferred from each system into the electrochemical cell to serve as the electrolyte for the EIS analysis [6]. EIS were made at open circuit potential using a 10 mV amplitude sinusoidal signal over frequencies ranging from 5 mHz to 100 kHz. Tafel plots were measured with a scan rate of 0.5 mVS^{-1} and were obtained by scanning from the open circuit potential (E_{corr}) towards 200 mV anodically and −200 mV cathodically using duplicate coupons. A scanning electron microscope coupled with energy-dispersive X-ray spectroscopy analysis (SEM-EDX) microscope (JEOL, model JSM-5600) with a beam voltage of 15 kV was used to visualize the morphology of the biofilm. Surface topography of the coupons and its elemental composition were characterized after removal of corrosion product, using SEM-EDX analyses. The corrosion product on the copper metal was removed by immersion in 50% concentrated hydrochloric acid for 1–3 min (ASTM G1-90). To examine the deposition of EPS on the coupons and its chemical composition, the coupons were analysed using FTIR and SEM as described earlier [48]. An FTIR microscope (Bio-Rad Model: FTS 135) equipped with a narrow band of liquid nitrogen-cooled HgCdTe (MCT) detector was used for collecting the IR spectra. An atomic force microscopy (AFM) analysis was carried out to visualize the biofilm formed on the metal surface after the corrosion experiment. The coupon was removed from the media, lightly rinsed in sterile distilled water, and air-dried. The coupon was examined under an AFM (Bruker Dimension Icon Nanoscope V) in the tapping mode to capture the images of biofilms on the coupon surface. Silicon N-type cantilever nanoprobes with a spring constant of $k = 25–75$ N/m (App NANO) were used.

2.3. Bacterial Adhesion to Hydrocarbons (BATH) and Salt Aggregation Test (SAT). BATH assay was used to assess the hydrophobicity of the bacteria ARTYP3, with hexadecane being used as the carbon source. *Massilia timonae* was harvested from growing cultures by centrifugation at 8000 g for 10 min at 4°C, washed twice, and suspended in PBS buffer (pH 7.1; containing (g/L) K_2HPO_4, 22.2; KH_2PO_4, 7.26; urea, 1.8; $MgSO_4 \cdot 7H_2O$, 0.2) at an initial absorbance of 0.12–0.14 (at 400 nm). The detailed procedure for the BATH test and the calculation of the emulsifying capacity as evaluated by an emulsification index (*E*24) has been described earlier [2]. SAT test was used to measure the relative cell surface hydrophobicity of the bacteria. Sodium phosphate (0.002 M, pH 6.8) was used to dilute a solution of 4 M $(NH_4)_2SO_4$ at pH 6.8. Serial dilutions were made giving $(NH_4)_2SO_4$ concentration ranging from 4.0 to 0.2 M differing by 0.2 M per dilution and ranging from 0.20 to 0.02 M differing by 0.02 M [49]. A bacterial suspension of 25 μL in 0.002 M sodium phosphate buffer (pH 6.8) was mixed with an equal volume of salt solution into 24-well tissue culture tray. The bacterial/salt mixture was gently rocked for 2 min at 25°C and visual observation was made against a dark background. The results were expressed as the lowest molarity of ammonium sulphate that caused bacterial aggregation.

3. Results

3.1. Physicochemical Characterization and Bacteriological Analysis of Cooling Tower Water. Table 1 shows the physiochemical parameters of the water sample collected from the cooling tower. The concentration of chloride and total hardness in the cooling water sample were 296 mg/L and 150 mg/L, respectively. The total viable count in the supplemented WNM was 6.1 × 10^5 CFU/mL. Preliminary identification of the bacteria by biochemical test indicated that the isolates belong to the genera *Massilia* sp. and *Pseudomonas* sp. The phenotypic profiles of the species isolated from cooling

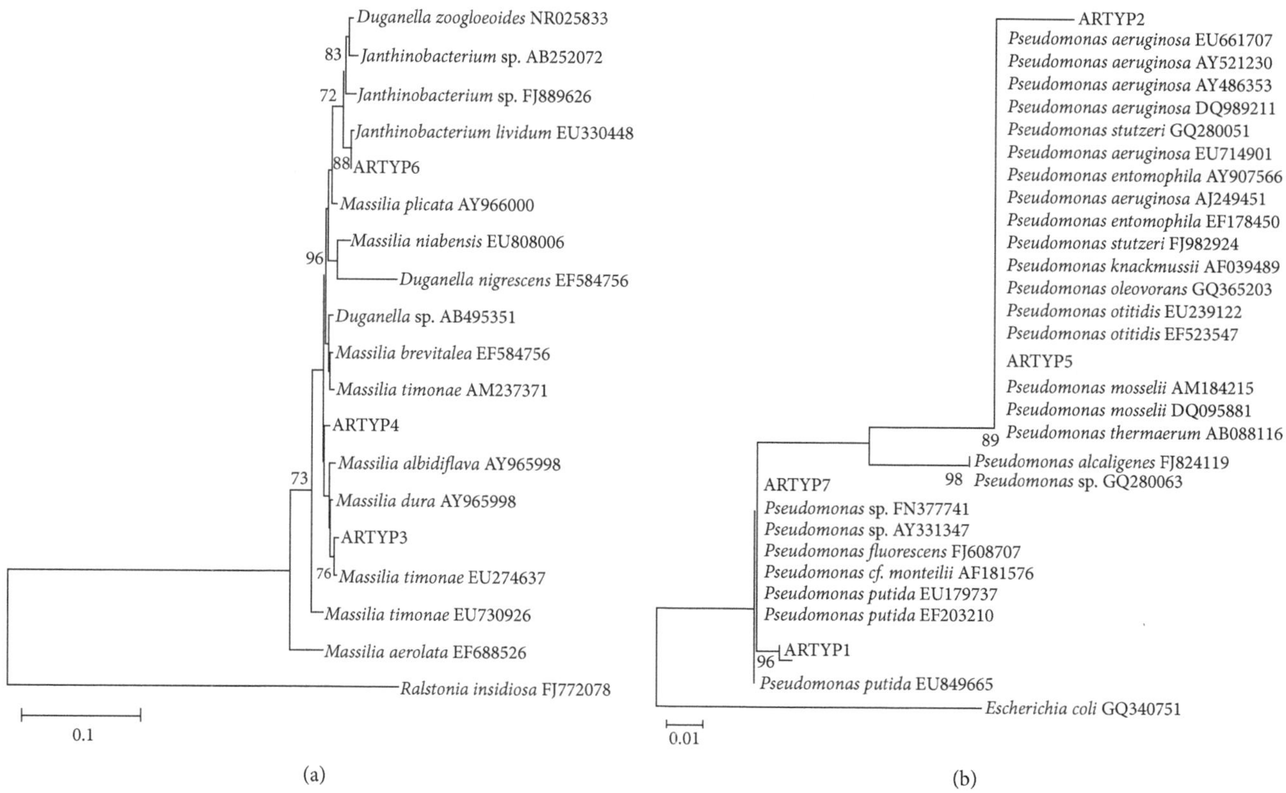

FIGURE 1: (a) Neighbor-joining tree based on 16S rRNA gene sequences, showing phylogenetic relationships between sequences of the *Massilia* related species. *Ralstonia insidiosa* was used as a bacterial out-group. Numbers at nodes indicate bootstrap values > 50% from 1,000 replicates. GenBank accession numbers are given in parentheses. The scale bar indicates sequence divergence. (b) Neighbor-joining tree based on 16S rRNA gene sequences, showing phylogenetic relationships between sequences of the *Pseudomonas* related species. *Escherichia coli* was used as a bacterial out-group. Numbers at nodes indicate bootstrap values > 50% from 1,000 replicates. GenBank accession numbers are given in parentheses. The scale bar indicates sequence divergence.

water strains are shown in Tables 2(a) and 2(b). The sequences obtained were submitted to a BLAST search to retrieve the corresponding phylogenetic relatives. The phylogenetic affiliations (Gammaproteobacteria and Betaproteobacteria) were confirmed by analyses of all related species recognized by the taxonomic and classification hierarchy done with the NCBI Taxonomy Homepage and Ribosomal Database Project-II Release 10. Two neighbor-joining phylogenetic trees were constructed for the *Massilia* related species (Figure 1(a)) and *Pseudomonas* related species (Figure 1(b)), to analyse the relationships among the sequences of the ribosomal library and related organisms from the GenBank database. The similarity and species identified with the phylogenetic analysis are given in Table 3. In the *Pseudomonas* genera tree, all the isolates belonging *to Pseudomonas* genus exhibited a high nucleotide sequence similarity of more than 99%; the corresponding bacterial isolates are, namely, *Pseudomonas putida* (99.2%), *Pseudomonas aeruginosa* (99%), *Pseudomonas mosselii* (99.5%), and *Pseudomonas* sp. (98.6%). In the class of *Massilia*, several isolates of the genus *Massilia* exhibited high sequence similarity (99%): *Massilia timonae* (99.5%) and *Massilia plicata* (99%), and one isolate, *Massilia albidiflava* had 98.5% similarity.

Phenotypic profile of strain ARTYP3 was shown in Table 2(a). Amplification of the gene encoding for small subunit ribosomal RNA of ARTYP3 was done using eubacterial 16S rDNA primers. The 16S rDNA amplicons derived from ARTYP3 are cloned in pTZ57R/T vectors. The recombinant plasmid (pARTYP3, harboring 16S rDNA insert) is partially sequenced. The sequence obtained is matched with the previously published sequences available in NCBI using BLAST. Sequence alignment and comparison revealed more than 99% similarity with *M. timonae*. *Massilia timonae* ARTYP3 has been deposited in GenBank with the accession number FJ755911 and also been submitted to the German Collection of Microorganisms and Cell Cultures (DSMZ).

3.2. Biocorrosion Studies and Surface Analysis

3.2.1. Weight Loss Study and Electrochemical Analysis. In the control system, the average weight loss of the metal coupons (0.0073 mg) was significantly less than in the presence of *M. timonae* ARTYP3 (0.065 mg). The average weight loss was lowest (0.0045 mg) in the presence of the biocide, that is, System 3. Figure 2 shows the Tafel polarization curves for copper in a cooling water system in the absence (System 1)

Table 2: (a) Biochemical characterization of *Massilia* related species isolates from cooling tower. (b) Biochemical characterization of *Pseudomonas* related species isolates from cooling tower.

(a)

Characteristics	FJ755911	FJ755912	FJ755914
Cell morphology			
Gram stain	Negative	Negative	Negative
Shape	Rod	Rod	Rod
Motility	+	+	+
Sporulation	−	−	−
Growth at			
20°C	−	−	−
30°C	+	+	+
40°C	+	+	+
Biochemical reaction			
Indole production test	−	−	+
Methyl red test	+	−	−
Voges-Proskauer test	−	−	−
Citrate utilization test	+	−	−
Oxidase test	+	+	+
Catalase test	+	+	+
Production of acid from			
Glucose	+	+	+
Galactose	−	−	−
Fructose	+	+	+
Sucrose	+	+	−
Mannitol	−	−	−
Lactose	+	+	−
Cellobiose	+	−	−
Adonitol	−	−	−
Arabinose	−	−	−
Raffinose	−	−	−
Inositol	−	−	−
Hydrolysis of			
Starch	−	+	−
Cellulose	−	+	−
Casein	−	+	−
Gelatin	−	−	−
Urea	+	−	+
Tween 80	−	+	−

(b)

Characteristics	FJ755909	FJ755910	FJ755913	FJ755915
Cell morphology				
Gram stain	Negative	Negative	Negative	Negative
Shape	Rod	Rod	Rod	Rod
Motility	+	+	−	+
Sporulation	−	−	−	−
Growth at				
20°C	+	+	+	+
30°C	+	+	+	+
40°C	+	+	+	+
Biochemical reaction				
Indole production test	−	+	+	+
Methyl red test	+	−	+	+
Voges-Proskauer test	−	−	−	−
Citrate utilization test	+	−	−	−
Oxidase test	+	−	−	−
Catalase test	+	+	+	+
Production of acid from				
Glucose	+	+	+	+
Galactose	−	−	−	−
Fructose	+	+	+	+
Sucrose	−	+	+	+
Mannitol	−	−	−	−
Lactose	+	−	−	−
Cellobiose	+	−	−	−
Adonitol	−	−	−	−
Arabinose	+	+	+	+
Raffinose	−	−	−	−
Inositol	−	−	−	−
Hydrolysis of				
Starch	+	+	+	+
Cellulose	−	+	+	+
Casein	−	+	+	+
Gelatin	+	−	−	−
Urea	−	+	+	+
Tween 80	−	+	+	+

and in the presence (System 2) of ARTYP3 and in the biocide system (System 3). The corresponding electrochemical data are presented in Table 4. The corrosion currents for System 1 and System 2 were 3.9×10^{-9} A/cm^2 and 5.5×10^{-7} A/cm^2, respectively, while, in presence of the biocide (System 3), the current was 1×10^{-9} A/cm^2. The open circuit potential (OCP) for the control system and in the presence of ARTYP3 and in the biocide systems were −255 mV, −225 mV, and −275 mV, respectively. Nyquist plots for the control and inoculated systems are given in Figures 3(a) and 3(b). The solution resistance (R_s) and the charge transfer resistance (R_{ct}) were derived from impedance measurements. In the control and in the presence of ARTYP3, the charge transfer resistance R_{ct} values were 12.7 Ω·cm^2 and 0.01 Ω·cm^2, respectively; R_{ct} increased significantly to 163 Ω·cm^2 in the presence of the biocide (Table 4).

3.2.2. Surface Analysis. Biofilm formed on the copper metal surface in System 2 and System 3 was observed under SEM and is presented in Figure 4. ARTYP3 colonized the metal surface and formed a thick biofilm (Figure 4(a)) while significantly less biofilm was formed when the biocide was present (Figure 4(b)). The cell density of ARTYP3 at the end of the biocorrosion study was about 2.3×10^7 CFU/mL and 1.2×10^2 CFU/mL in the absence and presence of

TABLE 3: 16S rRNA sequence analysis of the bacterial isolates.

Name of the isolates	Taxonomic phylum	Closest relationship in GenBank	Similarity (%)	Microbial group identified
ARTYP1	Gammaproteobacteria	*Pseudomonas putida* (EU849665)	99.2	*Pseudomonas putida* (FJ755909)
ARTYP2		*Pseudomonas aeruginosa* (EU661707)	99.0	*Pseudomonas aeruginosa* (FJ755910)
ARTYP5		*Pseudomonas mosselii* (AM184215)	99.5	*Pseudomonas mosselii* (FJ755913)
ARTYP7		*Pseudomonas* sp. (FN377741)	98.6	*Pseudomonas* sp. (FJ755915)
ARTYP3	Betaproteobacteria	*Massilia timonae* (EU274637)	99.5	*Massilia timonae* (FJ755911)
ARTYP4		*Massilia albidiflava* (AY965999)	98.5	*Massilia albidiflava* (FJ755912)
ARTYP6		*Massilia plicata* (AY966000)	99.0	*Massilia plicata* (FJ755914)

TABLE 4: Polarization and impedance parameters for copper in the presence/absence of bacterial isolates and biocide.

Systems	Polarization			Impedance data		
	E_{corr} (mV)	i_{corr} (A/cm^2)	Corrosion rate (mm/y)	R_s Ω	R_{ct} Ω	R_p Ω
System 1	−255	3.9×10^{-9}	4.5×10^{-5}	100	12.7	2×10^6
System 2	−225	5.5×10^{-7}	6.5×10^{-3}	148	0.01	1.4×10^3
System 3	−275	1×10^{-9}	1.1×10^{-5}	181	163	1.8×10^7

System 1: 250 mL sterilized cooling tower water with 1% of supplemented WNM medium. System 2: System 1 with 1 mL of *M. timonae* bacterial culture (about 10^5 CFU/mL). System 3: Systems 1 and 2 with the addition of 5 ppm biocide (2-bromo-2-nitropropane-1, 3-diol) (bronopol).

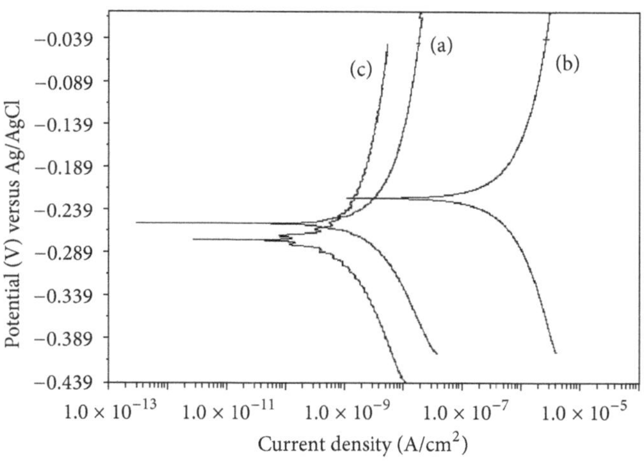

FIGURE 2: Tafel polarization curves of copper metal coupon in various systems after 10-day immersion in cooling water at 28°C: (a) control, (b) *M. timonae* ARTYP3, and (c) *M. timonae* ARTYP3 inoculated with biocide.

the biocide. AFM images (two- and three-dimensional view) of the ARTYP3 biofilm after exposure to the supplemented WNM medium are shown in Figure 4(c) where the thick biofilm on the copper metal surface is evident (Figure 4(c)). Figure 5 shows the SEM topography of the copper coupons after 10 days of exposure in the uninoculated system (Figure 5(a)), in the presence of the isolate (Figure 5(b)), and in the system with the isolate and the biocide (Figure 5(c)), after the removal of corrosion products and biofilm. Pitting corrosion was evident on the surface of the copper metal in the bacteria inoculated system but, in presence of the biocide, significantly less corrosion was observed. FTIR spectra of the ARTYP3 biofilm on the copper metal coupons exposed to the media with and without bronopol are shown in Figure 6. The distinct narrow peaks (1140 cm^{-1}) observed in the spectrum (Figure 6(a)) may be ascribed to the C–O–C group of polysaccharides. A broad band was noted in the range of 3000–3500 cm^{-1}, which may be assigned to adsorbed water molecule OH/NH group. The protein amide I (C=O stretch weakly coupled with C–N stretch and N–H bending) and amide II (C–N stretch strongly coupled with N–H bending) regions are found at approximately 1650 cm^{-1} and 1550 cm^{-1}, respectively. The covalent property of the C–O–Cu bond results in a strong band due to the C–O–C stretching vibration for ester in the range of 1675–1620 cm^{-1}. This region reflects vibrational modes of carbohydrate C–O–C and the broad phosphate-oxygen stretch. The band at 1242–1254 cm^{-1} is due to the -OH group. In the presence of bronopol, no significant peaks were observed on the copper metal surface in the FTIR spectrum (Figure 6(b)).

The bacteria cell surface hydrophobicity was assessed using BATH and SAT assays. Grown in *n*-hexadecane containing medium, ARTYP3 was found to be hydrophobic. The emulsifying capacity of ARTYP3 evaluated using the emulsification index (*E*) at different time periods, 24, 48, and 72 h, was about E_{24} 18%, E_{48} 22%, and E_{72} 22% [50]. Aggregation of *M. timonae* ARTYP3 with $(NH_4)_2SO_4$ was noted visually (against a dark background) even at the lowest calt concentration (0.002 M). This observation confirms the hydrophobic nature of the cell surface.

4. Discussion

As reported by Gallego et al. [51] *Massilia timonae* is one of five species under the genus *Massilia* belonging to the family of Oxalobacteraceae (Betaproteobacteria). Subsequently, other species of the genus *Massilia* were also successfully isolated from blood and environmental samples [52, 53]. *M. timonae* is identified and reported for the first time in water from a cooling tower. The bacteria *M. timonae* are Gram-negative, motile, and rod-shaped. Catalase and oxidase were present in all the *Massilia* sp. strains (Table 2(a)). The isolates ARTYP1–ARTYP7 from the cooling water deposited ferric hydroxide precipitate around the colony and formed as rust-red colour. As shown in the *Massilia* genera phylogenetic tree (Figure 1(a)), the isolates exhibited high sequence similarity with *M. timonae, M. albidiflava,* and *M. plicata* (>98.5%). The involvement of *M. timonae* is unexpected as this species has not previously been reported to be associated with MIC in cooling tower systems.

Pseudomonas species, the most prevalent in industrial water and seawater, have been reported in the corrosion of mild steel, stainless steel, and aluminum alloys in marine and industrial habitats [54, 55]. Aerobic *Pseudomonas* species are recognized to be the pioneer colonizer in biofilm formation; their primary role appears to be to create an oxygen-free environment to harbor other corrosive bacteria (such as the sulphate reducing bacteria). It has been reported that these aerobic slime-formers often grow in a patchy distribution over the metal surface and exclude oxygen via respiration, thereby creating oxygen concentration cells or ion concentration cells [56]. *Pseudomonas* has also been shown to enhance the corrosion of metals and cause pitting corrosion on the metal surface. Pedersen et al. [56] reported that *Pseudomonas* sp. facilitated the breakdown of passivity by excreting organic acids, thus resulting in an increase in the corrosion rates of metals. Busalmen et al. [57] attributed the acceleration in the corrosion rate to the cathodic reduction by the catalase secreted by *Pseudomonas*. The *Pseudomonas* species isolated from the cooling tower also produce the catalase enzyme. In the *Pseudomonas* phylogenetic tree, the isolates exhibited a high nucleotide sequence similarity with *P. aeruginosa, P. mosselii, P. putida,* and *Pseudomonas* sp. with high similarity values (>98%). The phylogenetic analysis of the isolates showed the dominance of *Pseudomonas* sp. among Gammaproteobacteria, and *Massilia* sp. among Betaproteobacteria (Table 3).

The effect of the bacteria on the metal surface and its induced corrosion process is manifested through electrochemical analysis. The data obtained from the polarization study in the presence/absence of ARTY3 and the biocide are presented in Table 4. The open circuit potential (OCP) in the presence of the biocide was negative (−275 mV) when compared to control (−225 mV), thus indicating that the biocide bronopol acts as a mixed inhibitor and suppresses both anodic and cathodic current significantly. The corrosion rate was higher in the presence of ARTYP3 (6.5×10^{-3} mm/y), when compared to the control (4.5×10^{-5} mm/y). In the presence of the biocide, however, the corrosion rate and current density were reduced (to about 1.1×10^{-5} mm/y and 1×10^{-9} A/cm^2). In the presence of ARTYP3, the corrosion current was the highest, thus indicating that ARTYP3 accelerates the corrosion of copper. Figure 2(b) shows that, in the presence of ARTYP3, both the anodic and cathodic curves shifted to the right when compared to the control, which indicated enhanced (anodic and cathodic) reactions, with the uptake of hydrogen and dissolution of copper on the metal surface. This led to pitting corrosion [58]. However, in the presence of the biocide (System 3), both anodic and cathodic reactions were suppressed significantly (i.e., Figure 2(c)). The suppression of corrosion current was due to the inhibition of the biofilm formation on the copper surface. Compared to the control, the resistances of the solution, R_s, in the bacterial cultures were higher, possibly due to bacterial exopolysaccharides in the medium. In the presence of the biocide (System 3), the charge transfer resistance, R_{ct}, was significantly higher (163 Ω) when compared to the control (12.7 Ω) and System 2 (0.01 Ω), possibly due to the compact nitrogen band with copper oxide layer on the metal surface which inhibits biocorrosion. In the presence of ARTYP3, the impedance curve (Figure 3) shows an activation control process due to defects and pores in the adsorbed film on the metal surface [59]. However, in the presence of the biocide, the curve shows a large and well-defined semicircle capacitive loop (Figure 3(b)). The capacitive loop shows the formation of the intact part of the adsorbed film on the metal surface and is large compared to the control. This observation corroborates the polarization studies.

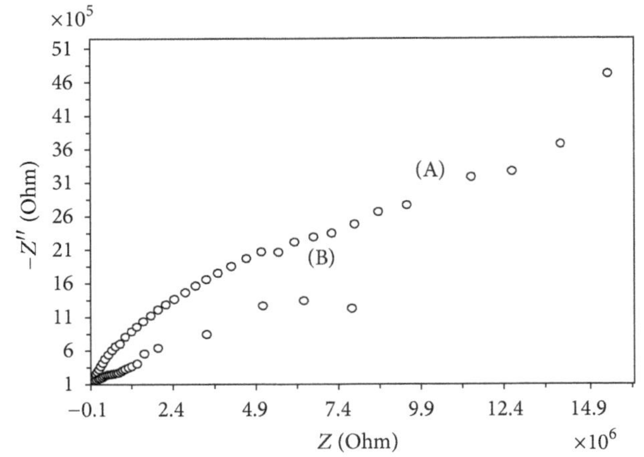

(a)

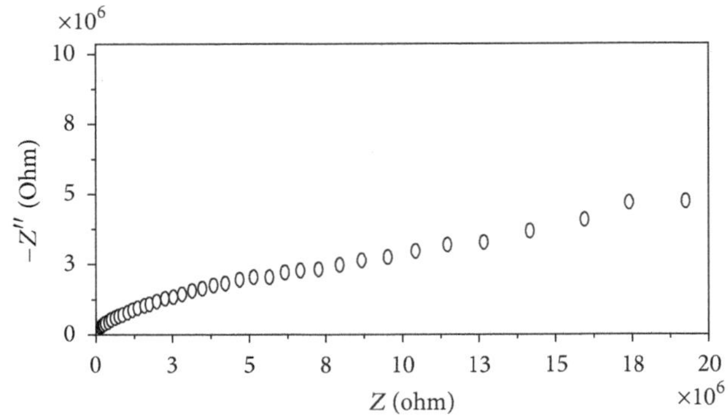

(b)

FIGURE 3: (a) Impedance curves of copper metal coupon in various systems after 10-day immersion in cooling water at 28°C: (A) control system and (B) *M. timonae* ARTYP3. (b) Impedance curves of copper metal coupon in *M. timonae* ARTYP3 inoculated with biocide after 10-day immersion in cooling water at 28°C.

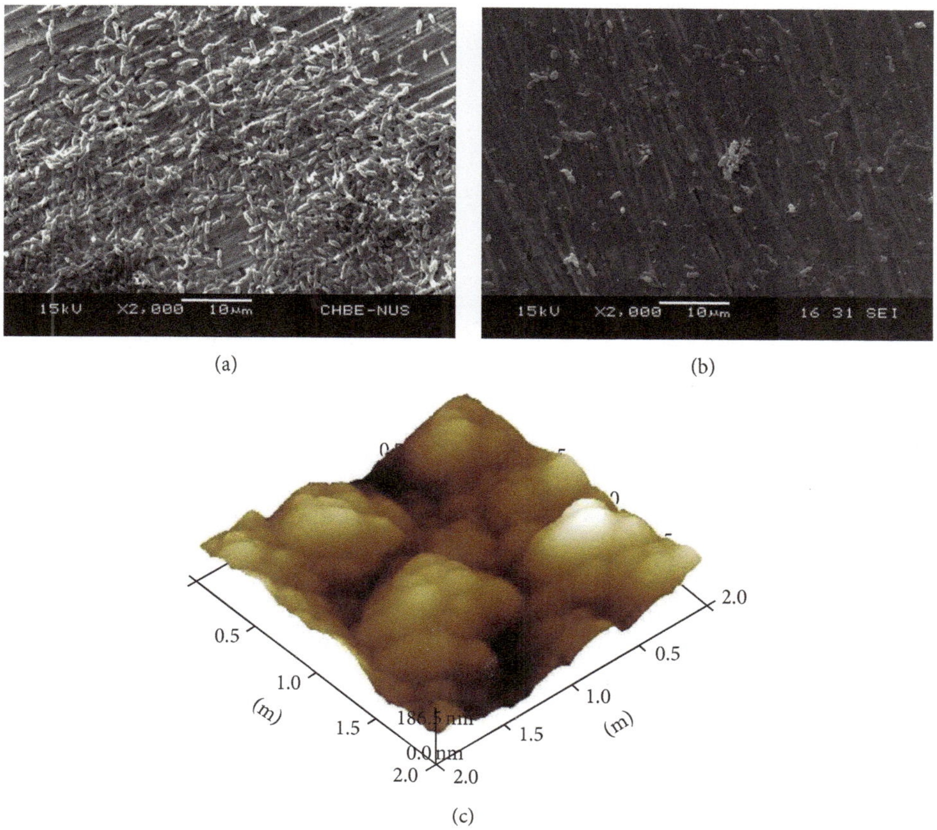

FIGURE 4: SEM and AFM micrographs of the copper surface coupon after immersion in cooling water systems (a) SEM: *M. timonae* ARTYP3 and (b) SEM: *M. timonae* ARTYP3 inoculated with biocide (c) AFM (three-dimensional view) *M. timonae* biofilm.

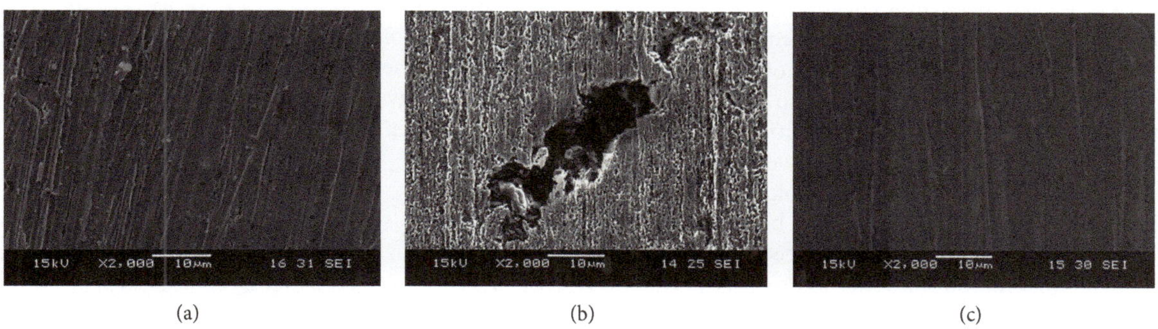

FIGURE 5: SEM micrograph of the copper surface coupon after immersion in cooling water systems (a). Copper surface exposed to the medium without bacteria and biocide. (b) Copper surface exposed to *M. Timonae*. (c) Copper surface exposed to *M. timonae* with the biocide bronopol.

Weight loss and corrosion rate were also monitored in the presence or absence of *M. timonae* and the biocide. Weight loss increased in the presence of the ARTYP3 (0.065 mg) compared to the uninoculated system (0.0073 mg) but was significantly reduced to about 0.0045 mg in the presence of the biocide. The corrosion rate, at 0.0177, 0.1573, and 0.0109 mm/year in Systems 1, 2, and 3, respectively, showed that bronopol inhibits the pitting corrosion of copper metal. Unlike the control system where uniform corrosion was noted, severe pitting attacks were observed over the surface of steel in the presence of ARTYP (Figure 5). In the presence of the biocide (System 3), no obvious corrosion pit was observed (Figure 5), thus showing that the biocide significantly inhibits biofilm formation (Figure 4) by ARTYP3 on the metal surface and minimizes corrosion. *M. timonae* and *Pseudomonas* sp. contribute to biofilm formation by producing exopolysaccharides and facilitating the attachment of other microorganisms, hence accelerating the corrosion process [60]. *Massilia* sp. and *Pseudomonas* sp. are capable of oxidizing copper ions as electron donors and gaining energy from the oxidation of

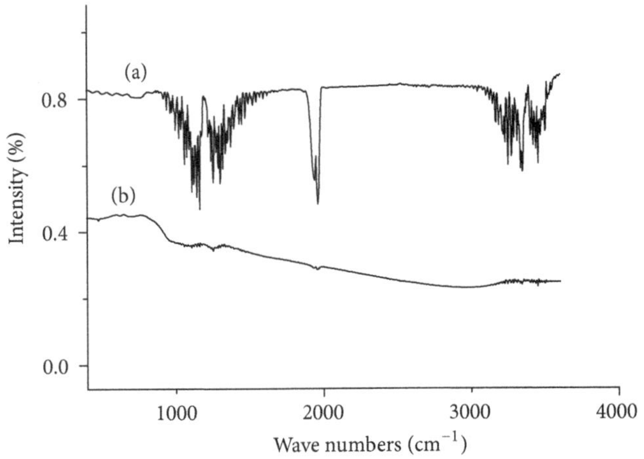

FIGURE 6: FTIR spectrum of copper surface coupon after immersion of cooling water systems. (a) *M. timonae* ARTYP3 and (b) *M. timonae* ARTYP3 inoculated with biocide.

copper to copper oxides. The bacteria promote the formation of copper oxides at low pH (5.0) which leads to pitting corrosion.

Surface hydrophobicity of microorganisms is recognized as a dominant factor that influences its adhesion on surfaces and hence surface corrosion. In biological systems, hydrophobic interactions are usually the strongest of all long-range noncovalent interactions and can be defined as the attraction between apolar or slightly polar molecules, particles, or cells, when immersed in water [61]. The hydrophobic nature of ARTYP3 has been established, and the emulsification index (E) was found to increase with the time of exposure [50]. This suggests the production of biosurfactant (rhamnolipid) by ARTYP3 and thus leads to higher uptake of hexadecane as the carbon source. In addition, aggregates of ARTYP3 were formed, even at the lowest dilution of ammonium sulphate, due to the hydrophobic nature of the cells. A higher hydrophobic behavior of the bacterial isolate induces greater adhesion on the metal surface [50, 62].

EPS comprises macromolecules such as proteins, polysaccharides, nucleic acids, and phospholipids. The capacity of EPS to bind metal ions is an important factor in determining the MIC behavior of the bacterial isolates [48, 60, 63]. Figure 4(a) shows typical rod-shaped cells (about 1 μm) on the metal surface (Figure 4(a)). In the presence of the biocide, significantly less bacteria were observed (Figure 4(b)). A thick layer of biofilm with clusters of microbial cells and EPS was also noted on the metal surface (Figure 4(c)). FTIR (Figure 6(b)) confirms the adhesion of EPS (COO^-, amide groups) secreted by the bacteria on the metal surface [63]. The addition of bronopol resulted in significant reduction in all the peak intensity, due to the inhibition of biofilm formation. The antibacterial mechanism of bronopol has been attributed to its interaction with essential thiols within the cell [32, 33]. Shepherd et al. [36] have also reported that bronopol catalyses the oxidation of thiol groups to disulfides with the consumption of oxygen. Such catalytic reaction also results in the generation of free radicals which subsequently inhibits the growth of the bacteria.

5. Conclusion

In this study, a corrosive bacterial community in water samples collected from a cooling tower was cultivated using traditional cultivation techniques and identified using 16S rRNA gene sequence. *M. timonae* was found to be the dominant corrosive bacteria amongst the culturable organisms and accelerated severe pitting corrosion in copper metal. The genus *Massilia* is identified for the first time in water from a cooling tower. The biosurfactant produced by the ARTYP3 contributes to an increase in the cell surface hydrophobicity of the bacteria and enhanced the bacterial adhesion on the copper metal surface. SEM-EDAX, AFM, and FTIR analyses of the copper metal coupons with bacterial biofilm developed after exposure to cooling water confirmed that EPS accumulate with exposure time and revealed that biofilms were formed as microcolonies, which subsequently caused pitting corrosion. Bronopol was shown to be an effective biocide in inhibiting both biofilm formation on the copper metal surface and the subsequent biocorrosion.

Conflict of Interests

The authors declare that there is no conflict of interests.

Acknowledgments

Rajasekar Aruliah acknowledges the award of a Research Fellowship supported under a Research Grant (R- 279-000-227-112) from the National University of Singapore, Singapore. Rajasekar Aruliah is thankful to the Department of Biotechnology (Government of India) for continuing his research career in India through the Ramalingaswami Fellow Award 2013.

References

[1] C.-G. Peng, J. K. Park, and R. W. Patenaude, "Statistics-based classification of microbially influenced corrosion in freshwater systems," *Water Research*, vol. 28, no. 4, pp. 951–959, 1994.

[2] A. Rajasekar and Y.-P. Ting, "Microbial corrosion of aluminum 2024 aeronautical alloy by hydrocarbon degrading bacteria *Bacillus cereus* ACE4 and *Serratia marcescens* ACE2," *Industrial & Engineering Chemistry Research*, vol. 49, no. 13, pp. 6054–6061, 2010.

[3] H. A. Videla, "Prevention and control of biocorrosion," *International Biodeterioration & Biodegradation*, vol. 49, no. 4, pp. 259–270, 2002.

[4] A. Rajasekar, S. Maruthamuthu, N. Palaniswamy, and A. Rajendran, "Biodegradation of corrosion inhibitors and their influence on petroleum product pipeline," *Microbiological Research*, vol. 162, no. 4, pp. 355–368, 2007.

[5] A. Rajasekar, T. G. Babu, S. K. Pandian, S. Maruthamuthu, N. Palaniswamy, and A. Rajendran, "Biodegradation and corrosion behavior of manganese oxidizer *Bacillus cereus* ACE4 in

diesel transporting pipeline," *Corrosion Science*, vol. 49, no. 6, pp. 2694–2710, 2007.

[6] A. Rajasekar, B. Anandkumar, S. Maruthamuthu, Y.-P. Ting, and P. K. S. M. Rahman, "Characterization of corrosive bacterial consortia isolated from petroleum-product-transporting pipelines," *Applied Microbiology and Biotechnology*, vol. 85, no. 4, pp. 1175–1188, 2010.

[7] A. Rajasekar and Y.-P. Ting, "Role of inorganic and organic medium in the corrosion behavior of *Bacillus megaterium* and *Pseudomonas* sp. in stainless steel SS 304," *Industrial & Engineering Chemistry Research*, vol. 50, no. 22, pp. 12534–12541, 2011.

[8] J. Elguindi, S. Moffitt, H. Hasman, C. Andrade, S. Raghavan, and C. Rensing, "Metallic copper corrosion rates, moisture content, and growth medium influence survival of copper ion-resistant bacteria," *Applied Microbiology and Biotechnology*, vol. 89, no. 6, pp. 1963–1970, 2011.

[9] R. MacDonald and V. S. Brözel, "Community analysis of bacterial biofilms in a simulated recirculating cooling-water system by fluorescent *in situ* hybridization with rRNA-targeted oligonucleotide probes," *Water Research*, vol. 34, no. 9, pp. 2439–2446, 2000.

[10] P. J. Bremer and G. G. Geesey, "Laboratory-based model of microbiologically induced corrosion of copper," *Applied and Environmental Microbiology*, vol. 57, no. 7, pp. 1956–1962, 1991.

[11] S. W. Borenstein, *Microbiologically Influenced Corrosion Handbook*, Woodhead, Cambridge, UK, 1994.

[12] J. M. Kusnetsovi, P. J. Martikainen, H. R. Jousimies-Somer et al., "Physical, chemical and microbiological water characteristics associated with the occurrence of *Legionella* in cooling tower systems," *Water Research*, vol. 27, no. 1, pp. 85–90, 1993.

[13] H. Yamamoto, M. Sugiura, S. Kusunoki, T. Ezaki, M. Ikedo, and E. Yabuuchi, "Factors stimulating propagation of *Legionella* in cooling tower water," *Applied and Environmental Microbiology*, vol. 58, no. 4, pp. 1394–1397, 1992.

[14] P. Wagner and B. Little, "Impact of alloying on microbiologically influenced corrosion—a review," *Materials Performance*, vol. 32, no. 9, pp. 65–68, 1993.

[15] G. G. Geesey and P. J. Bremer, "Evaluation of copper corrosion under bacterial biofilms," in *Proceedings of the Annual Meeting National Association of Corrosion Engineers*, paper 111, NACE, Cincinnati, Ohio, USA, 1991.

[16] H.-Y. Cheung, G. K.-L. Chan, S.-H. Cheung, S.-Q. Sun, and W.-F. Fong, "Morphological and chemical changes in the attached cells of *Pseudomonas aeruginosa* as primary biofilms develop on aluminium and CaF_2 plates," *Journal of Applied Microbiology*, vol. 102, no. 3, pp. 701–710, 2007.

[17] P. Angell and A. H. L. Chamberlain, "The role of extracellular products in copper colonisation," *International Biodeterioration*, vol. 27, no. 2, pp. 135–143, 1991.

[18] J. Li, J. Li, W. Yuan, and Y. Du, "Biocorrosion characteristics of the copper alloys BFe30-1-1 and HSn70-1AB by SRB using Atomic Force Microscopy and Scanning Electron Microscopy," *International Biodeterioration & Biodegradation*, vol. 64, no. 5, pp. 363–370, 2010.

[19] M. W. Mittelman and G. G. Geesey, "Copper-binding characteristics of exopolymers from a freshwater-sediment bacterium," *Applied and Environmental Microbiology*, vol. 49, no. 4, pp. 846–851, 1985.

[20] I. B. Beech and J. Sunner, "Biocorrosion: towards understanding interactions between biofilms and metals," *Current Opinion in Biotechnology*, vol. 15, no. 3, pp. 181–186, 2004.

[21] G. Blunn, "Biological fouling of copper and copper alloys," in *Biodeterioration*, S. Barry, D. R. Hozughton, G. C. Llewellyn, and C. E. O'Rear, Eds., vol. 6, pp. 567–575, CAB International, Slough, UK, 1986.

[22] N. Ceyhan and G. Ozdemir, "Extracellular polysaccharides produced by cooling water tower biofilm bacteria and their possible degradation," *Biofouling*, vol. 24, no. 2, pp. 129–135, 2008.

[23] G. D. Drysdale, H. C. Kasan, and F. Bux, "Denitrification by heterotrophic bacteria during activated sludge treatment," *Water SA*, vol. 25, no. 3, pp. 357–362, 1999.

[24] G. T. Huang, K.-Y. Chan, and H. H. P. Fang, "Microbiologically induced corrosion of 70Cu-30Ni alloy in anaerobic seawater," *Journal of the Electrochemical Society*, vol. 151, no. 7, pp. B434–B439, 2004.

[25] J. M. Kusnetsov, A. I. Tulkki, H. E. Ahonen, and P. J. Martikainen, "Efficacy of three prevention strategies against legionella in cooling water systems," *Journal of Applied Microbiology*, vol. 82, no. 6, pp. 763–768, 1997.

[26] S. Okabe, W. L. Jones, W. Lee, and W. G. Characklis, "Anaerobic SRB biofilms in industrial water systems: a process analysis," in *Biofouling and Biocorrosion in Industrial Water Systems*, G. G. Geesy, Z. Lewandowsky, and H. C. Flemming, Eds., pp. 189–204, Lewis, Boca Raton, Fla, USA, 1994.

[27] C. U. Schwermer, G. Lavik, R. M. Abed et al., "Impact of nitrate on the structure and function of bacterial biofilm communities in pipelines used for injection of seawater into oil fields," *Applied and Environmental Microbiology*, vol. 74, no. 9, pp. 2841–2851, 2008.

[28] J. Wen, K. Zhao, T. Gu, and I. I. Raad, "A green biocide enhancer for the treatment of sulfate-reducing bacteria (SRB) biofilms on carbon steel surfaces using glutaraldehyde," *International Biodeterioration & Biodegradation*, vol. 63, no. 8, pp. 1102–1106, 2009.

[29] S. A. Haveman, E. A. Greene, C. P. Stilwell, J. K. Voordouw, and G. Voordouw, "Physiological and gene expression analysis of inhibition of *Desulfovibrio vulgaris* Hildenborough by nitrite," *Journal of Bacteriology*, vol. 186, no. 23, pp. 7944–7950, 2004.

[30] R. Zuo, D. Örnek, B. C. Syrett et al., "Inhibiting mild steel corrosion from sulfate-reducing bacteria using antimicrobial-producing biofilms in Three-Mile-Island process water," *Applied Microbiology and Biotechnology*, vol. 64, no. 2, pp. 275–283, 2004.

[31] R. Frey, "Award-winning biocides are lean, mean, and green," *Today's Chemist at Work*, vol. 7, no. 6, pp. 34–35, 37–38, 1998.

[32] E. A. Greene, C. Hubert, M. Nemati, G. E. Jenneman, and G. Voordouw, "Nitrite reductase activity of sulphate-reducing bacteria prevents their inhibition by nitrate-reducing, sulphide-oxidizing bacteria," *Environmental Microbiology*, vol. 5, no. 7, pp. 607–617, 2003.

[33] E. A. Greene, V. Brunelle, G. E. Jenneman, and G. Voordouw, "Synergistic inhibition of microbial sulfide production by combinations of the metabolic inhibitor nitrite and biocides," *Applied and Environmental Microbiology*, vol. 72, no. 12, pp. 7897–7901, 2006.

[34] A. J. Telang, S. Ebert, J. M. Foght et al., "Effect of nitrate injection on the microbial community in an oil field as monitored by reverse sample genomeprobing," *Applied and Environmental Microbiology*, vol. 63, no. 5, pp. 1785–1793, 1997.

[35] B. Croshaw, M. J. Groves, and B. Lessel, "Some properties of bronopol, a new antimicrobial agent active against *Pseudomonas aeruginosa*," *The Journal of Pharmacy and Pharmacology*, vol. 16, no. S1, pp. 127T–130T, 1964.

[36] J. A. Shepherd, R. D. Waigh, and P. Gilbert, "Antibacterial action of 2-bromo-2-nitropropane-1,3-diol (bronopol)," *Antimicrobial Agents and Chemotherapy*, vol. 32, no. 11, pp. 1693–1698, 1988.

[37] D. Starosvetsky, R. Armon, J. Yahalom, and J. Starosvetsky, "Pitting corrosion of carbon steel caused by iron bacteria," *International Biodeterioration & Biodegradation*, vol. 47, no. 2, pp. 79–87, 2001.

[38] J. G. Holt, N. R. Kreig, P. H. A. Sneath, J. T. Stanely, and S. T. Williams, Eds., *Bergey's Manual of Determinative Bacteriology*, Williams & Wilkins, Baltimore, Md, USA, 1994.

[39] F. M. Ausubel, R. Brent, R. E. Kingston, D. D. Moore, J. G. Seidelman, and K. E. Struhl, *Current Protocols in Molecular Biology*, vol. 2, John Wiley & Sons, New York, NY, USA, 1988.

[40] W. G. Weisburg, S. M. Barns, D. A. Pelletier, and D. J. Lane, "16S ribosomal DNA amplification for phylogenetic study," *Journal of Bacteriology*, vol. 173, no. 2, pp. 697–703, 1991.

[41] S. F. Altschul, W. Gish, W. Miller, E. W. Myers, and D. J. Lipman, "Basic local alignment search tool," *Journal of Molecular Biology*, vol. 215, no. 3, pp. 403–410, 1990.

[42] M. A. Larkin, G. Blackshields, N. P. Brown et al., "Clustal W and Clustal X version 2.0," *Bioinformatics*, vol. 23, no. 21, pp. 2947–2948, 2007.

[43] M. Kimura, "A simple method for estimating evolutionary rates of base substitutions through comparative studies of nucleotide sequences," *Journal of Molecular Evolution*, vol. 16, no. 2, pp. 111–120, 1980.

[44] K. Tamura, J. Dudley, M. Nei, and S. Kumar, "MEGA4: Molecular Evolutionary Genetics Analysis (MEGA) software version 4.0," *Molecular Biology and Evolution*, vol. 24, no. 8, pp. 1596–1599, 2007.

[45] D. M. Hillis and J. J. Bull, "An empirical test of bootstrapping as a method for assessing confidence in phylogenetic analysis," *Systematic Biology*, vol. 42, no. 2, pp. 182–192, 1993.

[46] J. J. Campanella, L. Bitincka, and J. Smalley, "MatGAT: an application that generates similarity/identity matrices using protein or DNA sequences," *BMC Bioinformatics*, vol. 4, article 29, 2003.

[47] J.-Y. Maillard, "Bacterial target sites for biocide action," *Journal of Applied Microbiology Symposium Supplement*, vol. 92, no. S1, pp. 16S–27S, 2002.

[48] X. Sheng, Y.-P. Ting, and S. O. Pehkonen, "Evaluation of an organic corrosion inhibitor on abiotic corrosion and microbiologically influenced corrosion of mild steel," *Industrial and Engineering Chemistry Research*, vol. 46, no. 22, pp. 7117–7125, 2007.

[49] V. Pruthi and S. S. Cameotra, "Rapid identification of biosurfactant-producing bacterial strains using a cell surface hydrophobicity technique," *Biotechnology Techniques*, vol. 11, no. 9, pp. 671–674, 1997.

[50] A. Harimawan, A. Rajasekar, and Y.-P. Ting, "Bacteria attachment to surfaces—AFM force spectroscopy and physicochemical analyses," *Journal of Colloid and Interface Science*, vol. 364, no. 1, pp. 213–218, 2011.

[51] V. Gallego, C. Sánchez-Porro, M. T. García, and A. Ventosa, "*Massilia aurea* sp. nov., isolated from drinking water," *International Journal of Systematic and Evolutionary Microbiology*, vol. 56, no. 10, pp. 2449–2453, 2006.

[52] B. La Scola, R. J. Birtles, M.-N. Mallet, and D. Raoult, "*Massilia timonae* gen. nov., sp. nov., isolated from blood of an immunocompromised patient with cerebellar lesions," *Journal of Clinical Microbiology*, vol. 36, no. 10, pp. 2847–2852, 1998.

[53] Y.-Q. Zhang, W.-J. Li, K.-Y. Zhang et al., "*Massilia dura* sp. nov., *Massilia albidiflava* sp. nov., *Massilia plicata* sp. nov. and *Massilia lutea* sp. nov., isolated from soils in China," *International Journal of Systematic and Evolutionary Microbiology*, vol. 56, no. 2, pp. 459–463, 2006.

[54] D. A. Moreno, J. R. Ibars, I. B. Beech, and C. C. Gaylarde, "Biofilm formation on mild steel coupons by *Pseudomonas* and *Desulfovibrio*," *Biofouling*, vol. 7, no. 2, pp. 129–139, 1993.

[55] R. Gubner and I. B. Beech, "The effect of extracellular polymeric substances on the attachment of *Pseudomonas* NCIMB 2021 to AISI 304 and 316 stainless steel," *Biofouling*, vol. 15, no. 1–3, pp. 25–36, 2000.

[56] A. Pedersen, S. Kjelleberg, and M. Hermansson, "A screening method for bacterial corrosion of metals," *Journal of Microbiological Methods*, vol. 8, no. 4, pp. 191–198, 1988.

[57] J. P. Busalmen, M. Vázquez, and S. R. De Sánchez, "New evidences on the catalase mechanism of microbial corrosion," *Electrochimica Acta*, vol. 47, no. 12, pp. 1857–1865, 2002.

[58] S. Maruthamuthu, P. Dhandapani, S. Ponmariappan, J.-H. Bae, N. Palaniswamy, and P. K. S. M. Rahman, "Impact of ammonia producing *Bacillus* sp. on corrosion of cupronickel alloy 90:10," *Metals and Materials International*, vol. 15, no. 3, pp. 409–419, 2009.

[59] H. Ashassi-Sorkhabi and S. A. Nabavi-Amri, "Polarization and impedance methods in corrosion inhibition study of carbon steel by amines in petroleum-water mixtures," *Electrochimica Acta*, vol. 47, no. 13-14, pp. 2239–2244, 2002.

[60] J. F. Batista, R. F. C. Pereira, J. M. Lopes, M. F. M. Carvalho, M. J. Feio, and M. A. M. Reis, "*In situ* corrosion control in industrial water systems," *Biodegradation*, vol. 11, no. 6, pp. 441–448, 2000.

[61] C. J. van Oss, "Hydrophobicity and hydrophilicity of biosurfaces," *Current Opinion Colloid & Interface Science*, vol. 2, no. 5, pp. 503–512, 1997.

[62] L. Mei, H. C. Van Der Mei, Y. Ren, W. Norde, and H. J. Busscher, "Poisson analysis of streptococcal bond strengthening on stainless steel with and without a salivary conditioning film," *Langmuir*, vol. 25, no. 11, pp. 6227–6231, 2009.

[63] G. G. Geesey, M. W. Mittelman, T. Iwaoka, and P. R. Griffiths, "Role of bacterial exopolymers in the deterioration of metallic copper surfaces," *Materials Performance*, vol. 25, no. 2, pp. 37–40, 1986.

5

Corrosion Inhibition of Carbon Steel in 0.5 M H_2SO_4 by Phtalocyanine Blue

J. C. Valle-Quitana,[1] G. F. Dominguez-Patiño,[2] and J. G. Gonzalez-Rodriguez[1]

[1] *Universidad Autonoma del Estado de Morelos, CIICAP, Avenida Universidad 1001, 62209 Cuernavaca, MOR, Mexico*
[2] *Facultad de Ciencias Biologicas, Universidad Autonoma del Estado de Morelos, Avenida Universidad 1001, 62209 Cuernavaca, MOR, Mexico*

Correspondence should be addressed to J. G. Gonzalez-Rodriguez; ggonzalez@uaem.mx

Academic Editors: A. Hermann, I. Obot, and R. Salghi

Phthalocyanine blue dye has been investigated as a carbon steel corrosion inhibitor in 0.5 M sulfuric acid by using polarization curves, electrochemical impedance spectroscopy, and gravimetric tests. Dye concentrations included 0, 100, 200, 400, 600, 800, and 1000 ppm, whereas testing temperatures were 25, 40, and 60°C. Results indicated that phtalocyanine blue is a good corrosion inhibitor with its efficiency increasing with the concentration up to 40°C, but it increases at 60°C. Inhibitor improves the passive film properties and it forms an adherent, compact, protective film, acting, therefore, as an anodic-type inhibitor. At 25 and 40°C the corrosion process was under charge transfer, whereas at 60°C the adsorption/desorption of some species from the metal surface controlled the corrosion process.

1. Introduction

Organic compounds containing atoms such as nitrogen, sulphur, and oxygen are commonly used to reduce the corrosion attack on steel in acidic media. These compounds adsorb on the metal surface, block the active sites on the surface, and thereby reduce the corrosion process [1–6]. The inhibiting action of such compounds is based on the adsorption ability of their molecules, where the resulting adsorption film isolates the metal from the aggressive media. The corrosion inhibition of organic compounds is related to their adsorption properties. Adsorption depends on the nature and the state of the metal surface, on the type of corrosive environment, and on the chemical structure of the inhibitor [7]. Studies report that the adsorption of organic inhibitors mainly depends on some physicochemical properties of the molecule, related to its functional groups, to the possible steric effects and electronic density of donor atoms; adsorption is suppose, also to depend on the possible interaction of the π-orbitals of the inhibitor with the d-orbitals of the surface atoms, which induce greater adsorption of the inhibitor molecules onto the surface of metal, leading to the formation of a corrosion protection film [8].

Recently, there has been a lot of attention paid to the use of organic dyes as potential corrosion inhibitors for different metalsenvironments [9–24]. Thus, Oguzie evaluated the use of Congo red [9], methylene blue [10], indigo blue [11], safranin-o, thymol blue, and fluorescein [12] as corrosion inhibitors for mild steel in acidic media, although in some cases they evaluated these dyes for aluminum in hydrochloric acid [13]. Alizarin yellow and Bromophenol blue have been used also but now studying the synergistic effect of iodide ions for the corrosion inhibition of carbon steel in acidic media [14, 15]. Additionally, alizarin violet [16], 3H-phenothizzin-3-one, 7-dimethylamin [17], phenylazo-pyrazolones [18], acid violet 6B [19], and methyl violet [20] have also been studied as corrosion inhibitors for cold rolled steel in acidic environments. Finally, some traditional analytical indicators, for example, methyl red [21], cresol red [22], methyl orange [23], and fast green [24], behave as good corrosion inhibitors for steel in acid media. Thus, the aim of this work is to evaluate the corrosion inhibitory properties

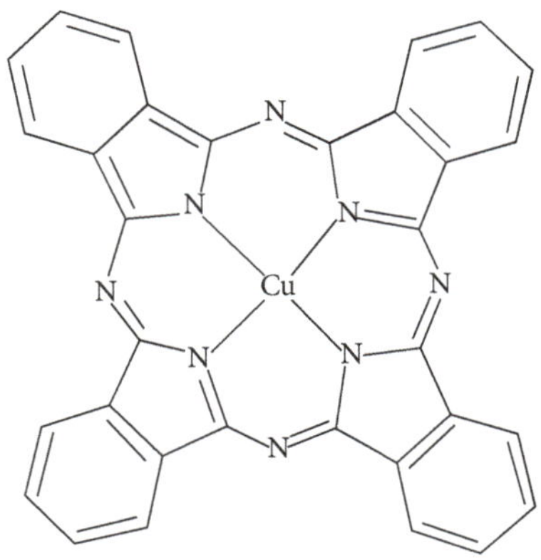

FIGURE 1: Chemical structure of phtalocyaine blue.

of phtalocyanine blue for carbon steel in acidic environment, an organic dye with good solubility in water, whose chemical structure is given in Figure 1 and possesses N and Cu atoms.

2. Experimental Procedure

Corrosion tests were performed on coupons prepared from 1018 carbon steel rods containing 0.14% C, 0.90% Mn, 0.30% S, 0.030% P, and as balance Fe, encapsulated in commercial epoxic resin with an exposed area of 1.0 cm^2. The organic dye phtalocyanine blue was obtained from a local manufacturer. The aggressive solution, 0.5 M H_2SO_4, was prepared by dilution of analytical grade H_2SO_4 with double distilled water. Weight loss experiments were carried out with steel rods 2.5 cm length and 0.6 cm diameter abraded with fine emery paper until 1200 grade, rinsed with acetone, and exposed to the aggressive solution during 72 h. After being corroded, specimens were taken out, washed with distilled water, degreased with acetone, dried, and weighed accurately. Tests were performed by triplicate at room temperature (25°C), 40 and 60°C by using a hot plate. Corrosion rates, in terms of weight loss measurements, ΔW, was calculated as follows:

$$\Delta W = \frac{m_1 - m_2}{A}, \quad (1)$$

where m_1 is the mass of the specimen before corrosion, m_2 the mass of the specimen after corrosion, and A the exposed area of the specimen. For the weight loss tests, inhibitor efficiency, IE, were calculated as follows:

$$\text{IE } (\%) = \frac{100\left(\Delta W_1 - \Delta W_2\right)}{\Delta W_1}, \quad (2)$$

where ΔW_1 is the weight loss without inhibitor and ΔW_2 the weight loss with inhibitor. Specimens were removed, rinsed in water and in acetone, dried in warm air, and stored in a dessicator. Specimens were weighed in an analytical balance with a precision of 0.1 mg. Electrochemical techniques employed included potentiodynamic polarization curves and electrochemical impedance spectroscopy measurements, EIS. In all experiments, the carbon steel electrode was allowed to reach a stable open circuit potential value, E_{corr}. Polarization curves were recorded at a constant sweep rate of 1 mV/s at the interval from −1000 to +1500 mV with respect to the E_{corr} value. Measurements were obtained by using a conventional three electrodes glass cell with two graphite electrodes symmetrically distributed and a saturated calomel electrode (SCE) as reference with a Lugging capillary bridge. Corrosion current density values, i_{corr}, were obtained by using Tafel extrapolation. Inhibitor efficiency, IE, was calculated as follows:

$$\text{IE } (\%) = \frac{100\left(i_{corr1} - i_{corr2}\right)}{i_{corr1}}, \quad (3)$$

where i_{corr1} is the corrosion current density value without inhibitor and i_{corr2} the corrosion current density value with inhibitor. EIS tests were carried out at E_{corr} by using a signal with amplitude of 10 mV in a frequency interval of 100 mHz–100 KHz. An ACM potentiostat controlled by a desk top computer was used for the polarization curves, whereas for the EIS measurements, a model PC4 300 Gamry potentiostat was used.

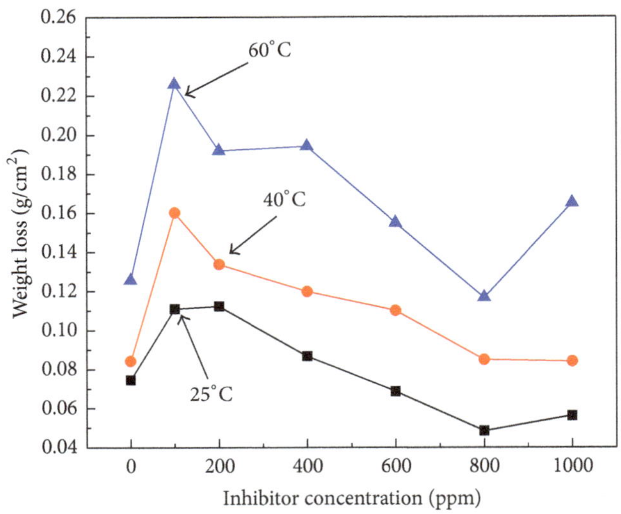

FIGURE 2: Effect of inhibitor concentration in the weight loss for 1018 carbon steel in 0.5 M H_2SO_4 at different temperatures.

3. Results and Discussion

Figure 2 shows the corrosion rate results, in terms of the weight loss, for carbon steel in 0.5 M H_2SO_4 at different inhibitor concentrations at 25–60°C. It can be seen that when 100 ppm of inhibitor is added the corrosion rate decreases compared to that for uninhibited solution, at the three different temperatures, but with a further increase in the inhibitor concentration the corrosion rate decreases, reaching its lowest value more or less around 800 ppm, increasing again at 1000 ppm. The fact that the corrosion rate decreases with the inhibitor concentration is due to the fact that the inhibitor adsorption coverage on the steel surface starts to increase [25]. It can also be seen that the corrosion rate increases with an increase in the temperature in both

TABLE 1: Inhibitor efficiency values obtained from the weight loss tests at different concentrations and temperatures.

C_{inh} (ppm)	I.E. (%)		
	25°C	40°C	60°C
0	—	—	—
100	−48.806	−79.781	−89.513
200	−50.759	−52.564	−58.637
400	−16.549	−2.8457	−54.426
600	7.942	−23.521	−30.119
800	35.052	−0.7194	7.1672
1000	24.6	0.2282	−31.134

TABLE 2: Potentiodynamical polarization parameters for the corrosion of 1018 carbon steel in 0.5 M H_2SO_4 at 25°C.

C_{inh} (ppm)	E_{corr} (mV)	i_{corr} (mA/cm^2)	E_{pas} (mV)	i_{pas} (mA/cm^2)	I.E. (%)
0	−503	0.3	725	6.2	—
100	−456	0.2	535	3.7	40
200	−457	0.15	645	5.3	15
400	−460	0.13	530	2.8	55
600	−462	0.12	535	1.7	67
800	−459	0.11	286	1.1	82
1000	−461	0.1	207	4.9	26

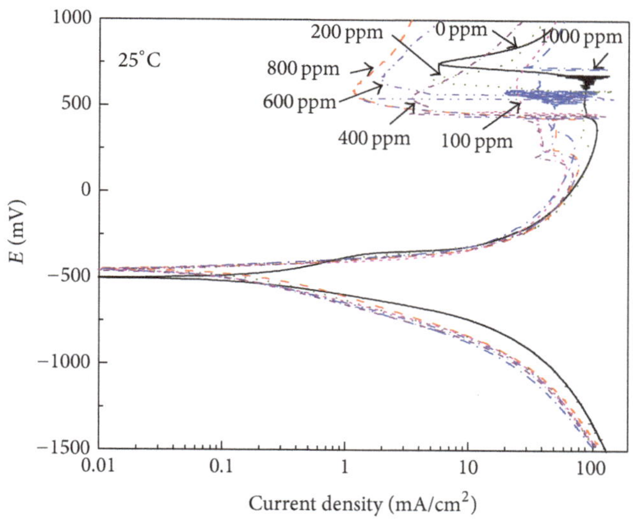

FIGURE 3: Effect of inhibitor concentration in the polarization curves for 1018 carbon steel in 0.5 M H_2SO_4 at 25°C.

uninhibited and inhibited solutions. Inhibitor efficiency at the different concentrations and temperatures is given in Table 1, indicating that the highest efficiency values are found between 800 and 1000 ppm. It should be noted that the inhibitor efficiency decreases with increasing the temperature up to 40°C but it increases with a further increase in the temperature up to 60°C. This is because inhibitor molecules are physically adsorbed on the metal surface, whereas a chemical adsorption implies an increase in the inhibitor efficiency with an increase in the temperature [8–13]. It has been reported that inhibitor can form coordination bonds between the unshared electron pair of N atom and the empty d iron electron [8, 22]. It means that coordinate bond could be formed by donating the lone electron pairs of N atoms to the unfilled orbits of Fe. Blue phtalocyanine can be protonated in acidic solution [24]; SO_4^{2-} ions could accumulate gradually close to the metal/solution interface, being specifically adsorbed; they create an excess negative charge towards the solution and favor more adsorption of the cations; then the protonated inhibitor may adsorb through electrostatic interactions between the positively charged molecules and the negatively charged metal surface. When protonated blue phtalocyanine adsorbed on metal surface, coordinate bond may be formed by partial transference of electrons from the N polar atom to the metal surface.

The effect of inhibitor concentration in the polarization curves for carbon steel in 0.5 M H_2SO_4 at 25°C is shown in Figure 3, whereas the electrochemical parameters are given in Table 2. It can be seen that the steel exhibits an active-passive behavior. The addition of the inhibitor did not affect very much neither the E_{corr} nor the i_{corr} values, but it improves the passive current density and passivation potential values, i_{pas} and E_{pas}, respectively. Thus, the E_{corr} value for the uninhibited solution was −503 mV, and with the addition of inhibitor this value was around −460 mV approximately regardless of the inhibitor concentration. The i_{corr} slightly decreased from 0.3 down to 0.1 mA/cm^2, and the lowest value obtained was with the addition of 1000 ppm. On the other side, as can be seen in Table 2, the passivation potential value, E_{pas}, for the uninhibited solution was 725 mV, but it decreased with the addition of inhibitor, obtaining the lowest value, 286 mV, with the addition of 800 ppm. Similarly, the passivation current density value, i_{pas}, was 6.2 mA/cm^2 in absence of inhibitor, but it decreased as the inhibitor concentration increased, obtaining the lowest value, 1.1 mA/cm^2 by adding 800 ppm of inhibitor; a further increase in the inhibitor concentration to 1000 ppm increased the i_{pas} value, similar to the behavior of the weight loss results shown in Figure 2. Thus, it could be concluded that the addition of phtalocyanine blue decreases the corrosion rate of carbon steel in acidic media by improving the passive film properties. In Table 2, inhibitor efficiency values were calculated by using the i_{pas} values instead of the i_{corr} ones, since the inhibitor effect is much clearer in the former than in the later.

Polarization curves at 40°C, Figure 4, were not very different to those obtained at room temperature: the steel exhibited an active-passive behavior, neither the E_{corr} nor the i_{corr} values were drastically affected by the addition of the inhibitor, Table 3. Unlike this, the E_{pas} and i_{pas} values were improved by the addition of the inhibitor, reaching the best effect with 1000 ppm. Once again, in Table 3, inhibitor efficiency values were calculated by using the i_{pas} values instead of the i_{corr} ones, since the inhibitor effect is much clearer in the former than in the later. This time, the highest efficiency value was obtained at 1000 ppm, and the efficiency values were lower than those obtained at 40°C. Finally, at 60°C, the steel did not show an active-passive behavior, only

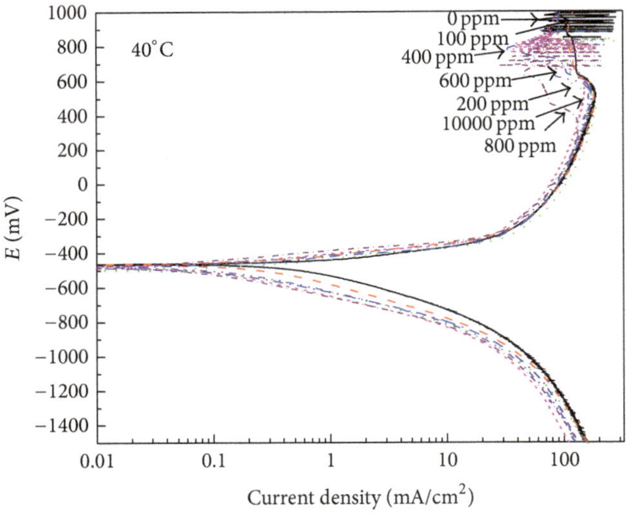

FIGURE 4: Effect of inhibitor concentration in the polarization curves for 1018 carbon steel in 0.5 M H_2SO_4 at 40°C.

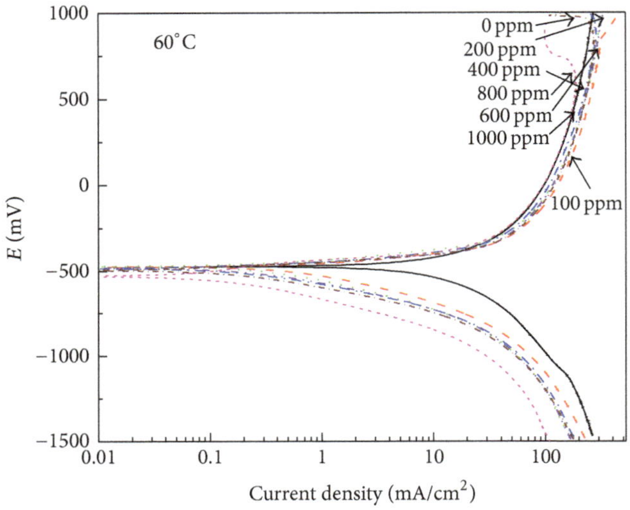

FIGURE 5: Effect of inhibitor concentration in the polarization curves for 1018 carbon steel in 0.5 M H_2SO_4 at 60°C.

anodic dissolution, Figure 5. The effect of adding inhibitor in the i_{corr} value at 60°C was somehow different, Table 4, since it increased with increasing the inhibitor concentration, reaching its lowest value with the addition of 1000 ppm. The anodic branch of the polarization curve was not affected by the inhibitor; instead the cathodic branch was affected, indicating that in this case the inhibitor acts as a cathodic-type of inhibitor. Additionally, the corrosion current density values increased with increasing the temperature. This time, efficiency values were higher than those obtained at both 25 and 40°C. The fact that the inhibitor does not affect the E_{corr} has been referred as it acts by a geometric blocking effect [22]; that is, the inhibition effect comes from the reduction of the reaction area on the surface of the corroding metal. However, it has been shown in this work that the main effect of the inhibitor is to improve the passive film properties,

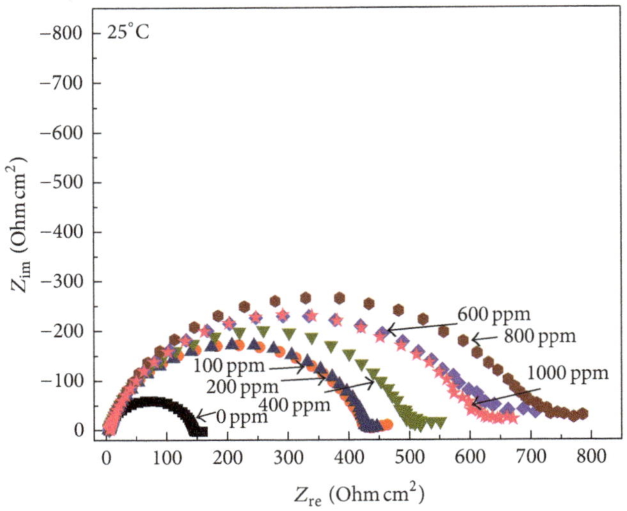

FIGURE 6: Nyquist plots for 1018 carbon steel in 0.5 M H_2SO_4 at 25°C.

TABLE 3: Potentiodynamical polarization parameters for the corrosion of 1018 carbon steel in 0.5 M H_2SO_4 at 40°C.

C_{inh} (ppm)	E_{corr} (mV)	i_{corr} (mA/cm^2)	E_{pas} (mV)	i_{pas} (mA/cm^2)	I.E. (%)
0	−480	0.8	650	133	—
100	−470	0.46	660	122	8
200	−500	0.2	640	110	17
400	−490	0.17	610	85	36
600	−483	0.12	630	77	42
800	−483	0.1	650	72	46
1000	−480	0.08	450	45	66

which is an anodic process, except at 60°C, where the cathodic slopes were affected. At 25 and 40°C the cathodic slopes were slightly affected; thus, the addition of the inhibitor implies the inhibitor molecules adsorption on both anodic and cathodic sites, resulting in an inhibition of both anodic and cathodic reduction reactions.

Impedance data in the Nyquist format for carbon steel corroded in 0.5 M H_2SO_4 at 25°C is given in Figure 6. Data describe a single, capacitive-like depressed semicircle with their centers in the real axis and with their diameters increasing with the inhibitor concentration up to 800 ppm, after which the diameter decreases with a further increase in the inhibitor concentration up to 1000 ppm. The single semicircle indicates that the corrosion process is under charge transfer control from the metal surface to the electrolyte through the double electrochemical layer and that the presence of the inhibitor does not change the mechanism of metal dissolution. Additionally, these impedance diagrams are not perfect semicircles which are related to the frequency dispersion as a result of the roughness and inhomogeneous of electrode surface [18]. Even more, the real axis intercepts at high and low frequencies in presence of the inhibitor is bigger in presence of the inhibitor than that in absence of the inhibitor, meaning that the impedance of the steel

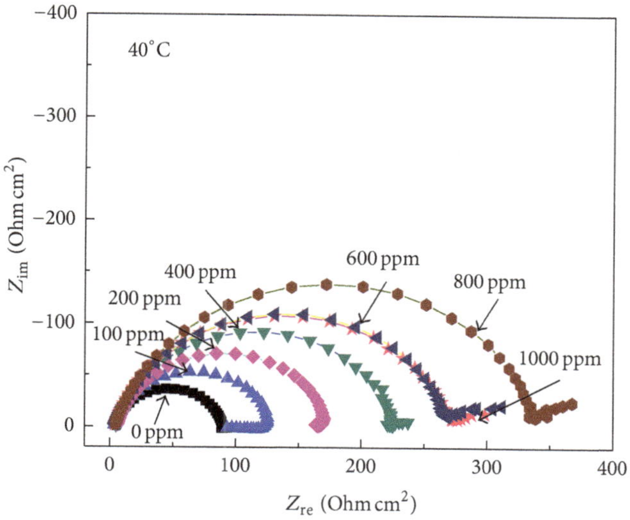

FIGURE 7: Nyquist plots for 1018 carbon steel in 0.5 M H_2SO_4 at 40°C.

TABLE 4: Potentiodynamical polarization parameters for the corrosion of 1018 carbon steel in 0.5 M H_2SO_4 at 60°C.

C_{inh} (ppm)	E_{corr} (mV)	i_{corr} (mA/cm^2)	I.E. (%)
0	−508	22	—
100	−503	11	50
200	−487	7.5	65
400	−490	5.6	75
600	−501	4.1	81
800	−503	3.5	84
1000	−500	2.9	87

in the inhibited environment increases with the inhibitor concentration, decreasing, thus, the metal corrosion rate. Thus, according to data in Figure 6, the highest inhibitor efficiency (81%) is reached with the addition of 800 ppm.

When the temperature is increased up to 40°C, Figure 7, Nyquist data also describe a single, depressed, capacitive-like semicircle with its centre increasing with increasing the inhibitor concentration, obtaining its biggest diameter value at 800 ppm of inhibitor. Thus, just like the tests at 25°C, the corrosion process is under charge transfer from the metal to the electrolyte and does not change with the presence of the inhibitor. The impedance values obtained at 40°C, however, are smaller than those obtained at 25°C, which implies that the corrosion rate increases with the temperature. The highest inhibitor efficiency, obtained with the addition of 800 ppm, was 75%, smaller than that obtained at 25°C. However, with a further increase in the testing temperature up to 60°C, Figure 8, Nyquist data described a capacitive-like, depressed semicircle at high and intermediate frequency values followed by an inductive semicircle at low frequencies. The capacitive semicircle diameter increased with increasing the inhibitor concentration reaching its maximum value when 800 ppm of inhibitor is added, with an efficiency value of 65%. The presence of an inductive loop at low frequencies means that the corrosion process is not any

TABLE 5: EIS Parameters for the corrosion of 1018 carbon steel in 0.5 M H_2SO_4 at 25°C.

C_{inh} (ppm)	R_{ct} (Ohm cm^2)	C_{dl} (μF/cm^2)	IE (%) 25°C	40°C	60°C
0	148	156	—	—	—
100	421	59	62	28	90
200	430	56	65	42	96
400	530	50	72	60	97
600	702	30	78	66	97
800	790	12	81	73	98
1000	671	22	78	67	96

longer under charge transfer control but under the control of the adsorption/desorption process of aggressive species or the inhibitor from metal surface.

EIS data can be calculated by using equivalent electric circuits as that shown in Figure 9. In this figure, R_s represents the solution resistance, R_{ct} the charge transfer resistance, C_{dl} the capacitance of the double electrochemical layer, and R_f represents the resistance of the corrosion products film and C_f its capacitance. However, the double-layer capacitance, C_{dl}, can also be calculated from the equation bellow:

$$C_{dl} = \frac{1}{2\pi f_{max} R_{ct}}, \tag{4}$$

where f_{max} is the frequency value at which the imaginary component of the impedance is maximal. For the uninhibited solution, a C_{dl} value of 156 μF/cm^2 was found. Table 5 gives the results for the R_{ct} and C_{dl} for the solution with and without inhibitor together with the inhibitor efficiency values at the three testing temperatures. It is important to note that by increasing the inhibitor concentration brings an increase in the charge transfer resistance value and a decrease in the double-layer capacitance. The decrease in the C_{dl} value can be interpreted as due to the adsorption of the inhibitor onto the electrode surface [20]. The double-layer formed at the metal-solution interface is considered as an electric capacitor, whose capacitance decreases due to the displacement of water molecules and other ions originally adsorbed on the electrode by the inhibitor molecules, forming a protective film. The thickness of the film formed increases with increasing concentration of the inhibitor, since more inhibitor adsorbs on the surface, resulting in lower C_{dl} values. It has to be noted that, according to Table 5, inhibitor efficiency values decrease when the temperature increases to 40°C; then they increase with a further increase in the temperature up to 60°C, similar to the reported values by using polarization curves, Tables 2–4.

It has been reported that the natural logarithm of the corrosion rate is a linear function with $1/T$, where T is the absolute temperature, for the corrosion of carbon steel in acidic conditions [2, 6, 9]. The apparent activation energy, E_a, associated with 1018 carbon steel in uninhibited and inhibited

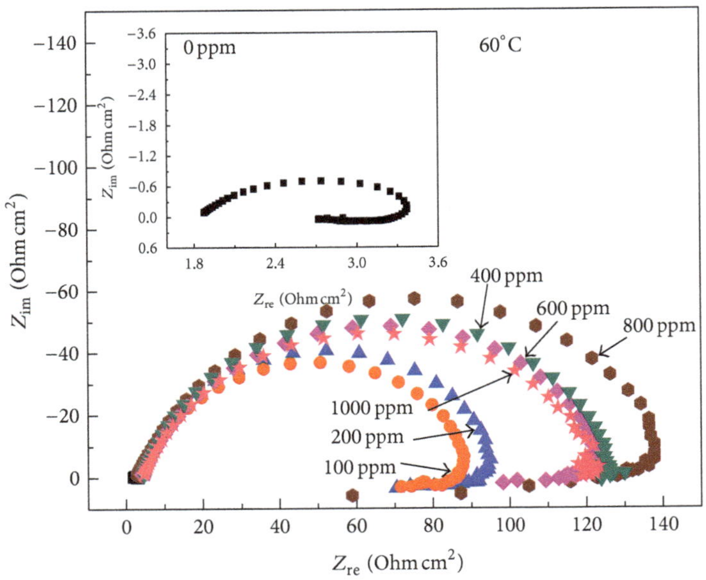

FIGURE 8: Nyquist plots for 1018 carbon steel in 0.5 M H_2SO_4 at 60°C.

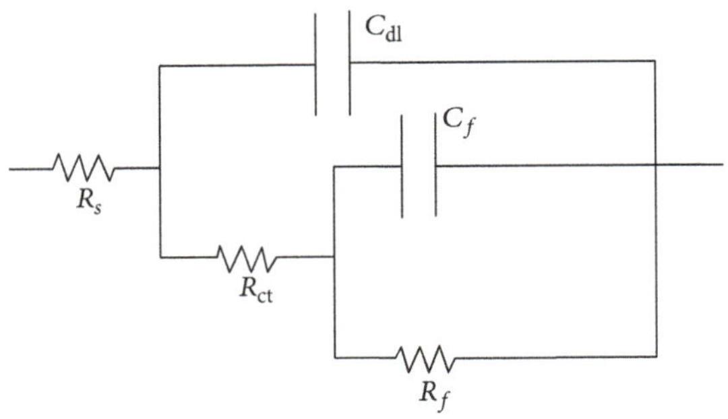

FIGURE 9: Electric circuit used to simulate EIS data.

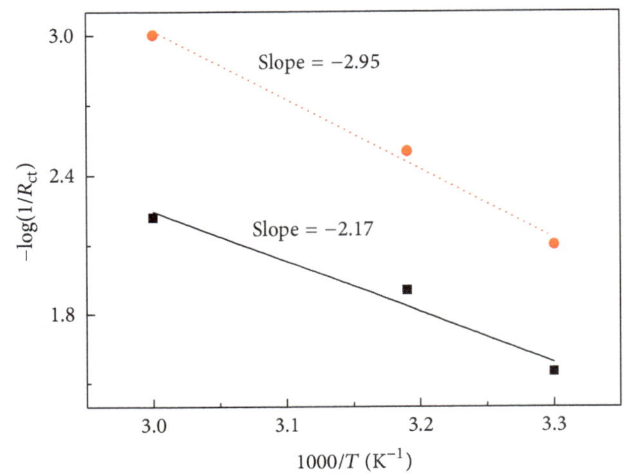

FIGURE 10: Arrhenius plot for the $1/R_{ct}$ values without and with 800 ppm of phtalocyanine blue.

acid solution was determined by using an Arrhenius-type plot according to the following equation:

$$\ln \mathrm{CR} = \frac{-E_a}{RT} + \ln F, \quad (5)$$

where CR is the corrosion rate, R is the molar gas constant, T is the absolute temperature, and F is the frequency factor. The corrosion rate can be given in terms of the weight loss, ΔW, corrosion current density, i_{corr}, or $1/R_{ct}$, where R_{ct} is the charge transfer resistance, which is the semicircles diameter of the Nyquist data. Arrhenius plots of ln $1/R_{ct}$ against T^{-1} for 1018 carbon steel in 0.5 M H_2SO_4 in absence and presence of inhibitor are shown in Figure 10. The apparent activation energy obtained for the corrosion process in the inhibitor-free, uninhibited acid solution was found to be 7.7 and 3.2 kJ mol^{-1} in presence of the inhibitor. The energy barrier of the corrosion reaction decreased in the presence of the inhibitor, which can be due to the adsorption of the inhibitor on the steel surface. Many studies [20, 21] showed that, in the presence of inhibitor, the decrease of E_a with respect to uninhibited solution is interpreted as physical adsorption.

Some micrographs of corroded specimens without and with the addition of inhibitor are shown in Figure 11. For the uninhibited solution, Figure 11(a), the steel surface shows very porous corrosion products, indicating the aggressive attack of the corroding environment on the steel surface. However, when the steel is corroded in the aggressive solution containing 800 ppm of phtalocyanine blue at 25°C Figure 11(b), very compact, rough, porous, or cracks-free, protective corrosion products are formed on top of the steel, which reduced the steel corrosion rate. When the steel is immersed in the inhibited solution but at higher temperatures, Figures 11(c) and 11(d), the corrosion products become more porous and some microcracks can be seen now, indicating that the aggressive environment can penetrate through them and corrode the underlying metal,

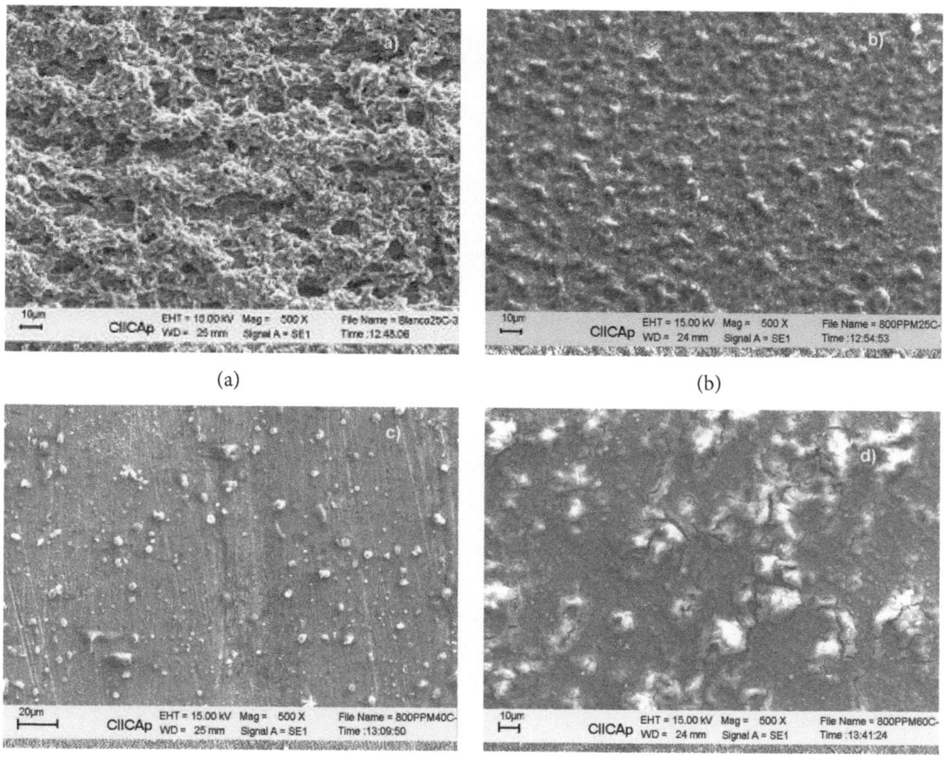

(a) (b) (c) (d)

Figure 11: Micrograph of corroded 1018 carbon steel at (a) 25°C with 0 ppm, and 800 ppm of phtalocyanine blue at (b) 25, (c) 40 and (d) 60°C.

increasing the corrosion rate and decreasing the inhibitor protectiveness.

4. Conclusions

A study of the use of phtalocyanine blue dye as corrosion inhibitor for 1018 carbon steel in 0.5 M H_2SO_4 has been carried out. It has been found that phtalocyanine blue acts as a good corrosion inhibitor with its efficiency increasing with the inhibitor concentration, reaching its highest value by adding 800 or 1000 ppm due to the formation of an adherent, compact, protective film. Additionally, inhibitor efficiency decreases with an increase in the temperature up to 40°C, but it increases at 60°C. Phtalocyanine blue dye improved the properties of the passive film formed on the steel surface, that is, decreases both the passivation current density and passivation potential values; therefore, it acts as an anodic-type inhibitor. At 25 and 40°C, Nyquist data displayed a single capacitive, depressed loop, indicating that the corrosion process is under charge transfer control, but at 60°C there appears an inductive loop, indicating that the corrosion process is now under adsorption/desorption control.

Conflict of Interests

The authors declare that there is no conflict of interests regarding the publication of this paper.

References

[1] L. R. Chauhan and G. Gunasekaran, "Corrosion inhibition of mild steel by plant extract in dilute HCl medium," *Corrosion Science*, vol. 49, no. 3, pp. 1143–1161, 2007.

[2] P. C. Okafor, M. E. Ikpi, I. E. Uwah, E. E. Ebenso, U. J. Ekpe, and S. A. Umoren, "Inhibitory action of *Phyllanthus amarus* extracts on the corrosion of mild steel in acidic media," *Corrosion Science*, vol. 50, no. 8, pp. 2310–2317, 2008.

[3] K. F. Khaled, "Inhibition of copper corrosion in 3.5% NaCl solutions by a new pyrimidine derivative: electrochemical and computer simulation techniques," *Journal of Solid State Electrochemistry*, vol. 11, pp. 1743–1749, 2007.

[4] M. Z. A. Rafiquee, S. Khan, N. Saxena, and M. A. Quraishi, "Investigation of some oleochemicals as green inhibitors on mild steel corrosion in sulfuric acid," *Journal of Applied Electrochemistry*, vol. 39, no. 8, pp. 1409–1417, 2009.

[5] H. Ashassi-Sorkhabi, B. Masoumi, P. Ejbari, and E. Asghari, "Corrosion inhibition of mild steel in acidic media by Basic yellow 13 dye," *Journal of Applied Electrochemistry*, vol. 39, no. 9, pp. 1497–1501, 2009.

[6] M. A. Quraishi, A. Singh, V. K. Singh, D. K. Yadav, and A. K. Singh, "Green approach to corrosion inhibition of mild steel in hydrochloric acid and sulphuric acid solutions by the extract of *Murraya koenigii* leaves," *Materials Chemistry and Physics*, vol. 122, no. 1, pp. 114–122, 2010.

[7] F. Bentiss, M. Traisnel, and M. Lagrenee, "Influence of 2,5-bis(4-dimethylaminophenyl)-1,3,4-thiadiazole on corrosion inhibition of mild steel in acidic media," *Journal of Applied Electrochemistry*, vol. 31, no. 1, pp. 41–48, 2001.

[8] I. A. Akpan and N. O. Offiong, "Inhibition of mild steel corrosion in HCl solution by ciprofloxacin drug," *International Journal of Corrosion*, vol. 2013, Article ID 301689, 5 pages, 2013.

[9] E. E. Oguzie, "Influence of halide ions on the inhibitive effect of congo red dye on the corrosion of mild steel in sulphuric acid solution," *Materials Chemistry and Physics*, vol. 87, no. 1, pp. 212–217, 2004.

[10] E. E. Oguzie, "Corrosion inhibition of mild steel in hydrochloric acid solution by methylene blue dye," *Materials Letters*, vol. 59, no. 8-9, pp. 1076–1079, 2005.

[11] E. E. Oguzie, C. Unaegbu, C. N. Ogukwe, B. N. Okolue, and A. I. Onuchukwu, "Inhibition of mild steel corrosion in sulphuric acid using indigo dye and synergistic halide additives," *Materials Chemistry and Physics*, vol. 84, no. 2-3, pp. 363–368, 2004.

[12] E. E. Ebenso and E. E. Oguzie, "Corrosion inhibition of mild steel in acidic media by some organic dyes," *Materials Letters*, vol. 59, no. 17, pp. 2163–2165, 2005.

[13] E. E. Oguzie, B. N. Okolue, E. E. Ebenso, G. N. Onuoha, and A. I. Onuchukwu, "Evaluation of the inhibitory effect of methylene blue dye on the corrosion of aluminium in hydrochloric acid," *Materials Chemistry and Physics*, vol. 84, pp. 394–401, 2004.

[14] E. E. Ebenso, H. Alemu, S. A. Umoren, and I. B. Obot, "Inhibition of mild steel corrosion in sulphuric acid using alizarin yellow GG dye and synergistic Iodide Additive," *International Journal of Electrochemical Science*, vol. 3, pp. 1325–1339, 2008.

[15] A. I. Onen, O. N. Maitera, J. Joseph, and E. E. Ebenso, "Corrosion inhibition potential and adsorption behaviour of bromophenol blue and thymol blue dyes on mild steel in acidic medium," *International Journal of Electrochemical Science*, vol. 6, no. 7, pp. 2884–2897, 2011.

[16] S. Deng, X. Li, and H. Fu, "Alizarin violet 3B as a novel corrosion inhibitor for steel in HCl, H_2SO_4 solutions," *Corrosion Science*, vol. 53, no. 11, pp. 3596–3602, 2011.

[17] H. Ashassi-Sorkhabi, D. Seifzadeh, and M. G. Hosseini, "EN, EIS and polarization studies to evaluate the inhibition effect of 3H-phenothiazin-3-one, 7-dimethylamin on mild steel corrosion in 1 M HCl solution," *Corrosion Science*, vol. 50, no. 12, pp. 3363–3370, 2008.

[18] M. A. Elmorsi and A. M. Hassanein, "Corrosion inhibition of copper by heterocyclic compounds," *Corrosion Science*, vol. 41, no. 12, pp. 2337–2352, 1999.

[19] S. Deng, X. Li, and H. Fu, "Acid violet 6B as a novel corrosion inhibitor for cold rolled steel in hydrochloric acid solution," *Corrosion Science*, vol. 53, no. 2, pp. 760–768, 2011.

[20] X. Li, S. Deng, and H. Fu, "Inhibition effect of methyl violet on the corrosion of cold rolled steel in 1.0 M HCl solution," *Corrosion Science*, vol. 52, no. 10, pp. 3413–3420, 2010.

[21] R. Q. Liu, L. Xiang, and X. G. Zhang, "Inhibition effect of methyl red on corrosion of steel in acidic solutions," *Corrosion Protection*, vol. 22, pp. 98–99, 2001.

[22] X. Li, S. Deng, H. Fu, N. Zhao, Y. Li, and G. Mu, "Corrosion inhibition of cresol red for cold rolled steel in phosphoric acid solution," *Corrosion Science and Protection Technology*, vol. 21, no. 4, pp. 354–357, 2009.

[23] J. Sathiyabama, S. Rajendran, J. A. Selvi, and A. J. Amalraj, "Methyl orange as corrosion inhibitor for carbon steel in well water," *Indian Journal of Chemical Technology*, vol. 15, no. 5, pp. 462–466, 2008.

[24] R. A. Prabhu, T. V. Venkatesha, and A. V. Shanbhag, "Carmine and fast green as corrosion inhibitors for mild steel in hydrochloric acid solution," *Journal of the Iranian Chemical Society*, vol. 6, no. 2, pp. 353–363, 2009.

[25] X. Li and G. Mu, "Tween-40 as corrosion inhibitor for cold rolled steel in sulphuric acid: weight loss study, electrochemical characterization, and AFM," *Applied Surface Science*, vol. 252, no. 5, pp. 1254–1265, 2005.

6

Development and Performance Evaluation of Corrosion Resistance Self-Healing Coating

Akshya Kumar Guin, Suryakanta Nayak, Manish Kumar Bhadu, Veena Singh, and Tapan Kumar Rout

Research and Development, Tata Steel Ltd., Jamshedpur 831001, India

Correspondence should be addressed to Akshya Kumar Guin; akshya.guin@tatasteel.com

Academic Editors: M. Aparicio and A. Hermann

Polymer based nanocapsule was developed using core-cell approach, where the core material was methyl diphenyl diisocyanate and the cell material was urea-formaldehyde. The synthesized capsules of 100 to 800 nm size were incorporated into sol-gel matrix to prepare a final coating for steel protection. This coating was found protecting the steel at the damage or crack locations in 3.5% NaCl solution. SEM micrographs confirmed healing of the coating at the damage or crack points.

1. Introduction

Having been inspired by nature, coating designers are trying to convert the traditional coating into the "smart or intelligent" coating. Self-healing is one of the examples of smart coating [1]. When coatings get damaged in the form of cracks induced by mechanical, thermal, and chemical means, regardless of the application, the integrity of the structure is significantly compromised. Since the damage deep inside material is difficult to perceive and to repair in particular, the material would be motivating to have ability of self-repairing through various stimuli response based on physical, chemical, mechanical, or thermal means which trigger the healing process. These progresses have now opened an age of new intelligent materials. Continuous efforts are made to mimic natural materials and to integrate self-repairing capability into polymers and polymer composites. Keeping all this in mind, self-healing concept has drawn considerable research interest in the last few years. Like pharmaceutical industry microcapsule technique is being used in corrosion prevention coatings. Healing is achieved by incorporating healing agents and a catalytic chemicals trigger within a polymer matrix. When coating is damaged by mishandling or misuse of the coated structure microcapsules rupture and active healing agent come out from the shell and with the help of catalyst healing the damage area.

In short, self-healing material provides the following advantages over the conventional coating:

(i) automatic repairing process;

(ii) barrier protection of underlying substrates;

(iii) autopreservation of aesthetics of surface appearance of coatings, plastic, and films;

(iv) restoration of mechanical integrity of load bearing materials as in composites [2–4].

Now researchers are focusing on the development of smart coatings, which are new generation functional coatings with mimicking nature. Self-healing coating is one of the examples of smart coating [5–7]. Conventional barrier protective coating when gets damaged during transportation or fabrication process, cracks or pin hole develops on the damaged area and localized corrosion initiates at that area and gradually propagates to other areas. These types of difficulties in repairing damaged area were well taken care of by scientists, by designing intelligent coatings using core-cell approach. The selection of a core material in a nanocapsule is important because this decides which stimuli response is needed to heal the damage. Self-healing concepts have drawn considerable research interest for pharmaceutical industries and in development of corrosion resistance coatings. When

Table 1: Composition of self-healing coating.

Sr. number	Ingredient	Composition by wt %
1	Titania-precursor	60–63
2	Coupling agent—methyl hydrogen silicone	36–39
3	Nano capsule	0.5–2
4	Zn-Al pigment	1

coating is damaged by mishandling or misuse of the coated structure, microcapsules rupture and active healing agent come out from the core through the cell. The core material reacts with the catalyst either in the coating matrix or by moisture to heal the damaged area [8–12].

Present research work emphasized on the development of a methyl diphenyl diisocyanate based nanosize capsule and incorporation of nanocapsule and corrosion resistance pigments like nanolamellar zinc flake into a titania based sol-gel coating matrix to generate the self-healing and corrosion resistance coating on steel substrate (see Table 1).

2. Experiment

2.1. Preparation of Nanocapsules. Capsules were prepared by in situ polymerization in an oil-in-water emulsion as reported by White, Jinglei et al. [2], Trask et al. [3], and Blaiszik et al. [4]. At room temperature (20–25°C), 5 mL of 5.0 wt % polyvinyl alcohol (PVA), urea, formaldehyde, and deionised water were mixed in a 250 mL three neck round bottomed flask connected to a reflux condenser and equipped with a mechanical stirrer. After the urea dissolved pH of the solution was adjusted to 3.5 by slow addition of 1 N NaOH and the temperature was kept at 60–70°C for one hour and then urea-formaldehyde prepolymer solution was obtained.

Under agitation, a slow stream of methyl diphenyl diisocyanate was added to the prepared prepolymer solution to form an oil-in-water emulsion. The tapered 3.2 mm tip sonicator horn of a 750 W ultrasonicator was placed in the solution for 3 minutes at 40% intensity with continuous mixing at 2000 rpm in room temperature. This sonication step changes the emulsion from slightly cloudy to opaque white. 12.67 g formalin (37% formaldehyde) was then added to obtain a 1 : 1.9 molar ratio of formaldehyde to urea. The emulsion was covered and heated at a rate of 1°C/m up to 55°C. After 4 h of continuous agitation at 800–2000 rpm the mixture was allowed to cool at ambient temperature. After cooling, the suspension was filtered by vacuum filtration method. The nanocapsules were properly rinsed with deionised water and air-dried for about 48 h.

2.2. Coating Formulation. The titania sol was prepared by mixing Ti-isopropoxide and 2-methoxy glycol as outlined by our previous work [6, 7]. After 24 h of aging, N-phenyl-3-aminopropyl triethoxy silane was added slowly into the titanium sol in 1 : 1(V/V) mixing ratio at room temperature; this solution is treated as base sol-gel solution. Self-healing coating formulation was prepared by incorporating previously synthesized 1% (by W/V) nanocapsules and 0.5% nanozinc-aluminium laminar alloy pigments (4 : 7 zinc to aluminium ratio by weight percentage, density is approximately 1.1 kg/L, and dimension is 20–50 nm dia and 10–15 micron length) and 1% into the above sol-gel polymer matrix. The commercially available dispersing agent like BYK 3441 (1% of total solid) was used for easy dispersion of pigments in sol-gel matrix. The cold rolled and annealed (CRCA) steel sheets were used for present study; these sheets were cleaned to remove oil and grease and subjected to phosphating in tricationic phosphate base solution to obtain a thin coating of <0.5 micron thickness (2-3 gm/m^2). The synthesized sol was applied on the substrate by dipping process, where dipping time was 1 min at room temperature (RT), followed by drying at 150°C for 1 hour.

The probable reaction between titania precursor and coupling agent N-phenyl-3-aminopropyl triethoxy silane is given below (Figure 1(a)).

2.3. Chemical Reaction in Nanocapsules Synthesis. Urea-formaldehyde based nanocapsules were synthesized as described in the experimental part. During the process, core chemicals like methyl diphenyl diisocyanate were covered with urea-formaldehyde (UF) resin. This reaction was carried out by in situ polymerization in an oil-in-water emulsion, where PVA is used as an emulsifying agent. The synthesis of nanocapsule in urea and formaldehyde backbone is a two-step process. First step is the methylation where formaldehyde reacts with urea, as shown in Figure 1. In the second step, the condensation reaction between the various methylol species and amines takes place. At low pH, the produced UF resins cure irreversibly to form a cross-linked network and this reaction can be induced by addition of further ammonium chloride. Resorcinol was used to improve the cocondensation reaction products.

The core chemical of nanocapsule contains isocyanate group. During the initiation of any crack, this isocyanate group reacts with water and moisture present in atmosphere and heals the damaged area.

3. Results and Discussion

3.1. Nanocapsule Characterization. A series of characterisation tests were carried out for capsules to understand the capsules morphology, physical properties, and stability. Cell wall integrity, aggregation, and nanocapsule sizes were observed by scanning electron microscope (SEM).

3.1.1. Capsule Size Analysis. Capsule size analysis was performed by SEM; the capsule's size fall in a wide range of 20–250 micrometer without the application of ultrasonicator. Microcapsule sizes decrease extensively to an order of 200–800 nm by using ultrasonicator treatment for about 3 min. The reason for this, is that the fluid flows away from the propeller, many large microeddies exist, and in the vicinity of the propeller blades many small microeddies exist, which results in a wide length scale. The capsule size can be controlled by adjusting the agitation rate. In the present study at 2000 rpm with 3 min ultrasonicator, the average diameter of the prepared capsule is found to be 200–800 nm.

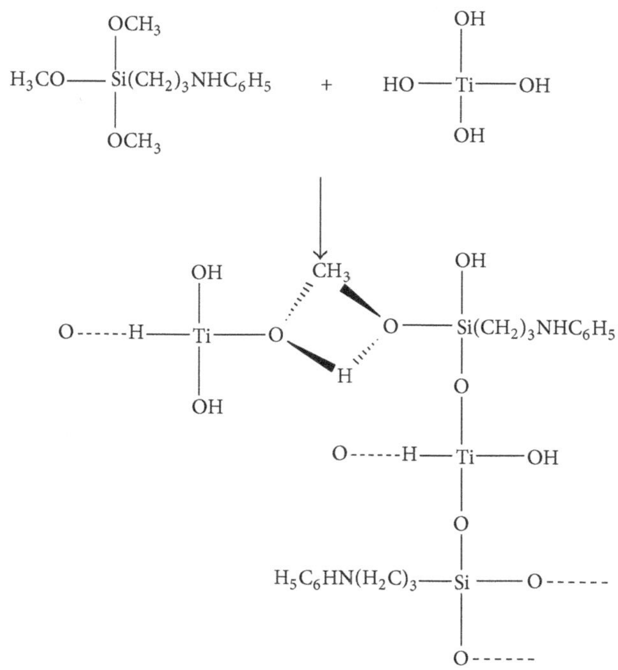

(a)

$$H_2NCONH_2 + CH_2O \rightleftharpoons NH_2CONHCH_2OH$$
Urea formaldehyde — Monomethylol urea derivative (prepolymer)

$$NH_2CONHCH_2OH + CH_2O \rightleftharpoons HOCH_2NHCONHCH_2OH$$
Prepolymer — Dimethylol urea derivative (prepolymer)

$$\text{-NH-CH}_2\text{OH} + \text{H-NH-CO-} \longrightarrow \text{-NH-CH}_2\text{-NH-CO-} + H_2O$$
Urea derivative urea derivative — Poly(urea-formaldehyde) (PUF) water

(b)

$$\text{-N=C=O} + H_2O \longrightarrow \text{-NH-CO-NH-} + CO_2$$
$$\text{-N=C=O} + \text{HO-} \longrightarrow \text{-NH-CO-O}^-$$

(c)

FIGURE 1: (a) Probable reaction mechanism of Ti-sol and coupling agent. (b) Reaction scheme of the formation of U-F prepolymer and PUF. (c) Reaction scheme of isocyanate derivative with water.

3.1.2. Nanocapsule Surface Morphology. Figure 2 shows the SEM morphology of nanocapsule. It was observed that the shape of nanocapsule is spherical and the surface of nanocapsule is rough and shaggy. There is a possibility of uniform distribution of nanocapsules, which covered the substrate fully for better corrosion and self-healing property.

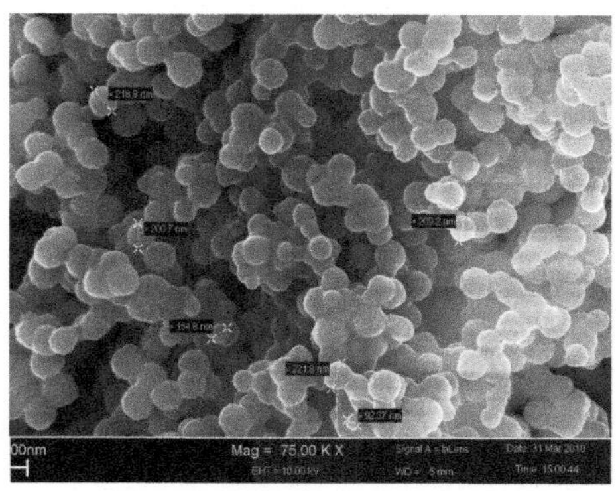

FIGURE 2: SEM image of microcapsule with 100–200 nm dia.

3.1.3. Thermal Resistance of Nanocapsules. The thermal stability of nanocapsules plays an important role in their application in self-healing composites. Figure 3 shows TGA diagram of nanocapsule with UF wall shell, weight loss near 100°C is mainly due to the removal of entrapped residual water, and the elimination of free formaldehyde. The weight loss at temperature between 200 and 300°C is mainly due to the decomposition of the PUF shell wall [13, 14]. The residue undergoes extensive fragmentation near 400°C and the prepared nanocapsules have a good thermal stability.

3.1.4. Characterization of Nanocapsule by FTIR. It is seen from the FTIR spectra in Figure 4, of urea-formaldehyde coated nanocapsule, core chemical, and nanocapsule rupture with water, that all are matching at characteristic peaks of a N–H vibration at 1571 cm^{-1} a C=O stretching vibration at 1650 cm^{-1} and a C–H vibration at 1460 cm^{-1}. C–N stretching vibrations are shown at 1286 and 1142 cm^{-1}. This spectrum confirms that the cell material is made of urea-formaldehyde polymer.

When the broken capsules are allowed to react with water it shows sharp absorption peaks at 1694 and 1562 cm^{-1}, as shown in Figure 4, which represents the presence of urethane linkage only, which means that core materials are active

Table 2: Salt spray test result.

Uncoated steel	Self healing coating	Sol-gel coating (with out self healing ingredient)

2 h	120 h	100 h

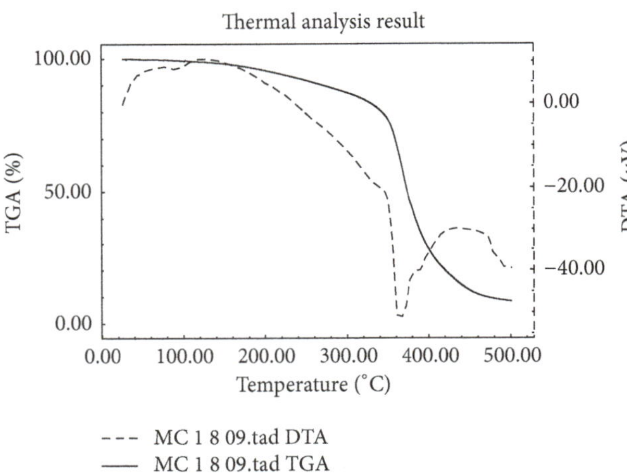

Figure 3: TGA-DTA graph of urethane capsule.

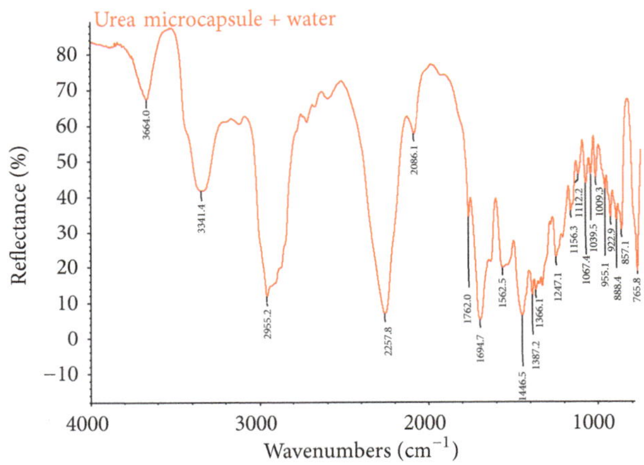

Figure 4: FTIR spectra of broken capsule in water.

enough and are reacting with water to form a urethane linkage.

3.2. Self-Healing Ability. After successful preparation, the nanocapsule was embedded in a titania based polymer matrix for achieving self-healing ability. In the present study, 1% of prepared nanocapsules along with 0.5% zinc-aluminium laminar alloy pigments was incorporated in sol-gel matrix and the final coating solution was applied on steel surface. After seven days of aging at room temperature, coated panels were punctured and investigated for self-healing ability. Under this investigation, the punctured coated steel substrates were exposed in 3.5% NaCl solution to check the propagation or healing of punctured area. SEM-EDX techniques were used for evaluation of self-healing property of coating. It was observed from the SEM-EDX analysis, as shown in Figure 5, that self-healing ability was observed within 8 h of damage. Incorporation of nanozinc-aluminium lamellar pigment and nanocapsule enhanced the self-healing property of coated steel. SEM-EDX analysis confirmed the healing ability of coated sample by synergic effect of nanocapsule and zinc-aluminium alloy based sol-gel coating. This self-healing ability can be occurred by two ways; firstly, by the capillary flow of core chemical from the capsule rupture and subsequent interaction with moisture (as shown in Figure 1); secondly by the interaction of zinc-aluminium pigment with moisture, which leads to the formation of zinc and aluminium oxide on the scratch surface (as shown in SEM-EDX, Figure 5).

3.3. Corrosion Resistance Study by ASTM B 117 Salt Spray. Salt spray test was conducted to check the coating delamination and blister formation across the inscribed area of coated sample with 6–10 micron dry film thickness and the result is shown in Table 2. It was observed that, due to strong

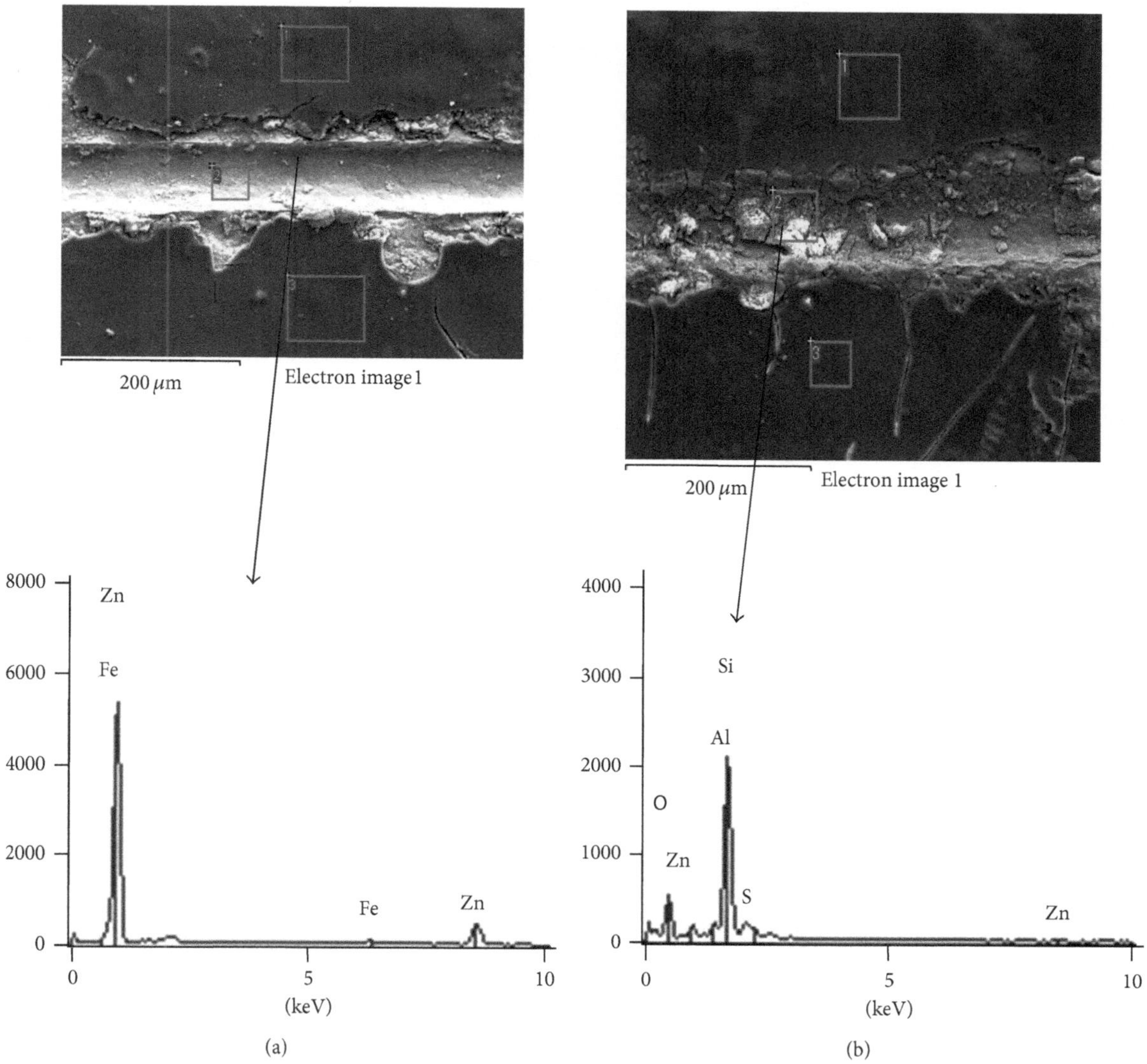

FIGURE 5: (a) SEM-EDX of initial cut sample. (b) SEM-EDX of self-healing coating after 8 h of cut.

adhesion of self-healing coating with steel surface, there was no trace of osmotic blister formation on coating surface, even after 100 h of exposed in salt spray chamber. After 120 h of exposure, coated panel developed medium size blister which was converted to dance blister after 150 h. The coating showed good barrier and corrosion resistance properties as against uncoated steel sheet which corrode after 2 h of exposure in ASTM B 117 test chamber and 100 h for sol-gel coating without having any self-healing ingredients. The reasonable salt spray resistance of the developed coating might be due to the formation of bonds between polymer molecule and metallic surface and due to presence of laminar structure of Zn-aluminum nanopigment, diffusion of water, and other corrosive ingredients through polymer coating which was delayed. So penetrable damaging efficiency of localized corrosion was reduced by self-healing mechanism.

3.4. EIS Study. EIS is an excellent tool to study the mechanism and kinetics of the corrosion reaction. The change in impedance spectrum due to immersion in the corrosion environment helps to understand the protective ability of the developed coatings. The resistance obtained at low frequency region is crucial to explain the corrosion resistance performance and intactness of the coatings. In the present study, EIS study of coated samples was carried out between 100 kHz and 0.01 Hz at the Ecorr. Figures 6(a)–6(e) show the EIS graph of coated samples at different intervals of time. The physical values were fitted with the VERSATEC software through appropriate equivalent circuit as shown in Figure 7. The system is assumed to consist of three interfaces with different resistance and capacitance characteristics. The first interface is between sol-gel coating and NaCl electrolyte. Second is the coating and metal substrate and the third is the metal surface where corrosion activity encounters. The spectra revelled three time constants, Rcoat-Qcoat, Rox-Qox, and Rct-Qdl, corresponding to resistance and capacitance of the sol-gel film, intermediate oxide layer, and double layer, respectively. Gradual reduction in Rcoat value with time indicated the attack of coating surface by chloride ions.

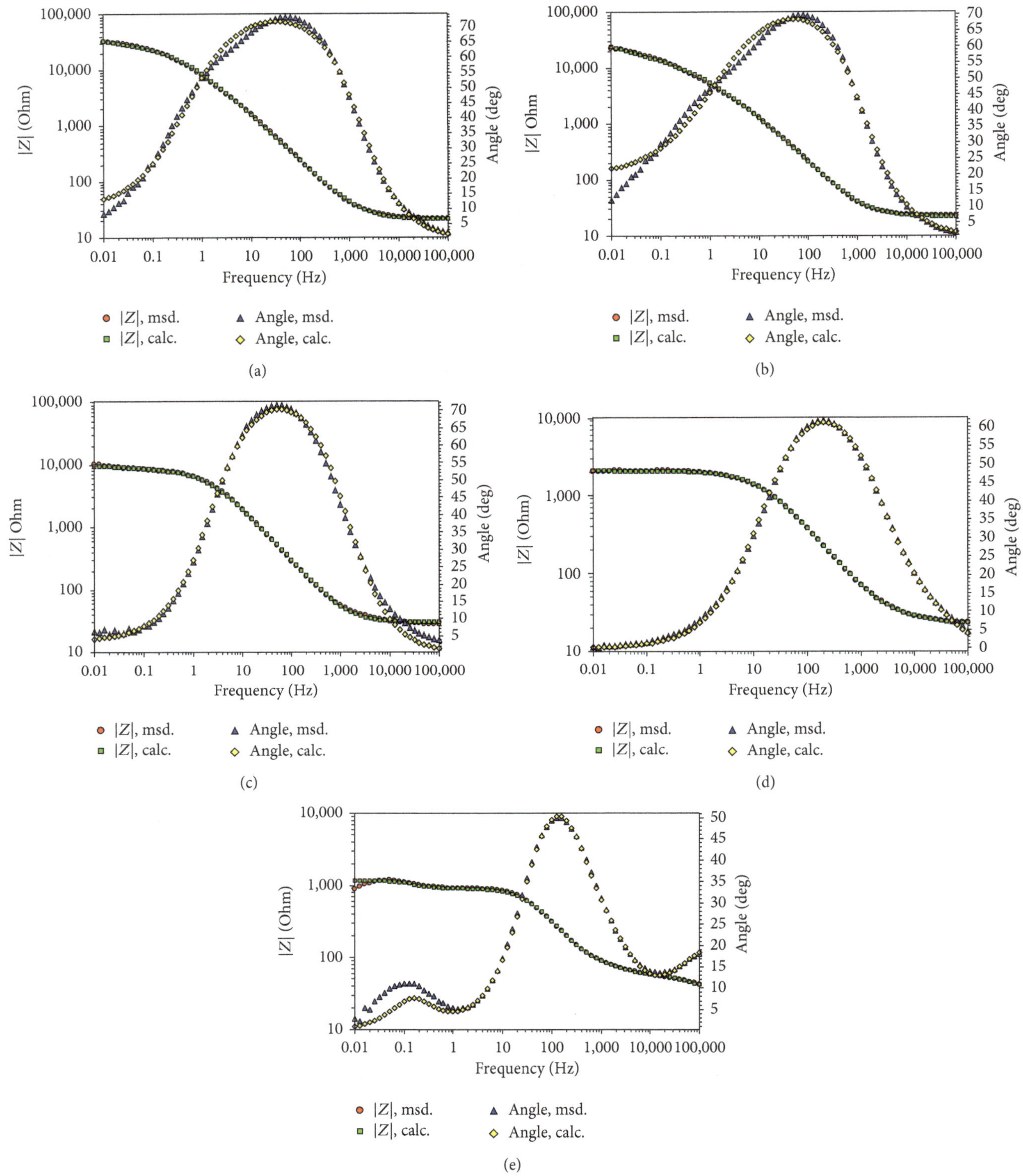

FIGURE 6: (a) Initial EIS result of self-healing coated sample. (b) EIS result after 48 h of exposure of self-healing coated sample in 3.5% NaCl solution. (c) EIS result after 96 h of exposure of self-healing coated sample in 3.5% NaCl solution. (d) Initial EIS result of sol-gel coated sample (without any capsule). (e) EIS result after 96 h of exposure of sol-gel coated sample (without any capsule) in 3.5% NaCl solution.

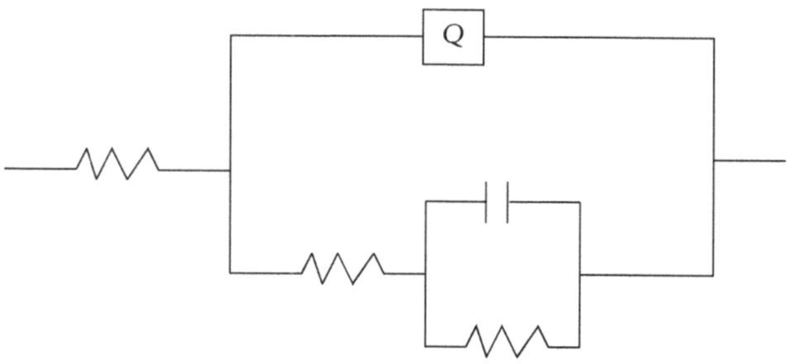

FIGURE 7: Curve fitting circuit of EIS diagram.

From the EIS graph, as shown in Figures 6(a)–6(e), it has been observed that sample coated with sol-gel based self-healing coating provides good barrier and corrosion resistance property than sample coated with only sol-gel solution (without self-healing capsule) in saline environment. This may be due to strong adhesion of coating with steel and uniform dispersion and coverage of nanocapsules in sol-gel coating. After 48 h of immersion, coating resistance decreased from 43625 ohm/cm^2 to 24875 ohm/cm^2, whereas corrosion resistance value dropped below 1000 ohm/cm^2 for sample coated with only sol-gel solution containing no self-healing capsules. More interestingly there is no additional time constant observed even after 96 h immersion in 3.5% NaCl solution and coating resistance was around 10823 ohm/cm^2; this indicates that sample coated provides a good barrier protection and improves corrosion resistance of steel substrate.

4. Conclusions

Methyl diphenyl diisocyanate as core material containing nanosize capsules was prepared using an in situ UF encapsulation by sonication technique. The capsule was found active and remained thermally stable, till 200°C. The nanocapsules and nanozinc-aluminium alloy pigments were successfully incorporated in sol-gel matrix, which provided uniformly coated self-healing coating. Healing effect observed within 8 h of damage was initiated using optical and SEM techniques. This nanocapsule and zinc-aluminium pigments embedded self-healing coating and enhanced corrosion resistance of coated steel.

Conflict of Interests

The authors declare that there is no conflict of interests regarding the publication of this paper.

Acknowledgment

The authors would like to thank the Management of Tata Steel Limited for the permission to publish and encouragement to carry out this work.

References

[1] S. K. Ghosh, T. Jana, R. Kumar, and B. Ray, “Self-healing—coating’s perspective & recent advances,” *Paints India*, vol. 4, p. 85, 2010.

[2] Y. Jinglei, M. W. Keller, J. S. Moore, S. R. White, and N. R. Sottos, “Microencapsulation of isocyanates for self-healing polymers,” *Macromolecules*, vol. 41, no. 24, pp. 9650–9655, 2008.

[3] R. S. Trask, G. J. Williams, and I. P. Bond, “Bioinspired self-healing of advanced composite structures using hollow glass fibres,” *Journal of the Royal Society Interface*, vol. 4, no. 13, pp. 362–371, 2007.

[4] B. J. Blaiszik, N. R. Sottos, and S. R. White, “Nanocapsules for self-healing materials,” *Composites Science and Technology*, vol. 68, no. 3-4, pp. 978–986, 2008.

[5] V. Sauvant-Moynot, S. Gonzalez, and J. Kittel, “Self-healing coatings: an alternative route for anticorrosion protection,” *Progress in Organic Coatings*, vol. 63, no. 3, pp. 307–315, 2008.

[6] A. K. Guin, S. Nayak, T. K. Rout, N. Bandyopadhyay, and D. K. Sengupta, “Corrosion resistance nano-hybrid sol-gel coating on steel sheet,” *ISIJ International*, vol. 51, no. 3, pp. 435–440, 2011.

[7] A. Guin, S. K. Nayak, T. K. Rout et al., “Corrosion behavior of nanohybrid titania—silica composite coating on phosphated steel sheet,” *Journal of Coatings Technology and Research*, vol. 9, no. 1, pp. 97–106, 2012.

[8] A. Kumar, L. D. Stephenson, and J. N. Murray, “Self-healing coatings for steel,” *Progress in Organic Coatings*, vol. 55, no. 3, pp. 244–253, 2008.

[9] S. R. White, N. R. Sottos, P. H. Geubelle et al., “Autonomic healing of polymer composites,” *Nature*, vol. 409, no. 6822, pp. 794–797, 2001.

[10] J. M. Kamphaus, J. D. Rule, J. S. Moore, N. R. Sottos, and S. R. White, “A new self-healing epoxy with tungsten (VI) chloride catalyst,” *Journal of the Royal Society Interface*, vol. 5, no. 18, pp. 95–103, 2008.

[11] J. M. Kamphaus, J. D. Rule, J. S. Moore, N. R. Sottos, and S. R. White, “A new self-healing epoxy with tungsten (VI) chloride catalyst,” *Journal of the Royal Society Interface*, vol. 5, no. 18, pp. 95–103, 2008.

[12] S. H. Cho, H. M. Anderson, S. R. White, N. R. Sottos, and P. V. Braun, “Polydimethylsiloxane-based self-healing materials,” *Advanced Materials*, vol. 18, pp. 997–1000, 2006.

[13] X. He and X. Shi, “Self-repairing coating for corrosion protection of aluminum alloys,” *Progress in Organic Coatings*, vol. 65, no. 1, pp. 37–43, 2009.

[14] S. V. Lamaka, M. L. Zheludkevich, K. A. Yasakau, R. Serra, S. K. Poznyak, and M. G. S. Ferreira, “Nanoporous titania interlayer as reservoir of corrosion inhibitors for coatings with self-healing ability,” *Progress in Organic Coatings*, vol. 58, no. 2-3, pp. 127–135, 2007.

7

Synergistic Effect of Iodide Ion and 4-Hydroxy Coumarin on the Corrosion Inhibition of Mild Steel in Hydrochloric Acid

Shyamala R. Krishnamurthy and Meenakshi H. N. Parameswaran

Department of Chemistry, Avinashilingam Institute for Home Science and Higher Education for Women, Coimbatore, Tamil Nadu 641043, India

Correspondence should be addressed to Meenakshi H. N. Parameswaran; meenaparam75@gmail.com

Academic Editors: C. Gervasi and H. Miao

No discussion of the use of inhibitors in acid solution is complete without mentioning the phenomenon of synergism. Synergism also operates in corrosion protection where an enhanced inhibition may be related to interaction between inhibitor compounds. This effect has been observed since the earlier days of inhibitor technology and continues to be a potent tool in the development of acid inhibitors for specialized uses. Hence, in this paper, the corrosion inhibition behavior of mild steel (MS) in 1 M hydrochloric acid in the presence of 4-hydroxy coumarin (4HC) and potassium iodide (KI) has been investigated using the mass loss method and electrochemical techniques. The inhibitive performance of 4HC is considerably enhanced by the addition of 1×10^{-3} KI. The addition of KI to different concentrations of 4HC has intensified its efficiency through considerable reduction in the mass loss, corrosion current density I_{corr}, double layer capacitance C_{dl}, and increase in charge transfer resistance R_{ct}. The calculated synergism parameter "S" is greater than unity, thereby proving the fact that the improvement in inhibition efficiency of 4HC, generated by the addition of KI, is due to synergism.

1. Introduction

Synergism literally means joint action of different processes in producing an effect greater than the sum of their individual effects. This effect has been observed since the earlier days of inhibitor technology and continues to be a potent tool in the development of acid inhibitors for specialized uses. In fact, the inhibitive nature of most of the extracts of common plants used to combat corrosion of metals is known to be fortified through synergism of many organic constituents present. In modern practice, inhibitors are rarely used in the form of single compounds. A formulation of two or more organic compounds and/or inorganic compounds is usually employed. The synergistic action of anions on the inhibition of mild steel corrosion in acid solution containing organic compound has been reported by several authors [1–4]. In view of this, the enhancement of protection efficiency of the chosen coumarin inhibitor by iodide ions against mild steel corrosion in 1 M hydrochloric acid has been studied.

2. Experimental Parts

2.1. Synthesis of 4-Hydroxy Coumarin. The microwave assisted synthesis of 4-hydroxy coumarin (4HC) was carried out in an ordinary domestic microwave oven by the von Pechmann reaction of phenol, malonic acid fused $ZnCl_2$, and a few drops of $POCl_3$ for 60–40 seconds at 450 W. The mechanism of the reaction is shown in Scheme 1 [5].

2.2. Preparation of the Sample. MS specimens of area $5 \times 1\,cm^2$ having composition C = 0.049%, Mn = 0.346%, Si = 0.026%, P = 0.038%, S = 0.020%, Cr = 0.050%, Mo = 0.019%, Ni = 0.019%, and the reminder Fe were used for the entire study. MS specimens with an exposed area of 1 cm^2 were used for polarization and impedance studies. The specimens were degreased, cleaned with fine quality emery sheet, washed with distilled water, and stored in desiccators before they are used for the investigation.

TABLE 1: %IE in presence of 10^{-3} M KI and synergism parameter.

[Inh]		10^{-3} M KI	0.001	0.003	0.005	0.007	0.009	0.01
	I_1, I_2	0.1778	0.0500	0.1537	0.2583	0.3718	0.4329	0.5586
1 h	I'_{1+2}	—	0.3472	0.4966	0.6200	0.6687	0.7122	0.7920
	S	—	1.1970	1.3820	1.6650	1.5590	1.6200	1.7450
	I_1, I_2	0.0915	0.0996	0.2727	0.3402	0.4373	0.4460	0.5678
3 h	I'_{1+2}	—	0.2610	0.4537	0.4959	0.5820	0.6474	0.7537
	S	—	1.1070	1.2100	1.1890	1.2230	1.4270	1.5940
	I_1, I_2	0.2799	0.2665	0.3949	0.4164	0.4897	0.5742	0.6252
6 h	I'_{1+2}	—	0.5339	0.5814	0.6189	0.6699	0.7307	0.7589
	S	—	1.1330	1.0410	1.1030	1.1130	1.1390	1.1200
	I_1, I_2	0.1139	0.1483	0.3011	0.3731	0.4592	0.4611	0.4759
12 h	I'_{1+2}	—	0.1907	0.3495	0.4346	0.4795	0.5619	0.6764
	S	—	0.9330	0.9520	0.9830	0.9210	1.0900	1.4350

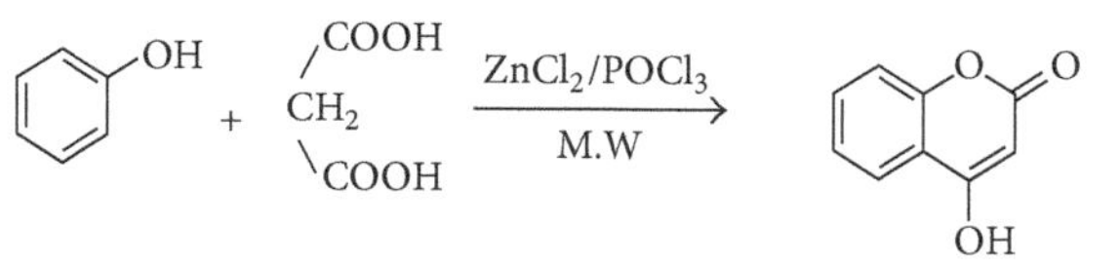

SCHEME 1: Synthesis of 4-hydroxy coumarin.

2.3. Weight Loss Method. MS specimens were accurately weighed and fully immersed in 100 mL of 1 M HCl in the absence and presence of the inhibitors, of various concentrations, removed after a particular period of immersion, dipped in sodium bicarbonate solution, (for neutralizing the remaining acid on the specimen), washed, dried, and reweighed. The loss in weight was determined in triplicate and the results were averaged. The experiment was performed for investigating the parameters such as inhibitor concentration (0.001 to 0.01 M), immersion (1 h, 3 h, 6 h, and 12 h), and synergism (presence of 10^{-3} M KI). The rates of corrosion and inhibition efficiency were calculated by the following formulae [6]:

$$\text{Corrosion rate (mpy)} = \frac{534 \times \text{weight loss in mgs}}{\text{Density} \times \text{Area in square inch} \times \text{Time in hrs}},$$

$$\text{IE\%} = (\text{Corrosion rate without inhibitor} - \text{Corrosion rate with inhibitor}) \times (\text{Corrosion rate without inhibitor})^{-1} \times 100. \quad (1)$$

2.4. Polarization Technique. A frequency response analyzer 1280B (Solartron) and an IBM personal computer were used for the polarization study. The cell for the polarization studies was a glass beaker containing the aerated unstirred test solution with a platinum electrode as the counter electrode, a saturated calomel electrode as the reference electrode, and the mild steel electrode as the working electrode. 100 mL of 1 M HCl with and without different concentrations of inhibitors was taken in an electrochemical cell. The polished electrode was then introduced. The electrode was placed at −1 mV cathodic to its open circuit potential. Thus, the potential was scanned at −0.1 mV/sec towards the anodic direction in Tafel extrapolation. Applied potential versus current was plotted and on extrapolation of linear portion to the corrosion potential gives the corrosion current. In anodic and cathodic plots, the slope of the linear portion gave Tafel constants "b_a" and "b_c" respectively. Inhibition efficiency was calculated by the following formula [7, 8]:

$$\text{IE\%} = (\text{Corrosion current without inhibitor} - \text{Corrosion current with inhibitor}) \times (\text{Corrosion current with inhibitor})^{-1} \times 100. \quad (2)$$

2.5. AC Impedance Method. The electrochemical impedance measurements were carried out using mild steel in acid medium with and without an inhibitor using the same potentiostat. A plot of real part of impedance (z') and imaginary part of impedance (z'') was made, and from this, the charge transfer resistance (R_{ct}), double layer capacitance (C_{dl}), and inhibition efficiency were calculated using ZView software. The inhibitor efficiency was obtained from the electrochemical parameters as below [9–11]

$$\%\text{IE} = \frac{\text{Rp}_{(i)} - \text{Rp}}{\text{Rp}_{(i)}} \times 100, \qquad \%\text{IE} = \frac{\text{Rct}_{(i)} - \text{Rct}}{\text{Rct}_{(i)}} \times 100, \quad (3)$$

where Rp and Rct are Polarization resistance and charge transfer resistance in the absence of the inhibitor, respectively, and $\text{Rp}_{(i)}$ and $\text{Rct}_{(i)}$ are Polarization resistance and charge transfer resistance in the presence of the inhibitor, respectively.

3. Results and Discussion

3.1. Mass Loss Method. The values of IE for the specific concentrations of 4HC and in combination with 10^{-3} M KI are given in Table 1. The interesting feature of the investigation is

TABLE 2: Electrode kinetic parameters for corrosion of mild steel in 1 M HCl with various concentrations of 4-hydroxy coumarin in the presence and absence of 10^{-3} M KI.

[Inh] M	[KI] M	$-E_{corr}$ mV	I_{corr} mA/cm^2	b_a mV/dec	b_c mV/dec	%IE	S
Blank	—	506	11.12	206	166	—	—
—	10^{-3}	503	8.39	199	162	24.55	—
0.001	—	509	10.04	230	180	9.73	1.126
—	10^{-3}	496	6.72	152	130	39.52	
0.007	—	483	6.54	199	143	41.19	1.197
—	10^{-3}	516	4.12	149	152	62.94	
0.009	—	492	6.40	180	143	43.17	1.167
—	10^{-3}	496	4.08	163	129	63.27	
0.010	—	495	4.41	174	154	60.81	1.128
—	10^{-3}	509	2.96	126	111	73.39	

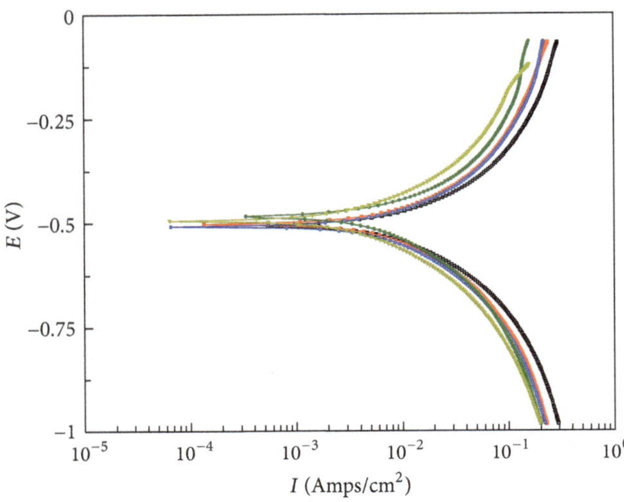

FIGURE 1: Tafel plot for corrosion of mild steel in 1 M HCl containing different concentrations of 4-hydroxy coumarin.

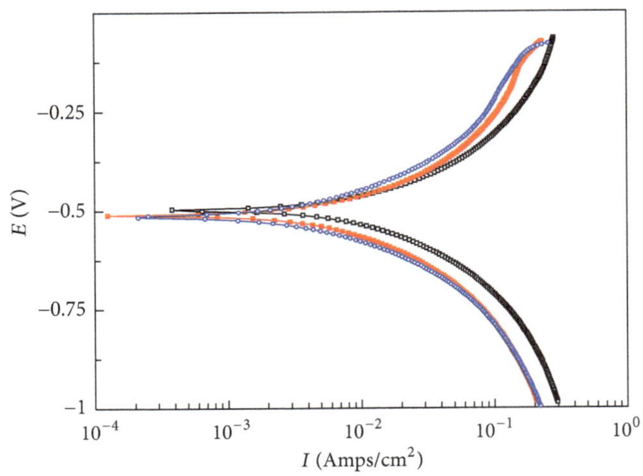

FIGURE 2: Tafel plot for corrosion of mild steel in 1 M HCl containing different concentrations of 4-hydroxy coumarin in the presence of 10^{-3} KI.

that the compound gave better inhibitive performance in the presence of 1×10^{-3} M potassium iodide. This enhancement in the IE on the addition of KI occurs due to synergistic effect.

The interaction of inhibitor molecules can be described by the introduction of synergism parameter; S was evaluated using the relationship given by Aramaki and Hackerman and reported elsewhere [12, 13]

$$S = \frac{1 - I_{1,2}}{1 - I_{1+2}}. \quad (4)$$

$I_{1,2}$ is the calculated inhibition effect supposing additivity (no interaction between the inhibitor compounds exists). $I_{1,2}$ can be expressed as $I_{1,2} = I_1 + I_2 - (I_1 I_2)$, I_1 is the measured inhibition effect of 1×10^{-3} M KI, I_2 is the measured inhibition effect of 4HC, and $I_{(1+2)}$ is the measured inhibition effect for the combination of 4HC and 1×10^{-3} M KI. The values of "S," the synergism parameter, for different concentrations of 4HC in 1 M HCl in combination with iodide ions are higher than unity suggesting the synergistic effect.

3.2. Polarization Measurements. Figures 1 and 2 represent the anodic and cathodic polarization curves of mild steel in 1 M HCl devoid of and containing different concentrations of 4HC and in the presence of 1×10^{-3} M KI. The electrochemical parameters of mild steel corrosion in free and inhibited 1 M HCl solutions were given in Table 2. It can be seen from the table that as the concentration of the coumarins is increased, there is little change in the corrosion potential. Moreover, the corrosion current (I_{corr}) decreases markedly in the presence of the inhibitor, and the magnitude of such an effect increases with increasing inhibitor concentration. These results showed the inhibitive action of the coumarin to corrosion of mild steel in the acidic medium. Also, it can be seen from the table, that the addition of 1×10^{-3} M KI further reduces the I_{corr} values exhibiting synergism.

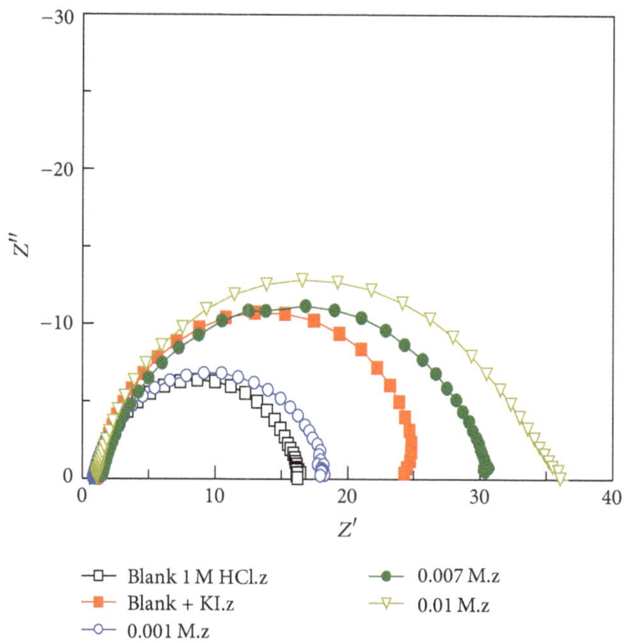

FIGURE 3: Complex-plane impedance of mild steel in 1 M HCl with different concentrations of 4-hydroxy coumarin.

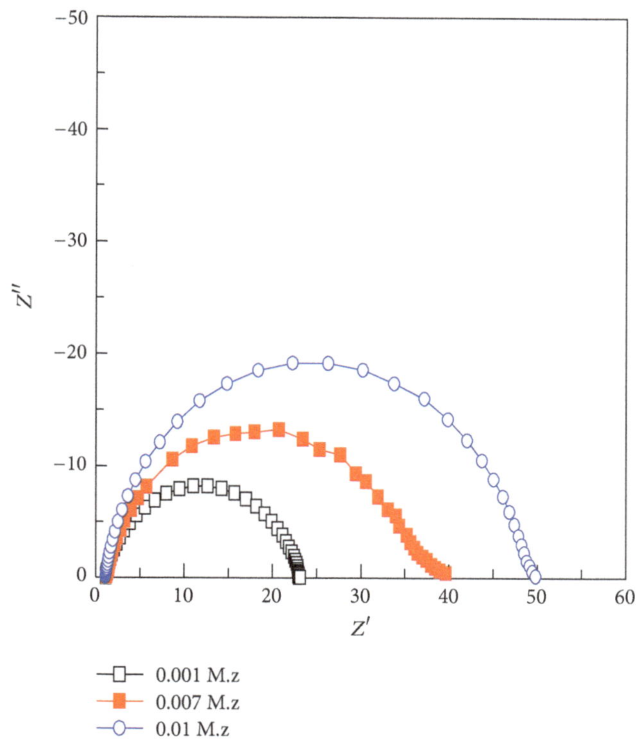

FIGURE 4: Complex-plane impedance of mild steel in 1 M HCl with different concentrations of 4-hydroxy coumarin in the presence of 10^{-3} M KI.

The values of inhibition efficiency increase with the increase in coumarin concentrations, reaching the maximum values of 60.81% at 0.01 M. The addition of KI has enhanced the percentage efficiency, which could be attributed to synergistic effect. The presence of synergism has been established by calculating the synergism parameter "S" which is recorded in Table 2. Almost all values are greater than unity, thereby proving the fact that the improvement in inhibition efficiency of 4HC generated by the addition of KI is due to synergism.

3.3. Electrochemical Impedance Spectroscopy. Figure 3 represents the Nyquist plot obtained for MS in 1 M HCl in absence and presence of various concentrations of the inhibitor. Impedance responses of acid solution containing various concentrations of the inhibitor in combination with 1×10^{-3} M KI solution are depicted in Figure 4.

The inhibition efficiency was calculated from C_{dl} values applying the following formula:

$$\%\text{IE} = 1 - \left[\frac{C_{dl(i)}}{C_{dl}}\right] \times 100, \tag{5}$$

where $C_{dl(i)}$ is the double layer capacitance value in the presence of the inhibitor and C_{dl} is the double layer capacitance value in the absence of the inhibitor.

The calculated inhibition efficiencies from the values of C_{dl} obtained from the impedance spectra was given in Table 3. The protecting capacity of 4HC was found to be increasing with the increase in its concentration. Due to synergism, the addition of KI has caused an enhancement in the protecting ability.

In order to correlate impedance and polarization methods, I_{corr} values were obtained from the Nyquist plots for different concentration of the inhibitor in the absence and presence of 10^{-3} M KI using the Stern-Geary equation [14]:

$$I_{corr} = \frac{b_a \times b_c}{2.3\,(b_a + b_c)\,R_{ct}}. \tag{6}$$

The obtained I_{corr} and %IE values with different concentrations of 4HC in the absence and presence of KI are given in Table 4.

The enhanced inhibition efficiency can be explained as due to joint adsorption of iodide ions and inhibitor cations or molecules. When irreversible chemisorption of iodide ions takes place and stable metal-iodide covalent bonds are formed, the iodide ions enter the metallic part of the double layer and the charge of the metal surface, which in turn attracts the positively charged inhibitor cations rendering the adsorption more effective.

Synergism can also happen according to a molecular mechanism. The coumarin cations created as a result of protonation exist only in the solution. On entering the double layer, they become discharged. The molecules produced are adsorbed at the expense of the unpaired electrons of the heteroatom.

4. Conclusions

(1) The inhibition efficiency of the inhibitor increased with the increase in its concentration at all immersion times.

TABLE 3: Impedance parameters for corrosion of mild steel in 1 M HCl for different concentration of 4-hydroxy coumarin in the presence and absence of 10^{-3} M KI.

[Inh] M	[KI] M	R_{ct} Ω cm^2	%IE	C_{dl} μF/cm^2	%IE
Blank	—	15.65	—	425	—
—	10^{-3}	22.00	28.86	328	22.82
0.001	—	18.03	13.20	385	9.41
—	10^{-3}	23.04	32.07	344	19.06
0.003	—	19.71	20.60	376	11.53
—	10^{-3}	23.96	34.68	318	25.18
0.007	—	30.12	48.04	280	34.12
—	10^{-3}	41.94	62.68	124	70.82
0.009	—	31.44	50.22	278	34.59
—	10^{-3}	44.31	64.68	121	71.53
0.010	—	37.48	58.24	149	64.94
—	10^{-3}	51.58	69.60	111	73.88

TABLE 4: Values of I_{corr} and %IE for corrosion of mild steel in 1 M HCl with different concentrations of 4-hydroxy coumarin in absence and presence of 10^{-3} M KI obtained by Stern-Geary equation.

[Inh] M	[KI] M	I_{corr} mA/cm^2	%IE
Blank	—	2.55	—
—	10^{-3}	1.76	30.98
0.001	—	2.43	4.71
—	10^{-3}	1.01	60.39
0.007	—	1.20	52.94
—	10^{-3}	0.78	69.41
0.009	—	1.15	54.90
—	10^{-3}	0.71	72.16
0.010	—	0.95	62.75
—	10^{-3}	0.50	80.39

(2) The presence of 10^{-3} M KI boosted the inhibitive efficiency of the coumarin.

(3) The synergism parameter S obtained by the Aramaki and Hackerman equation was found to be greater than unity. This suggests that the enhanced inhibition efficiency caused by the addition of iodide ion is due only to synergistic effect.

(4) The concentration dependence of the inhibition efficiency calculated from electrochemical studies has the same trend as that obtained from weight loss measurements.

(5) 4-Hydroxy coumarin covers both the anodic and cathodic regions through general adsorption. The adsorption occurs through the activated nonbonded electrons on the lactone oxygen, and the π electrons of the aromatic rings can adsorb onto the metal surface. Further, the –OH group at the 4th position shifts the electron flow towards the pyranone ring due to +M effect. Also, the oxonium ion formed in the acidic solution can be adsorbed on the metal surface through columbic interaction.

Acknowledgment

The authors would like to thank the authorities of the Avinashilingam Institute for Home Science and Higher Education for Women for providing necessary facilities to carry out this study.

References

[1] K. S. Sudhish, K. S. Ashish, C. M. Lutendo, M. K. Mwadham, and E. E. Eno, "Inhibitive effect of azorubine dye on the corrosion of mild steel in hydrochloric acid medium and synergistic iodide additive," *International Journal of Electrochemical Science*, vol. 7, pp. 5057–5068, 2012.

[2] M. Ajmal, A. S. Mideen, and M. A. Quraishi, "2-hydrazino-6-methyl-benzothiazole as an effective inhibitor for the corrosion of mild steel in acidic solutions," *Corrosion Science*, vol. 36, no. 1, pp. 79–84, 1994.

[3] A. Y. Musa, A. B. Mohamad, A. A. H. Kadhum, M. S. Takriff, and L. T. Tien, "Synergistic effect of potassium iodide with phthalazone on the corrosion inhibition of mild steel in 1.0M HCl," *Corrosion Science*, vol. 53, no. 11, pp. 3672–3677, 2011.

[4] E. E. Ebenso, A. Hailemichael, S. A. Umoren, and I. B. Obot, "Inhibition of mild steel corrosion in sulphuric acid using Alizarin yellow GG dye and synergistic iodide additive," *International Journal of Electrochemical Science*, vol. 3, pp. 1325–1339, 2008.

[5] V. K. Ahluwalia and R. Aggarwal, *Comprehensive Practical Organic Chemistry, Preparation and Quantitative Analysis*, Universities Press, Hyderabad, India, 2000.

[6] R. Saratha, M. Saranya Devi, H. N. Meenakshi, and R. Shyamala, "Enhanced corrosion resistance of Tecoma stans extract on mild steel in 0. 5M H_2SO_4 solution," *International Journal of Current Research*, vol. 2, pp. 92–96, 2011.

[7] H. Baeza, M. Guzmán, P. Ortega, and L. Vera, "Corrosion inhibition of copper in 0.5 M hydrochloric acid by 1,3,4-thiadiazole-2,5-dithiol," *Journal of the Chilean Chemical Society*, vol. 48, no. 3, pp. 23–26, 2003.

[8] W. A. W. E. Amira, A. A. Rahim, H. Osman, K. Awang, and P. B. Raja, "Corrosion inhibition of mild steel in 1 M HCl solution by Xylopia ferruginea leaves from different extract and partitions,"

International Journal of Electrochemical Science, vol. 6, no. 7, pp. 2998–3016, 2011.

[9] S. Ambrish, S. Ashish Kumar, and M. A. Quraishi, "Dapsone: a novel corrosion inhibitor for mild steel in acid media," *The Open Electrochemistry Journal*, vol. 2, pp. 43–51, 2010.

[10] M. Z. A. Rafiquee, S. Khan, N. Saxena, and M. A. Quraishi, "Influence of some thiadiazole derivatives on corrosion inhibition of mild steel in formic and acetic acid media," *Portugaliae Electrochimica Acta*, vol. 25, pp. 419–434, 2007.

[11] J.-Q. Xue, M. Wu, J.-X. Li, C.-B. Tang, and C.-X. Yin, "Inhibition behavior of a novel acidification inhibitor for steels at high temperature," *Materials Science Forum*, vol. 686, pp. 574–578, 2011.

[12] S. A. Umoren and E. E. Ebenso, "The synergistic effect of polyacrylamide and iodide ions on the corrosion inhibition of mild steel in H_2SO_4," *Materials Chemistry and Physics*, vol. 106, no. 2-3, pp. 387–393, 2007.

[13] S. A. Umoren, E. E. Ebenso, P. C. Okafor, U. J. Ekpe, and O. Ogbobe, "Effect of halide ions on the corrosion inhibition of aluminium in alkaline medium using polyvinyl alcohol," *Journal of Applied Polymer Science*, vol. 103, no. 5, pp. 2810–2816, 2007.

[14] L. Larabi, Y. Harek, M. Traisnel, and A. Mansri, "Synergistic influence of poly(4-vinylpyridine) and potassium iodide on inhibition of corrosion of mild steel in 1M HCl," *Journal of Applied Electrochemistry*, vol. 34, no. 8, pp. 833–839, 2004.

8

The Corrosion Resistance of Nickel Electrocomposite Coating Containing $BaFe_{12}O_{19}$ Particles

S. T. Aruna, G. Savitha, Jyothi Shedthi, and V. K. William Grips

National Aerospace Laboratories, Council of Scientific and Industrial Research, Post Bag No. 1779, Bangalore 560 017, India

Correspondence should be addressed to S. T. Aruna; staruna194@gmail.com

Academic Editors: G. El Mahdy, C. Gu, H. Miao, and C. Valentini

Electroplating composite coating is an effective method to prepare composite coating through the codeposition of metallic, nonmetallic, or polymer particles with metal to improve properties such as corrosion resistance, hardness, and wear performance. This paper reports the synthesis of a novel Ni-$BaFe_{12}O_{19}$ magnetic nanocomposite coating exhibiting improved corrosion resistance. In the present paper, $BaFe_{12}O_{19}$ particles were synthesized by a single-step solution combustion method and characterized for phase, particle size, and morphology. These particles were incorporated in a nickel metal matrix, and the properties of the coatings like nanohardness and corrosion resistance were investigated. The coating microstructure was also studied using field emission scanning electron microscope. A Vickers hardness of 777 HV was exhibited by Ni-$BaFe_{12}O_{19}$, and plain Ni coating exhibited a hardness of 517 HV. The Ni-$BaFe_{12}O_{19}$ composite coating exhibited improved corrosion resistance compared to plain Ni coating with an i_{corr} value of 0.034 $\mu A/cm^2$ compared to 0.361 $\mu A/cm^2$ for plain Ni. The Ni-$BaFe_{12}O_{19}$ coating also exhibited higher charge transfer and polarization resistance compared to plain Ni coating.

1. Introduction

In recent years, there is an increased interest in the synthesis and properties of particle reinforced metal matrix nanocomposite coatings with grain size of both matrix and dispersed particles less than 100 nm [1]. Particle-reinforced metal matrix composite (MMC) coatings have been widely used in various engineering applications due to their excellent mechanical, wear, and corrosion resistant properties compared to their components [2]. In composite materials, the metal matrix properties are modified by the addition of various insoluble substances like hard oxides (Al_2O_3, ZrO_2, SiO_2, TiO_2, and CeO_2); carbides (SiC and WC); diamond or solid lubricants (graphite, MoS_2, or PTFE) [2, 3]. Electrocodeposition is a coating protective technique which consists of introducing homogeneously dispersing inert particles of a different material in the metal matrix; this two-phase coating improves corrosion, wear, and mechanical resistance. The improvement of these properties depends on the combination of both particles and metal matrix. This method has the following advantages like low cost, low temperature and single-step process without additional thermal treatment [4]. The application of magnetic particles in electrodeposited composite coatings is less explored. Barium ferrite thin film media are attractive candidates for high-density overcoat free magnetic recording due to their large uniaxial anisotropy, high chemical stability, and corrosion resistance [5]. Thin films and micromagnets arrays of a composite matrix consisting of magnetic particles and a ferromagnetic alloy have been fabricated based on magnetic composite electroplating. Guan and Nelson have reported the fabrication of arrays of CoNiP-$BaFe_{12}O_{19}$ magnets with a variety of dimensions and features as small as 8 μm [6]. Qu and Jiao also reported the fabrication of CoNiMnP-$BaFe_{12}O_{19}$ magnetic composite coating by electrodeposition for applications in microelectromechanical systems actuators. This coating exhibited a maximum magnetic energy density of 17.68 kJm^{-3} and maximum coercivity of 1342 Oe [7]. However, to the best of our knowledge, there are no reports on the application of barium hexaferrite particles for improving the corrosion resistance of electrodeposited Ni-metal matrix composites. The main aim of the present work is to prepare barium hexaferrite magnetic particles and incorporate them in Ni-metal matrix and study the corrosion resistance and microstructure of the composite coating.

2. Experimental

2.1. Preparation and Characterization of Barium Hexaferrite Powder. The solution combustion method is a very versatile technique that has been used for the preparation of magnetic oxides and in particular ferrites [8]. Barium hexaferrite powder was prepared by the solution combustion method as follows: an aqueous redox mixture containing stoichiometric amounts of barium nitrate (5.39 g), ferric nitrate (100 g), and 46.33 g of oxalyl dihydrazide (ODH) was dissolved in minimum amounts of water and transferred into an alumina crucible and placed in a muffle furnace. The oxidizer-to-fuel ratio was maintained as unity. The temperature of the furnace was raised to 450°C and then switched off. The solution boils, undergoes frothing, and undergoes smoldering type combustion with low intensity sparks. This is followed by the evolution of copious amounts of gases and results in a brown voluminous mass. The obtained brown voluminous mass was the hard magnetic barium hexaferrite particles (Figure 1(a)). The as-prepared powder was magnetic in nature (Figure 1(b)).

Assuming complete combustion, the theoretical equation for the formation of barium ferrite using ODH can be written as follows:

$$Ba(NO_3)_{2(aq)} + 12\,Fe(NO_3)_{3(aq)} + 19\,C_2H_6N_4O_2 \longrightarrow BaFe_{12}O_{19} + 38\,CO_{2(g)} + 57\,H_2O + 57\,N_2 \quad (1)$$

For every mole of $BaFe_{12}O_{19}$ formed 152 moles of gases were evolved. The powder was characterized by powder X-ray diffractometer (Bruker D8, Advance), particle size analyzer (Mastersizer 2000, Malvern Instruments), and field emission scanning electron microscope (FESEM, Carl Zeiss Supra 40 VP).

2.2. Preparation of Ni-$BaFe_{12}O_{19}$ Composite Coatings. Nickel sulfamate plating bath was prepared by mixing 300 g L^{-1} of nickel sulfamate solution (50 g of nickel per liter), 10 g L^{-1} of nickel chloride, 30 g L^{-1} boric acid, and 0.2 g L^{-1} of sodium lauryl sulfate. The Ni-sulfamate plating bath (approximately 200 mL) containing $BaFe_{12}O_{19}$ particles (5 g/L) was taken in a glass beaker and was mechanically stirred overnight. For higher bath loadings, the electrolyte bath was saturated with powders because of its very fine nature, and there was no coating formation, and hence 5 g/L of barium hexaferrite particles was chosen. During electrodeposition, the bath was held at room temperature, and its pH was maintained at 4 by the addition of sulfamic acid and basic nickel carbonate. A pure nickel sheet (2.5 cm × 12 cm) and a brass substrate of the same dimension were used as anode and cathode of respectively. The polished brass substrate of area 2.5 cm × 3.75 cm was degreased with acetone followed by cathodic cleaning and acid dipping and finally washed with distilled water. In order to ensure uniform dispersion of the powder, the electrolyte bath containing particles was subjected to mechanical agitation for 15 hours before the deposition process, and during electrodeposition the particles were also mechanically stirred at a fixed rpm. The codeposition was carried out at 1.55 Adm^{-2} for 3 h such that the deposit thickness was approximately 40 μm based on Faraday's laws.

2.3. Characterization of Ni-$BaFe_{12}O_{19}$ Composite Coatings

2.3.1. XRD, Nanohardness, and Microstructural Studies of $BaFe_{12}O_{19}$ Composite Coatings. The XRD patterns of plain Ni and Ni-$BaFe_{12}O_{19}$ composite coatings electrodeposited at 1.55 A dm^{-2} were also recorded. The metallographic specimens for cross-sectional studies were prepared by sandwiching electrodeposited Ni-$BaFe_{12}O_{19}$ brass coupons with a copper backup in a Bakelite matrix followed by mechanical grinding and polishing with Al_2O_3 slurry, down to 0.05 μm. The nanohardness was obtained on electrodeposited Ni and Ni-$BaFe_{12}O_{19}$ cross-sections using CSM Instruments using the Oliver Pharr method. A Berkovich diamond indenter with 200 mN load was used for the measurements. The surface FESEM images of plain Ni and Ni-$BaFe_{12}O_{19}$ coatings were also recorded.

2.3.2. Corrosion Studies. Corrosion behavior of Ni and Ni-$BaFe_{12}O_{19}$ composite coatings electrodeposited at 1.55 A dm^{-2} for 45 min on mild steel coupons was conducted using CHI 604 2D electrochemical workstation. The test was carried out in deaerated 3.5 wt% (0.6 M) NaCl solution (200 ± 2 mL) using conventional three-electrode cell equipped with coated brass coupon with an active area of 1 cm^2 as working electrode and platinum foil and saturated calomel electrode (SCE) used as counter and reference electrodes, respectively. The reference electrode was connected to a Luggin capillary, and its tip was placed very close to the surface of the working electrode to minimize IR drop. The coupon was immersed in NaCl solution for an hour in order to establish the open circuit potential (EOCP). Electrochemical Impedance Spectroscopy (EIS) studies were carried out in the frequency range of 100 kHz–10 mHz. The amplitude of applied alternating potential was 10 mV on the EOCP. The impedance data was displayed as Nyquist and Bode plots, respectively. The Nyquist plot is a plot of real (Z') versus imaginary impedance (Z''). The Bode plot is a plot of $|Z|$ versus frequency and frequency versus phase angle (θ), where $|Z|$ is the absolute impedance. The acquired data were curve-fitted and analyzed using ZSimpWin program. After EIS measurements, the system was allowed to attain open circuit potential, then the upper and lower potential limits of linear sweep voltammetry were set at ±200 mV with respect to the EOCP. The sweep rate was 1 mV/s. The Tafel plots obtained have been represented as potential versus $\log i$ plot.

The corrosion potential E_{corr}, corrosion current density i_{corr}, and polarization resistance R_p were deduced from the

(a) (b)

FIGURE 1: Photographs showing combustion synthesized $BaFe_{12}O_{19}$ powder (a) after crushing and (b) attracted to magnetic stirring bar.

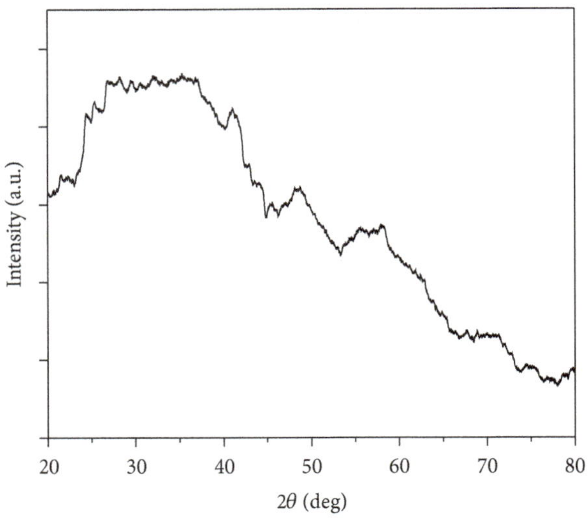

FIGURE 2: Powder XRD pattern of as-prepared $BaFe_{12}O_{19}$ powder.

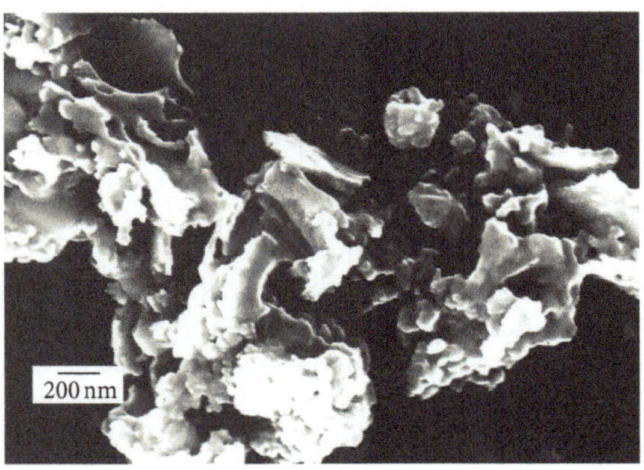

FIGURE 3: Field emission scanning electron microscope image of as-prepared barium ferrite particles.

Tafel plot. The polarization resistance was obtained using the Stern-Geary equation:

$$R_p = \left[\frac{b_a b_c}{2.303\,(b_a + b_c)}\right]\left[\frac{1}{i_{\text{corr}}}\right], \quad (2)$$

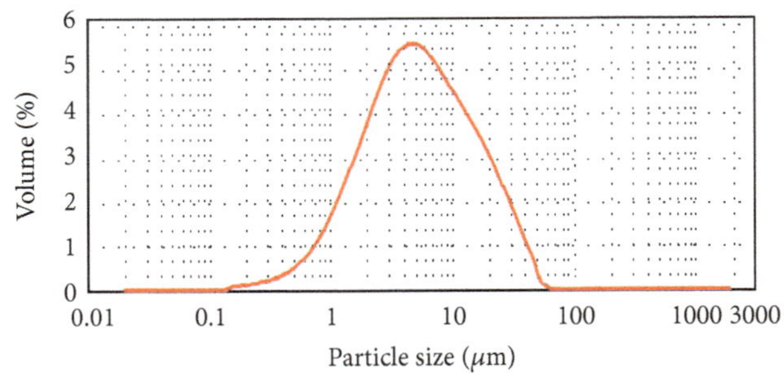

FIGURE 4: Particle size distribution of as-prepared $BaFe_{12}O_{19}$ powder.

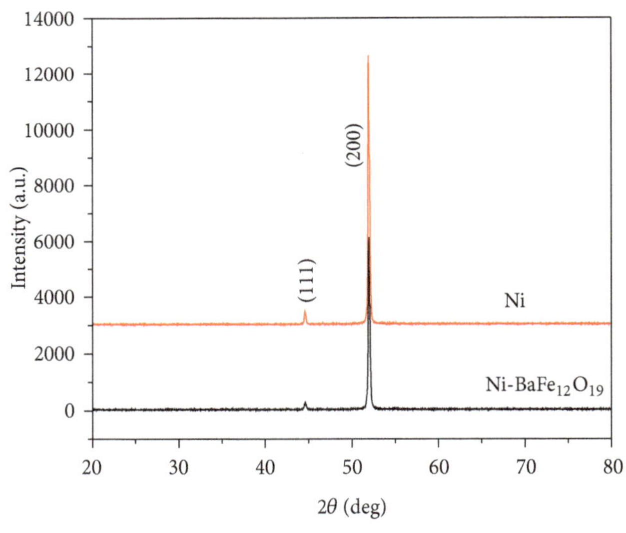

FIGURE 5: XRD patterns of electrodeposited (a) Ni and (b) Ni-$BaFe_{12}O_{19}$ composite coatings.

where b_a and b_c are the Tafel slopes or the Tafel constants, expressed in mV/dec.

Corrosion rates (CR) were calculated by the following equation:

$$\text{CR} = \frac{1.13\ i_{\text{corr}} \times \text{Eq. wt. of Ni}}{\text{Density of Ni}}. \quad (3)$$

Figure 6: Surface FESEM images of electrodeposited (a-b) Ni and (c-d) Ni-$BaFe_{12}O_{19}$ composite coatings.

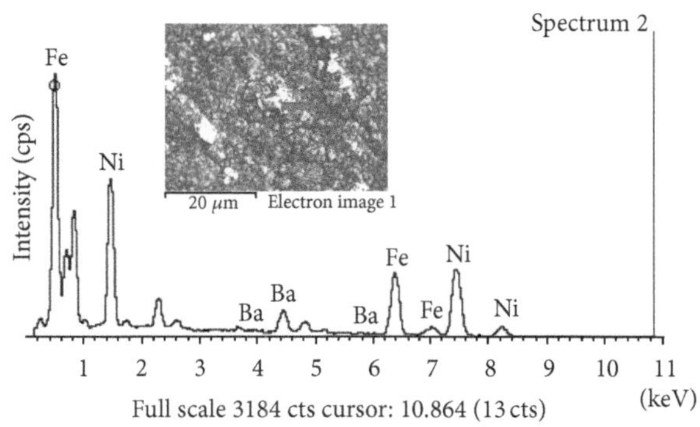

Figure 7: EDAX spectra recorded on the surface of electrodeposited Ni-$BaFe_{12}O_{19}$ composite coating.

3. Results and Discussion

3.1. Powder Characterization. The as-prepared $BaFe_{12}O_{19}$ powder was poorly crystalline as seen from the XRD pattern (Figure 2).

And hence it was not possible to determine the crystallite size. However, the phase purity of the sample was indirectly evident from the strong magnetic behavior of the as-prepared $BaFe_{12}O_{19}$ powder (Figure 1). Further calcinations of the barium ferrite will result in crystalline powder as reported in the literature [8].

The powder morphology of the combustion synthesized barium ferrite powder was analyzed by field emission scanning electron microscopy (FESEM) (Figure 3). From the FESEM image, it is evident that the particles are forming flakes and some hexagonal shaped particles mostly in nanosize were observed.

The particle size analysis of $BaFe_{12}O_{19}$ powder prepared by solution combustion method is shown in Figure 4. The particles have a tendency to agglomerate, and this is evident from the particle size analysis. The particle size distribution curve showed a Gaussian distribution with an average agglomerated particle size of 5.3 μm.

3.2. Ni-$BaFe_{12}O_{19}$ Composite Coating. The XRD patterns of Ni and Ni-$BaFe_{12}O_{19}$ electrocomposite coatings are shown in Figure 5, and the peaks were very sharp in case of plain Ni. The XRD patterns show the peaks corresponding to plain Ni, and no peaks corresponding to $BaFe_{12}O_{19}$ were observed as the percentage of particles incorporated is less than 5 wt% and also due to the poorly crystalline nature of the powder. Also, there was no change in the orientation of Ni peaks with the particle incorporation.

FESEM images of electrodeposited plain Ni and Ni-$BaFe_{12}O_{19}$ composite coating surfaces are shown in Figure 6. From the figure, it is evident that the barium ferrite particles are uniformly distributed throughout the Ni matrix (Figure 6(c)), and with the incorporation of the particles, there is a change in the shape of the nickel grains.

The energy dispersive X-ray analysis (EDAX) spectra recorded on the surface of the coatings confirmed the incorporation of barium hexaferrite particles in the coating (Figure 7). The weight percentage of Ni, Fe, Ba, and O were

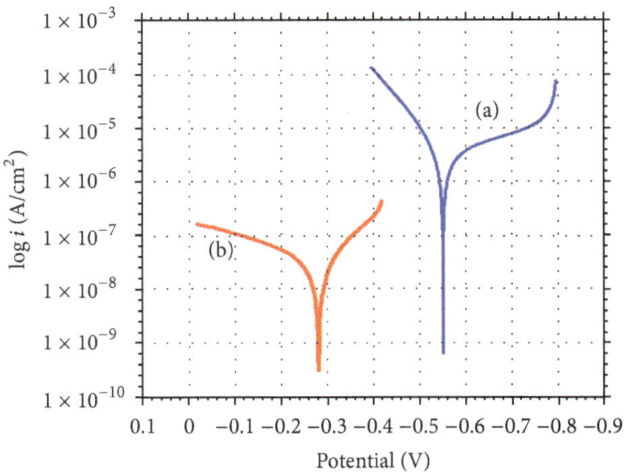

FIGURE 8: Tafel plots of electrodeposited (a) plain Ni and (b) Ni-$BaFe_{12}O_{19}$ coatings.

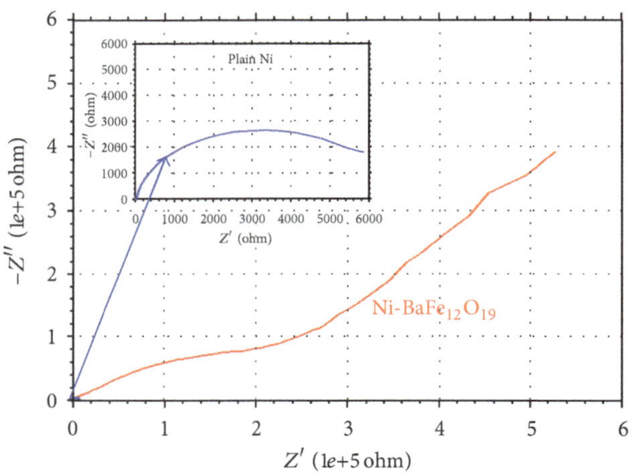

FIGURE 9: Nyquist plots of electrodeposited plain Ni (inset) and Ni-$BaFe_{12}O_{19}$ coatings.

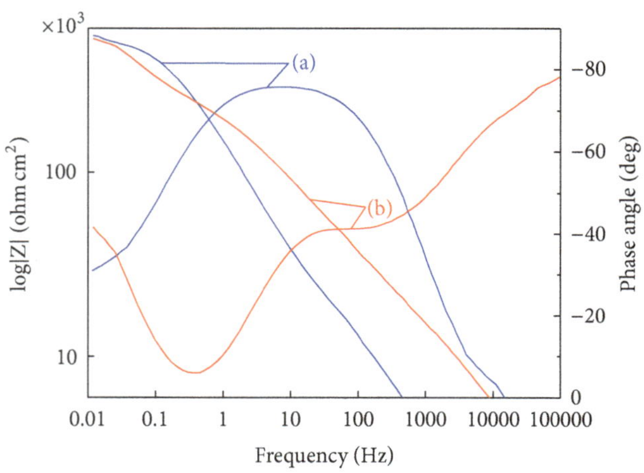

FIGURE 10: Bode plots of electrodeposited (a) Ni and (b) Ni-$BaFe_{12}O_{19}$ coatings.

31.71, 18.28, 9.41, and 40.59%, respectively, on a particular spot where white particles were seen.

The nanohardness values obtained from Oliver Pharr method on the cross-sections of the electrodeposited (1.55 A dm^{-2}) Ni and Ni-$BaFe_{12}O_{19}$ coating surface s were 517 ± 38 HV and 777 ± 16 HV, respectively. The improved hardness may be attributed to the dispersion strengthening of the Ni matrix due to the incorporation of barium hexaferrite particles.

The Tafel plots obtained for Ni and Ni-$BaFe_{12}O_{19}$ coatings in 3.5 wt% NaCl are shown in Figure 8. The corrosion potential (E_{corr}), corrosion current (i_{corr}), and polarization resistance (R_p) obtained from the Tafel plots are listed in Table 1. From the Tafel plots, it is clear that Ni-$BaFe_{12}O_{19}$ coating shows nobler behavior compared to plain Ni coating. Corrosion potential and corrosion current are lower for Ni-$BaFe_{12}O_{19}$ compared to Ni. This shows that, by incorporating $BaFe_{12}O_{19}$ particles in Ni matrix, the corrosion resistance of Ni can be enhanced (Table 1).

The Nyquist impedance plots for Ni and Ni-$BaFe_{12}O_{19}$ obtained in free-air (nonaerated) condition in 3.5% NaCl solution are shown in Figure 9. At higher frequencies, interception with the real axis is ascribed to the electrolyte bulk resistance (R_s), and at low frequencies an interphase appears whose interception with the real axis is ascribed to the charge transfer resistance (R_{ct}). In case of Ni-$BaFe_{12}O_{19}$, additional Warburg diffusion behavior was observed at low frequency region, which represents a kind of resistance to mass transfer. A Warburg diffusional phenomenon, which suggests diffusion-controlled mechanism in the corrosion reaction, was observed in the low frequency region for Ni-$BaFe_{12}O_{19}$ coating. The Warburg impedance is usually observed at low frequencies in electrochemical experiments due to the concentration polarization induced by a sluggish diffusion process [9]. The Warburg impedance can occur due to the following two reasons: (i) if there is some porosity/cracks in the coating the electrolyte can enter through the pores and attack the substrate, and in turn the corrosion products can block the pores and further offer resistance to the electrolyte; (ii) some loosely held particles may be dislodged in the electrolyte, and thereby exposing the substrate to the electrolyte. However, cracks and pores were not seen in the FESEM images of as-prepared coating surface.

The Bode plot of Ni showed a distinct single peak corresponding to the electrode/electrolyte interface compared to Ni-$BaFe_{12}O_{19}$ (Figure 10). In both equivalent circuits, the double layer capacitance C_{dl} is represented by a constant phase element (Q), which accounts for the deviation from the ideal dielectric behavior of the nonideal capacitor. The R_{ct} value of Ni was about 7 $k\Omega cm^{-2}$, and for Ni-$BaFe_{12}O_{19}$ it was 1081 $k\Omega cm^{-2}$. It is gratifying to note that there is a good correlation between the R_p and R_{ct} values. This shows that the corrosion resistance of Ni-$BaFe_{12}O_{19}$ is higher than that plain Ni. The higher R_{ct} value of Ni-$BaFe_{12}O_{19}$ indicates that the active area available for the corrosion attack is less when compared to Ni.

The phase angle versus log (frequency) shows that phase angle of both Ni and Ni-$BaFe_{12}O_{19}$ coatings was less than 90° indicating their lower capacitive behavior. The n value for Ni

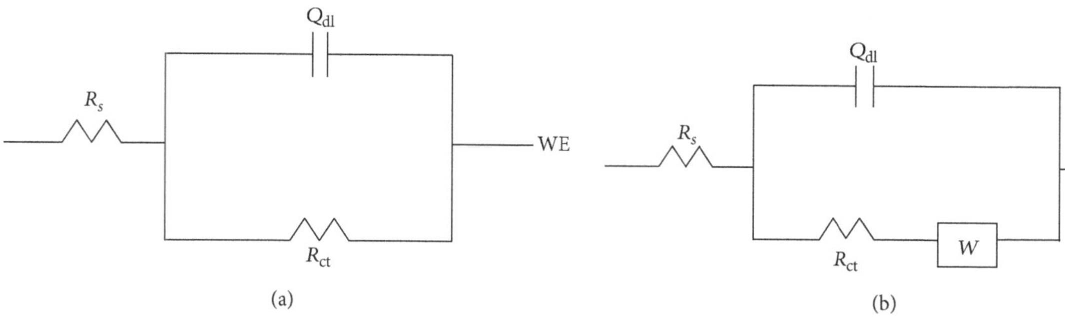

FIGURE 11: (a): Equivalent circuit used to fit the parameters of nickel (R_s: solution resistance; Q_{dl}: double layer capacitance; R_{ct}: charge transfer resistance; WE: working electrode). (b): Equivalent circuit used to fit the parameters of Ni-$BaFe_{12}O_{19}$ composite coating (R_s: solution resistance; Q_{dl}: double layer capacitance; R_{ct}: charge transfer resistance; W: Warburg impedance; WE: working electrode).

TABLE 1: Corrosion potential, corrosion rates, and Tafel slopes calculated from potentiodynamic diagrams for mild steel, pure Ni, and Ni-$BaFe_{12}O_{19}$.

Coating	i_{corr} (μA/cm^2)	E_{corr} (V)	b_c (V/dec)	b_a (V/dec)	R_P (kΩ cm^2)
Ni	3.61	−0.551	0.031	0.114	8.29
Ni-$BaFe_{12}O_{19}$	0.034	−0.279	0.2615	0.1262	1087.67

TABLE 2: Electrochemical impedance analysis data.

Sample	R_s (Ω m^2)	Q_{dl}-Y_0 (μF/cm^2)	n_{dl}	R_{ct} (kΩ cm^2)	W ($\mu\Omega$ cm^{-2} s$^{-1/2}$)
Ni	1.05	226.1	0.83	7.03	—
Ni-$BaFe_{12}O_{19}$	1.12	2.74	0.37	1081	9.9

coating was higher than that of Ni-$BaFe_{12}O_{19}$ which indicates a higher capacitive behavior. The corrosion rates as calculated from (3) were 1.34×10^{-6} and 1.2×10^{-7} g/h, respectively, for Ni and Ni-$BaFe_{12}O_{19}$ composite coatings.

The equivalent circuit shown in Figure 11(a) was used to fit the parameters of plain Ni, and the fitted values are tabulated in Table 2. The appropriate equivalent circuit was used for fitting the parameters of Ni-$BaFe_{12}O_{19}$, and the acquired data are shown in Figure 11(b).

4. Conclusions

Barium hexaferrite particles having magnetic property prepared by the solution combustion method were incorporated in the Ni-metal matrix. The bath loading beyond 5 g/L was not possible due to the very fine nature of the combustion synthesized powder. It was gratifying to note that the anticorrosion properties of the Ni matrix could be further improved by the incorporation of hard $BaFe_{12}O_{19}$ particles in the Ni matrix. Corrosion potential and corrosion current were lower for Ni-$BaFe_{12}O_{19}$ when compared to Ni. There was an increase in the nanohardness of the Ni-composite coating with the incorporation of $BaFe_{12}O_{19}$ particles. The R_{ct} value of Ni was about 7 kΩcm^{-2}, and for Ni-$BaFe_{12}O_{19}$ it was 1081 kΩcm^{-2}. This shows that the incorporation of barium hexaferrite particles enhances the corrosion resistance of Ni-$BaFe_{12}O_{19}$ coating.

Acknowledgments

The authors thank the Director of CSIR-NAL for his encouragement and SIP-SED-02 project for the funding. The authors thank Mr. Siju and Mr. Praveen for the FESEM and nanohardness measurements.

References

[1] I. Garcia, J. Fransaer, and J. P. Celis, "Electrodeposition and sliding wear resistance of nickel composite coatings containing micron and submicron SiC particles," *Surface and Coatings Technology*, vol. 148, no. 2-3, pp. 171–178, 2001.

[2] C. T. J. Low, R. G. A. Wills, and F. C. Walsh, "Electrodeposition of composite coatings containing nanoparticles in a metal deposit," *Surface and Coatings Technology*, vol. 201, no. 1-2, pp. 371–383, 2006.

[3] J. Niedbała, A. Budniok, and E. Łagiewka, "Hydrogen evolution on the polyethylene-modified Ni-Mo composite layers," *Thin Solid Films*, vol. 516, no. 18, pp. 6191–6196, 2008.

[4] J. K. Dennis and T. E. Such, *Nickel and Chromium Plating*, Woodhead Publishing, 3rd edition, 1993.

[5] X. Sui, M. Scherge, M. H. Knyder, J. E. Snyder, V. G. Harris, and N. C. Koon, "Barium ferrite thin-film recording media," *Journal of Magnetism and Magnetic Materials*, vol. 655, no. 1–3, pp. 132–139, 1996.

[6] S. Guan and B. J. Nelson, "Magnetic composite electroplating for depositing micromagnets," *Journal of Microelectromechanical Systems*, vol. 15, no. 2, pp. 330–337, 2006.

[7] N. S. Qu and F. Jiao, "Fabrication of CoNiMnP-$BaFe_{12}O_{19}$ nanocomposite coatings by electrodeposition," *Transactions of the Institute of Metal Finishing*, vol. 90, no. 2, pp. 92–97, 2012.

[8] K. Suresh and K. C. Patil, "A Recipe for an instant synthesis of fine particle oxide materials," in *Perspectives in Solid State Chemistry*, K. J. Rao, Ed., pp. 376–388, Narosa Publishing House, New Delhi, India, 1995.

[9] J. S. Yoo and S. M. Park, "Reply to comments on the article 'an electrochemical impedance measurement techniques employing fourier transform,'" *Analytical Chemistry*, vol. 73, no. 16, pp. 4060–4061, 2001.

Microstructural Characteristics and Mechanical Properties of Nanostructured and Conventional TiAlN and AlCrN Coatings on ASTM-SA210 Grade A-1 Boiler Steel

Vikas Chawla

D.A.V. College of Engineering and Technology, Kanina 123027, India

Correspondence should be addressed to Vikas Chawla; vikkydmt@gmail.com

Academic Editors: S. J. Lee, Q. Qu, and R. Solmaz

Nanostructured and conventional TiAlN and AlCrN coatings were deposited on ASTM-SA210 Grade A-1 boiler steel. The present work has been focused to characterize the conventional thick (by plasma spraying and gas nitrided) and nanostructured thin (by physical vapor deposition process) TiAlN and AlCrN coatings developed on ASTM-SA210 Grade A-1 boiler steel. The surface morphology of the coated samples was studied using FE-SEM with EDAX attachment, XRD analysis, AFM analysis, and the X-ray mapping analysis. The presence of metal nitride phases as identified by XRD analysis indicates that the desired coatings were successfully developed. The bond strength, surface roughness, and microhardness of the gas nitrided plasma sprayed conventional coatings were measured. A good adhesion of the conventional thick TiAlN and AlCrN coatings was evident from bond test results.

1. Introduction

The boiler components are subjected to various types of degradation phenomenon such as high-temperature corrosion, erosion corrosion, overheating, solid particle abrasion and wear. High-temperature oxidation and erosion caused by the impact of fly ashes and unburnt carbon particles are the main problems to solve in these applications. Therefore, the development of wear and high-temperature oxidation protection systems in industrial boilers is a very important topic from both engineering and an economic perspective [1, 2]. Present materials being capable of resisting erosive and corrosive environments are highly alloyed and thus expensive. In search for cost-effective solutions for erosion-corrosion problems, the effective method of erosion-corrosion prevention is to coat the alloy with a protective layer, which has been used in the current investigation.

The high-temperature protection system must meet several criteria: that is, it should provide adequate environmental resistance, be chemically and mechanically compatible with the substrate, and be practically applicable, reliable, and economically attractive [3]. The primary aim of the coating/surface treatment is the ability to produce a stable, slow-growing surface oxide providing a barrier between the coated alloy and the environment [4]. Recent studies show that 80% of the total cost for the protection of metals is related to coating application [5, 6].

According to DeMasi-Marcin and Gupta [7], physical vapor deposition and plasma spray (a thermal spray process) have been reported to be two major coating processing technologies which are used worldwide. The PVD processes have been successful in improving mechanical properties such as wear, friction, and hardness [8]. Their use as corrosion-resistant coatings is also becoming widespread. Among the thermal spray coating processes, plasma spraying is reported to be versatile technology that has been successful as a reliable cost-effective solution for many industrial problems [9, 10]. Since the commercialization of physical vapor deposited (PVD) TiN coatings in early 1980s, transition metal nitrides based hard coatings have been successfully used for the materials protection particularly to improve cutting tools lifetime [11].

In recent years, corrosion performance of nanostructured materials/coatings is a hot topic in corrosion field [12, 13]. Nanostructured materials as a new class of engineering materials indeed behave differently from their microscopic

TABLE 1: Chemical composition (wt%) of Grade A-1 boiler steel (ASTM-SA210 Grade A-1).

Elements	C	Mn	Si	S	P	Fe
Nominal	0.27	0.93	0.1	0.058	0.048	Bal.
Actual	0.2768	0.6725	0.1725	0.00517	0.093	Bal.

counterparts because their characteristic sizes are smaller than the characteristic length scales of physical phenomenon occurring in bulk materials [14]. As reported by Chawla et al. [15, 16], in the past decade, attractive properties associated with a nanostructure have been documented for bulk materials, where most of the research in the field of nanomaterials has been focused. Little work has been published on high-temperature corrosion behavior of nanostructured and conventional hard metal nitride coatings to the best of knowledge of the author. Therefore, these coatings were developed on the ASTM-SA210 Grade A-1 boiler steel, and their behavior was studied when exposed to high-temperature oxidation in air, molten salt environment, industrial environment, high temperature solid particle erosion, and simulated marine environment. The present work has been focused to characterize the conventional thick (by plasma spraying and gas nitrided) and nanostructured thin (by physical vapor deposition process) TiAlN and AlCrN coatings developed on ASTM-SA210 Grade A-1 boiler steel.

2. Experimental Details

2.1. Selection of Substrate and Sample Preparation. The boiler steel material "ASTM-SA210 Grade A-1" has been selected for the present study as the substrate material. Selection of candidate material for the study has been made after consultation with Guru Nanak Dev Thermal Plant, Bathinda (India). Grade A-1 has a wide range of applications in boilers, especially in the construction of their water walls. The actual chemical composition of the selected alloy has been determined with the help of Optical Emission Spectrometer (Thrmo Jarrel Ash, TJA181/81, USA), which is reported in Table 1 along with nominal chemical composition.

Specimens with dimensions of approximately 20 mm × 15 mm × 5 mm were cut from the alloy sheets. The specimens were polished using emery papers of 220, 400, 600 grit sizes and subsequently on 1/0, 2/0, 3/0, and 4/0 grades. Then the samples were mirror polished using cloth polishing wheel machine with 1 μm lavigated alumina powder suspension for deposition of nanostructured coatings and grit blasted with alumina powders (Grit 45) prior to the deposition of the conventional coatings by plasma spraying (a thermal spray process), for developing better adhesion between the substrates and the coatings. The specimens were prepared manually, and whole care was taken to avoid any structural changes in the specimens.

2.2. Development of Nanostructured Thin Coatings. The nanostructured thin TiAlN and AlCrN coatings, with a thickness around 4 μm, were deposited on the substrates at Oerlikon Balzers Coatings India Limited, Gurgaon, India.

TABLE 2: Summary of nanostructured thin coatings' deposition parameters.

Coating parameters	Nanostructured TiAlN	Nanostructured AlCrN
Machine used	Standard balzers rapid coating system (RCS) machine	Standard balzers rapid coating system (RCS) machine
Made in	Oerlikon Balzers, Swiss	Oerlikon Balzers, Swiss
Targets composition	Ti, $Ti_{50}Al_{50}$	$Al_{70}Cr_{30}$
Number of targets	Ti (02), $Ti_{50}Al_{50}$ (04)	$Al_{70}Cr_{30}$ (06)
Targets power	3.5 KW	3.5 KW
Reactive gas	Nitrogen	Nitrogen
Deposition pressure	3.5 Pa	3.5 Pa
Substrate bias voltage	−40 V to −170 V	−40 V to −170 V
Coating thickness	4 μm ± 1 μm	4 μm ± 1 μm

A front-loading *Balzer's rapid coating system* (RCS) *machine* (*made in Oerlikon Balzers, Swiss*) was used for the deposition of the coatings. The machine is equipped with 6 cathodic arc sources. In case of nanostructured thin TiAlN coatings, two of the six sources were used to deposit a thin, 0.3 μm thick TiN sub-layer to improve adhesion of coating. The remaining four sources were employed to deposit the main layer of the coatings, which was obtained using customized sintered targets. Whereas in case of nanostructured thin AlCrN coating, all the six targets sources were employed to deposit the main layer of the coatings. The Oerlikon Balzers Coatings India Limited, Gurgaon, India has optimized the coating parameters for all the substrates. The compositions of the targets used, coating thickness and the summary of the process parameters, are presented in Table 2. For all coatings, argon (Ar) and pure nitrogen atmosphere were used during deposition. Prior to deposition all the substrates were cleaned in two steps: firstly with Ultrasonic Pre-Cleaner (made in Imeco, Pune, India) and secondly with Ultrasonic Cleaning Machine with 9 Tanks including hot air dryer (make Oerlikon Balzers Ltd. India) for 1.5 Hrs.

2.3. Development of Conventional Thick Coatings. Two types of coating powders, namely, Ti-50Al (50 Ti-50 Al Wt%) and Al-30Cr (70 Al-30 Cr Wt%) were chosen for plasma spray deposition on the substrate. These powders were prepared in laboratory ball mill for 8 hrs to form a uniform mixture from three types of commercially available alloy powders, namely, titanium powder with minimum assay 98.0% (Art. 6322) supplied by Loba Chemie, Mumbai, India, aluminum powder with minimum assay 90.0% and size 200 mesh (Art. 37064) and chromium powder with minimum assay 99.0% and size 60 mesh (Art. 36054) supplied by S.D. Fine Chemical Limited, Mumbai, India. The coating work were carried out a commercial firm, namely, Anod Plasma Limited, Kanpur, India. They used 40 kW Miller Thermal (USA) plasma spray apparatus to apply the coatings. Argon was used as powder carrying and shielding gas. Standard spray parameters were

Table 3: Parameters of the argon shrouded plasma spray process.

Arc current (A)	750
Arc voltage (V)	45
Powder feed rate (rev/min)	5.2
Spraying distance (mm)	90–110
Plasma arc gas (argon) pressure	58 psi
Powder gas pressure	60 psi
AVX gas pressure	10 psi

Table 4: Parameters of the gas nitriding process.

Temperature	650°C
Duration	3 hrs For Ti-Al coating 5 hrs For Al-Cr coating
Discharge rate of nitrogen	0.3 L/min

designed by the previous firm for depositing the coatings in the present work. All the process parameters were kept constant throughout the coating process while spraying distance was maintained in a narrow range of 90–110 mm. The process parameters for the shrouded plasma spray process employed for applying the coatings are summarized in Table 3.

Gas nitriding is a case hardening process whereby nitrogen is introduced into the surface of an alloy by holding it at a suitable temperature in contact with a nitrogenous gas, that is, nitrogen or ammonia. After reviewing the literature critically, it was decided to do the gas nitriding of the plasma sprayed conventional thick Ti-Al and Al-Cr coatings in the lab. The apparatus used for the gas nitriding process is shown in Figure 1, which consists of a long steel tube placed in a silicon tube furnace. In the front view (Figure 1), the steel tube is connected to the nitrogen cylinder for nitrogen supply, and in the rear view, the outlet of the nitrogen gas is through water for slow and constant discharge rate. The gas nitriding parameters were optimized after conducting several trials on plasma sprayed coated specimens. The process parameters employed for the gas nitriding are as summarized in Table 4.

2.4. Characterization of the Coatings. Field emission scanning electron microscope (FESEM, FEI, Quanta 200F Company) with EDAX Genesis software attachment (made in Czech Republic) is used to characterize the surface morphology of the coatings. SEM micrographs along with EDS spectrum were taken with an electron beam energy of 20 keV. The surface morphology (2D and 3D) of the thin films was also characterized by AFM (Model: NTEGRA, NT-MDT, Ireland) to calculate the surface roughness and particle size.

The coated specimens were subjected to XRD analysis using Bruker AXS D-8 advance diffractometer (Germany) with Cu Kα radiation. The scan rate used was 2°/min and the scan range was from 20° to 120°. The grain size of the thin films was estimated from Scherrer formula, as given in (1). In this expression, the grain size D is along the surface normal direction, which is also the direction of the XRD diffraction vector:

$$D = \frac{0.9\lambda}{B\cos\theta}, \quad (1)$$

where B is the corrected full-width at half maximum (FWHM) of a Bragg peak, λ is the X-ray wavelength, and θ is the Bragg angle. B is obtained from the equation $B^2 = B^2_r - B^2_{\text{strain}} - C^2$, where B_r is the FWHM of a measured Bragg peak, $B_{\text{strain}} = \varepsilon \tan\theta$ is the lattice broadening from the residual strain ε measured by XRD using the $\cos^2\alpha \sin^2\psi$ method, and C is the instrumental line broadening. A Zeiss Axiovert 200 MAT inverted optical microscope, fitted with image software Zeiss Axiovision Release 4.1, was used for optical microscopy. The porosity measurements were made with image analyser, having software of Dewinter Materials Plus 1.01 based on ASTM B276. A PMP3 inverted metallurgical microscope was used to obtain the images.

The as-coated specimens were cut with a diamond cutter (Buehler's Precision Diamond Saw, Model ISOMET 1000, made in USA) across its cross-section and subsequently hot mounted in Buehler's transoptic powder (20-3400-080). Subsequently, the mounted specimens were polished manually using emery papers of 220, 400, and 600 grit sizes and subsequently on 1/0, 2/0, 3/0, and 4/0 grades. Finally, the specimens were mirror polished on a cloth polishing wheel machine with 0.05 μm alumina powder suspension. The specimens were washed thoroughly with flowing water and dried in hot air to remove any moisture.

The thickness of the coatings was monitored during the coating processes. In case of plasma sprayed conventional thick coatings, the thickness was monitored with a Minitest-2000 thin film thickness gauge (made in Germany, precision ±1 μm). Efforts were made to obtain the coatings with uniform thickness. The thickness of some of the as-coated specimens (nanostructured as well as conventional coatings) was further verified by cutting along the cross-section and mounted (as explained previously). A Field emission scanning electron microscope (FE-SEM)(FEI Quanta 200F, made in Czech Republic) was used to obtain the back scattered electron (BSE) images.

The bond strength of the gas nitrided plasma sprayed coatings was tested using the ASTM standard C633-01. This test method covers the determination of the degree of adhesion (bonding strength) of a coating to a substrate or the cohesion strength of the coating in a tension normal to the surface. The test consists of coating one face of a substrate fixture, bonding this coating to the face of a loading fixture, and subjecting this assembly of coating and fixtures to a tensile load normal to the plane of the coating. A data acquisition system has continuously recorded the tensile load exerted by the machine. The surface roughness (R_a) values of the gas nitrided plasma sprayed as coated specimens were measured using Surface Roughness Tester (Mitutoyo SJ-201, Japan). Each reported value of surface roughness (R_a) is the mean of five observations taken at different locations. The centre line average (CLA) method was used to obtain the R_a values.

X-ray mapping analysis of the samples was done on field emission scanning electron microscope (FEI, Quanta 200F Company) for image acquisition entailed a backscattered

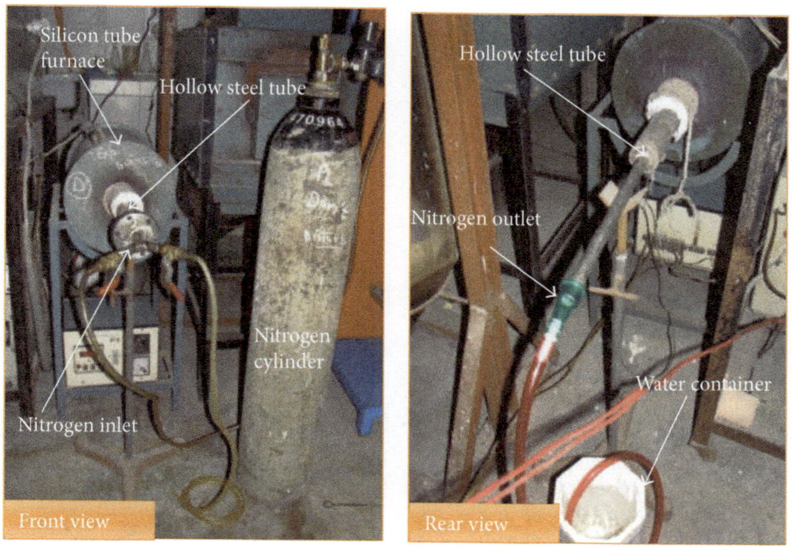

(a) (b)

FIGURE 1: Front and rear views of the setup used for gas nitriding of the plasma sprayed conventional thick Ti-Al and Al-Cr coatings in the laboratory.

electron image (BSEI) and secondary electron image (SEI) mode. An accelerating voltage of 20–25 kV, a working distance of 9–10 mm, and an image size of 1024∗884 pixels were used for getting quality images. Energy Dispersive X-ray Analysis (EDAX) and X-ray mapping were employed while imaging on FE-SEM was carried out to obtain elemental composition at different areas of the coating and surface morphology of the coatings, respectively. The specimen were silver pasted between samples and the stub in order to have conductivity, thereafter, gold-coated to facilitate elemental X-ray mapping analysis of the different elements present across the coating. The selected area has three regions, that is, substrate, coating, and some epoxy region. X-ray mappings were obtained for all the elements of the substrate and the coatings, but only those mappings are reported which indicates the presence of some element.

For obtaining microhardness of the gas nitrided plasma sprayed conventional coatings, the specimens were cut, mounted, and polished (as explained polished). The microhardness of the as sprayed coatings was measured by using Miniload 2 Microhardness Tester (Leitz, Germany) fitted with a Vickers pyramidal diamond indenter. A 15 g (147.1 mN) load was applied to the indenter for penetration of as sprayed coatings. Hardness value was calculated from the relation $Hv = 1854.4 \times F/d^2$ where "F" is the load in grams and "d" is the diameter of the indenter in micrometer. Each reported value of the microhardness is the average value of five measurements.

3. Observations and Results

3.1. Visual Observations. The nanostructured thin (by Physical Vapor Deposition process) and conventional thick (by plasma spraying and gas nitriding) TiAlN and AlCrN coatings have been formulated successfully on ASTM-SA210 Grade A-1 boiler steel. The macrographs of as-coated Grade A-1 boiler steel are shown in Figure 2. It can be observed from the macrographs that the nanostructured thin TiAlN coating is violet grey in color and nanostructured thin AlCrN coating is light grey in color. Visual observations indicate that the surface of nanostructured coatings is smooth, whereas the surface of conventional thick coatings is rough. The color of the thick coatings appeared as grey with some bluish shining. Also, optical micrographs of the nanostructured and conventional coatings are shown in Figure 3. The nanostructured coatings (Figures 3(a) and 3(b)) have uniform microstructure. It is evident from the microstructure that the coatings contain some pores and inclusions. In case of conventional thick coatings (Figures 3(c) and 3(d)), the massive microstructure can be observed with irregularly shaped grains.

3.2. Surface Analysis

3.2.1. X-Ray Diffraction Analysis (XRD). XRD diffractograms for nanostructured and conventional TiAlN and AlCrN coated ASTM-SA210 Grade A-1 boiler steel are depicted in Figure 4 on reduced scale. As indicated by the diffractograms in Figure 4; TiN and AlN are the main phases present in the nanostructured thin TiAlN coating. Further, in case of nanostructured AlCrN coating, the prominent phases are CrN and AlN. The grain size of the nanostructured thin coatings was estimated from Scherrer formula; which is reported in Table 5. The calculated grain size for nanostructured thin TiAlN and AlCrN coatings is 16 and 26 nm, respectively. Further, the main phases identified for the conventional thick TiAlN coating are Al_2O_3, TiN, Ti_3Al, AlN and small peaks of TiO_2 and Fe_2O_3. The phases identified in case of conventional thick AlCrN coating are CrN, AlN, and Al_2O_3 with minor peaks of Cr_2O_3.

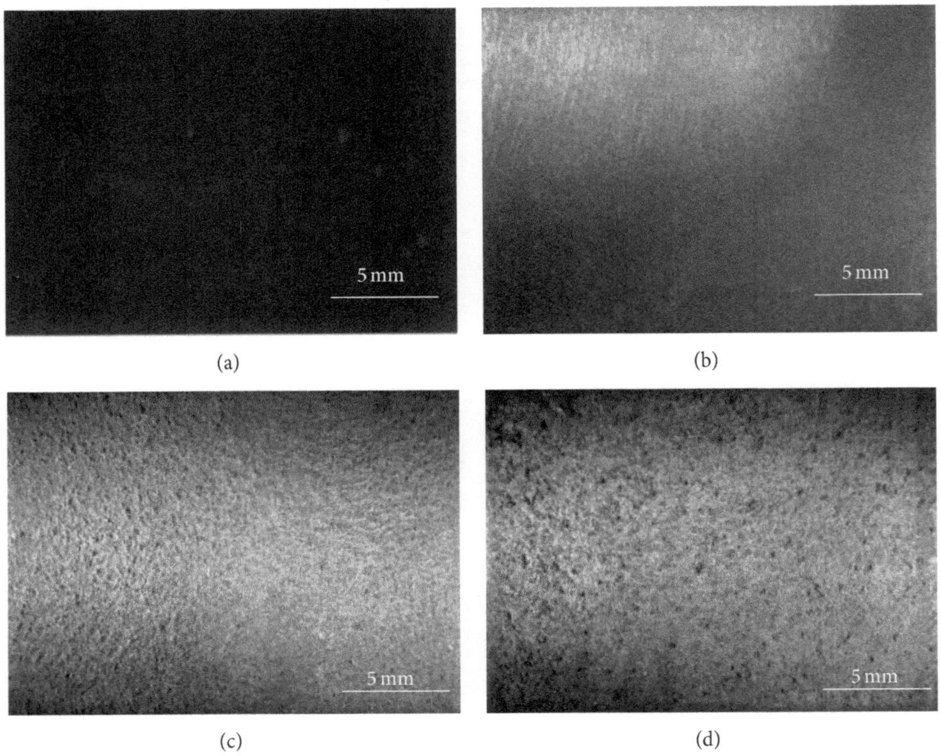

FIGURE 2: Surface macrographs for the as-coated ASTM-SA210 Grade A-1 boiler steel. (a) Nanostructured TiAlN coating, (b) nanostructured AlCrN coating, (c) conventional TiAlN coating, and (d) conventional AlCrN coating.

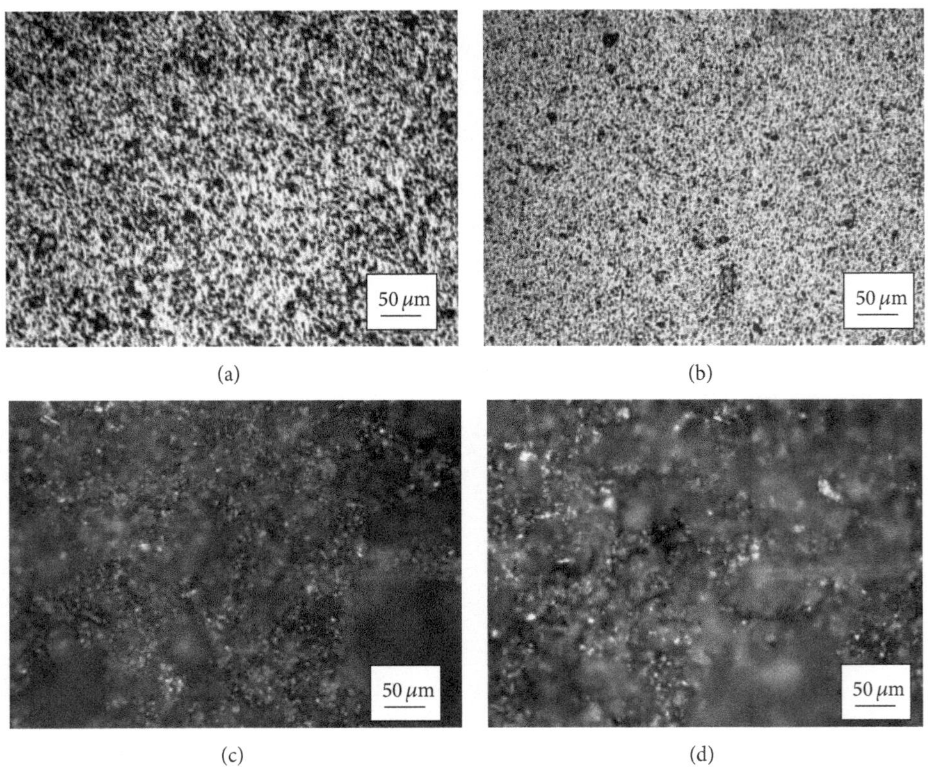

FIGURE 3: Optical micrograph (200x) of the surface of as-coated ASTM-SA210 Grade A-1 boiler steel. (a) Nanostructured TiAlN coating, (b) nanostructured AlCrN coating, (c) conventional TiAlN coating, and (d) conventional AlCrN coating.

TABLE 5: Microstructural and mechanical properties of nanostructured thin TiAlN and AlCrN coated ASTM-SA210 Grade A-1 boiler steel.

Coating	Surface roughness (nm)	Particle size (nm)		Porosity (% age)	Coating thickness (μm)	Coating color
		Scherrer formula	AFM analysis			
Nanostructured TiAlN coating	03.75	16	18	<0.5	6.1	Violet grey
Nanostructured AlCrN coating	06.10	26	28	<0.5	4.2	Light grey

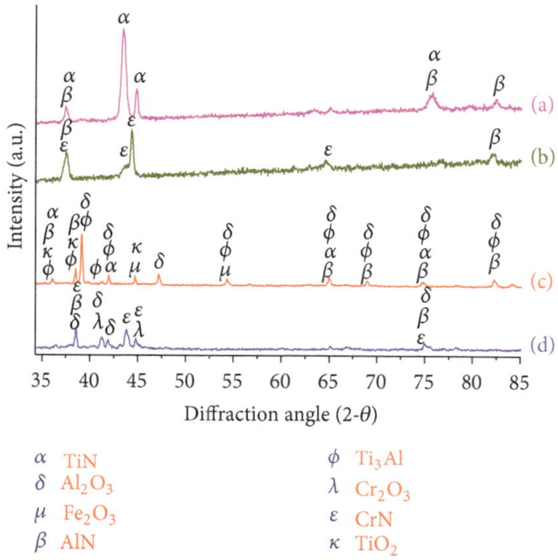

FIGURE 4: X-Ray Diffraction pattern of as-coated ASTM-SA210 Grade A-1 boiler steel. (a) Nanostructured TiAlN coating, (b) nanostructured AlCrN coating, (c) conventional TiAlN coating, and (d) conventional AlCrN coating.

3.2.2. Porosity Analysis. The porosity analysis is of prime importance in high temperature corrosion studies. The dense coatings are supposed to provide very good corrosion resistance as compared to porous coatings. The porosity measurements were made by PMP3 inverted metallurgical microscope with stereographic imaging. The porosity of nanostructured thin TiAlN and AlCrN coatings was found below 0.5% (Table 5). In case of conventional thick coatings; the porosity of as-sprayed Ti-Al (2.30–4.25%) and Al-Cr (2.00–4.30%) coatings was also evaluated, which is reported in Table 6. A considerable decrease in the porosity had been observed after gas nitriding, and it was found to be less than 0.6% (Table 6).

3.2.3. AFM Analysis of the As-Deposited Nanostructured Thin Coatings. The surface topography of the nanostructured thin TiAlN and AlCrN coatings was studied using Atomic Force Microscope (AFM: NT-MDT: NTEGRA Model) in semi-contact mode. Figure 5 shows the AFM surface morphology (2D and 3D) of the nanostructured thin TiAlN and AlCrN coatings deposited on ASTM-SA210 Grade A-1 boiler steel. The difference in the morphology between the two coatings can be inferred by comparing the 2D images in Figures 5(a) and 5(c); however, a clearer comparison of the coatings could be afforded by viewing 3D images in Figures 5(b) and 5(d). As the axis scale indicates, the overall roughness of the nanostructured TiAlN coating (Figure 5(b)) is less than that of AlCrN coating (Figure 5(d)). The surface roughness and particle size in the coatings are analyzed by AFM analysis, reported in Table 5. The surface roughness in case of nanostructured TiAlN coating was observed to be 3.75 nm and 6.10 nm in case of nanostructured AlCrN coating (Table 5). The nanostructured thin TiAlN coating had shown lesser particle size (18 nm) as compared to nanostructured thin AlCrN coating (28 nm).

3.2.4. Surface Roughness (R_a) Values of the Conventional Thick Coatings. The coating surface was very rough in case of conventional thick TiAlN and AlCrN coatings due to the presence of melted/partially melted particles and the roughness were found to be in the range of 10.35–15.13 μm and 11.84–15.23 μm, respectively. The centre line average (CLA) method was used to obtain the R_a values.

3.2.5. Evaluation of Microhardness and Bond Strength of the Conventional Thick Coatings. The hardness of the coatings has been measured along the cross-section of the conventional thick TiAlN and AlCrN coated Grade A-1 boiler steel. Figure 6 shows the microhardness profiles along the cross-section of the coatings as a function of distance from the coating-substrate interface. The critical hardness values of the substrate Grade A-1 boiler steel was found to be in the range 200–250 Hv. From the microhardness profiles (Figure 6) it is obvious that the conventional TiAlN coating has shown maximum microhardness of the order of 900–950 Hv. The conventional thick AlCrN coating has shown microhardness of the order of 600–700 Hv.

The bond strength of the conventional thick TiAlN and AlCrN coatings was measured on three specimens as per ASTM standard C633-01. The coatings failed at the substrate-coating interface while remaining attached to the adhesive (Figure 7). Average bond strength of 68.74 MPa and 54.69 MPa was observed in case of conventional TiAlN and AlCrN coatings, respectively (Table 6).

3.2.6. Surface Morphology of Coatings. FE-SEM micrographs along with EDS spectrum that reveal the surface morphology of the nanostructured and conventional TiAlN and AlCrN

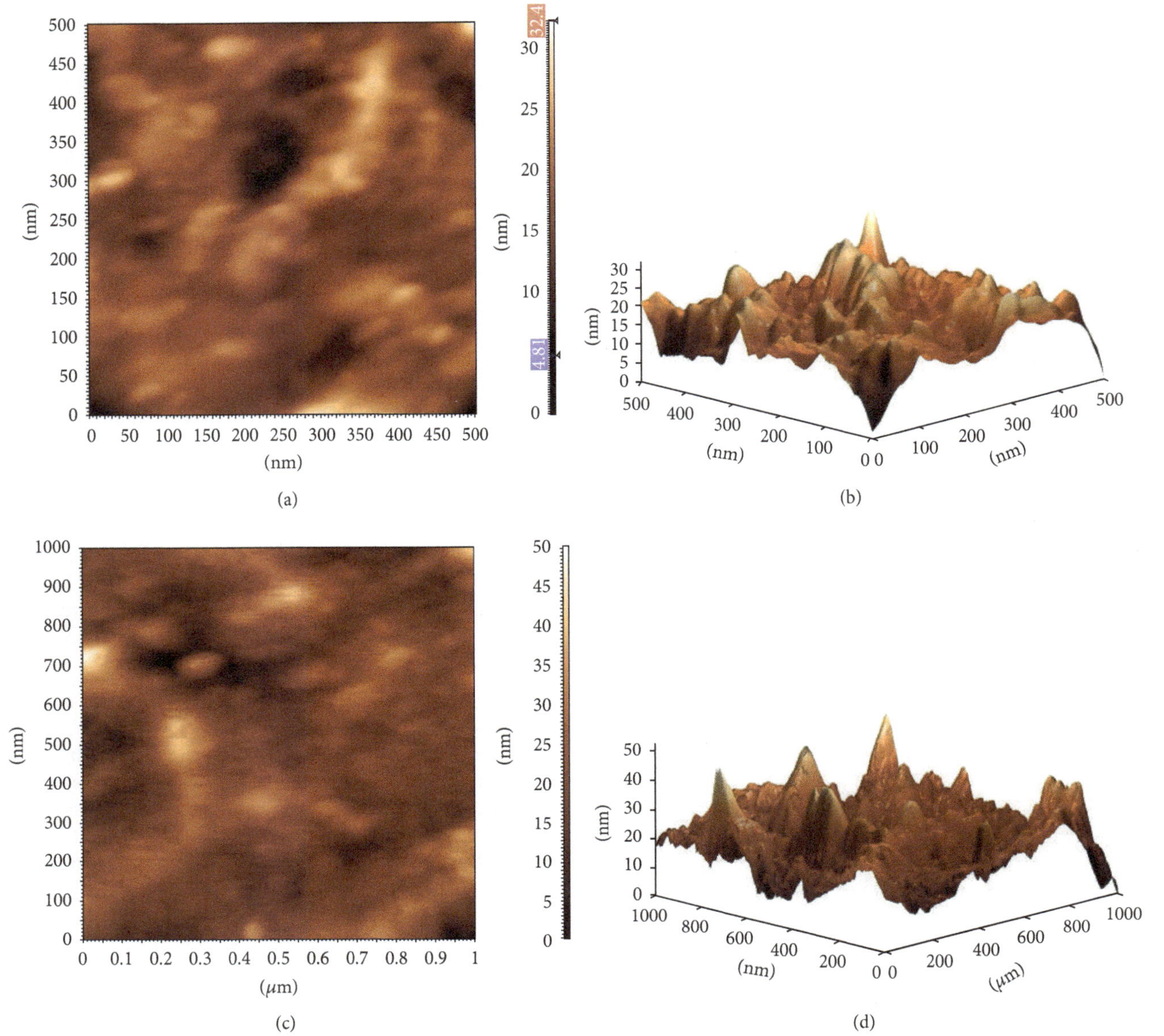

Figure 5: 2D and 3D AFM images for the as-coated ASTM-SA210 Grade A-1 boiler steel. (a) and (b) Nanostructured TiAlN coating, (c) and (d) nanostructured AlCrN coating.

coated ASTM-SA210 Grade A-1 boiler steel are shown in Figure 8. Micrograph (Figure 8(a)) for nanostructured thin TiAlN coating at higher magnification (10000x) indicates grey matrix with some black and white contrast regions. Dense structure with lower porosity is observed in the coating microstructure, and also it is free from cracks. EDAX analysis at point 1 and point 2 in Figure 8 indicates the presence of Ti, Al, and N with negligible amount of Fe. The black colored areas revealed the higher amount of Ti and less concentration of Al as compared to the white contrast region. In case of nanostructured thin AlCrN coated Grade A-1 boiler steel, the FE-SEM micrograph indicates dense grey colored coating with tiny dark grey particles dispersed in the matrix. EDAX point analysis (points 3 and 4 in Figure 8) shows the coating is rich in Al, Cr, and N along with small amount of Fe and oxygen (Figure 8(b)).

The micrographs in case of conventional plasma sprayed gas nitride TiAlN and AlCrN coatings are shown in Figures 8(c) and 8(d). In general microscopic features indicate that the conventional coatings are homogeneous and massive, free from cracks. Presence of some oxide stringers as well as open pores has been noticed in general in the conventional coatings. The EDAX point analysis (point 5 and 6 in Figure 8) in case of conventional TiAlN coating indicates the higher concentration of Ti and Al in the coating along with some amount of N and oxygen. The presence of Al and Cr along with N and oxygen is revealed by EDAX analysis in case of conventional AlCrN coating (points 7 and 8 in Figure 8).

TABLE 6: Microstructural and mechanical properties of conventional thick TiAlN and AlCrN coated ASTM-SA210 Grade A-1 boiler steel.

Coating	Surface roughness (μm)	Coating thickness (μm)	Porosity (% age)		Bond strength (MPa)	Coating color
			As-sprayed	After gas nitriding		
Conventional TiAlN coating	10.35–15.13	172	2.30–4.25	<0.6	68.74	Grey
Conventional AlCrN coating	11.84–15.23	166	2.00–4.30	<0.6	54.69	Bluish grey

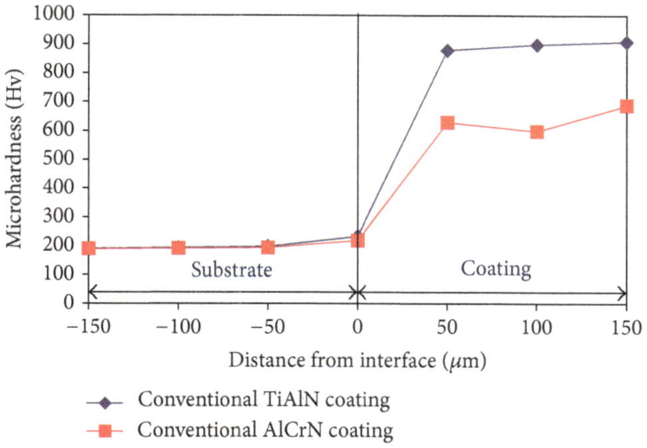

FIGURE 6: Microhardness profile across the cross-section for conventional TiAlN and conventional AlCrN coating on ASTM-SA210 Grade A-1 boiler steel.

3.3. Cross-Sectional Analysis

3.3.1. Coating Thickness. The as-coated samples were cut across the cross-section using Buehler Isomet 1000 precision saw and mounted in transoptic mounting resin and subsequently mirror-polished to obtain scanning electron back-scattered micrographs and X-ray mapping of different elements for coated and uncoated ASTM-SA210 Grade A-1 boiler steel. The coating thickness values were measured from SEM back-scattered micrographs as shown in Figure 9 and average coating thickness is reported in Table 6. The measured average coating thickness values for nanostructured thin and conventional thick TiAlN and AlCrN coatings are 6.1, 4.2, 172, and 166 and 122 μm, respectively.

3.3.2. X-Ray Mapping. BSEI and X-ray mapping analysis for nanostructured and conventional TiAlN and AlCrN coated ASTM-SA210 Grade A-1 boiler steel is shown in Figure 9. The BSEI and X-ray mapping analysis of the nanostructured TiAlN coated Grade A-1 boiler steel are presented in Figure 9(a). The X-ray mapping indicates presence of Al and Ti along with small amount of N in the coating, and no diffusion of Fe from the substrate has been observed. In case of nanostructured thin AlCrN coated Grade A-1 boiler steel, the BSEI and X-ray mapping are shown in Figure 9(b). The X-ray mapping indicates the presence of Al and Cr along with some concentration of N in the coating, and no diffusion of Fe and other elements from the substrate has been observed.

In case of conventional thick TiAlN coated Grade A-1 boiler steel, Figure 9(c), Al- and Ti-rich coating along with negligible amount of N and O has been observed. The diffusion of Fe from the substrate has taken place as indicated by X-ray mapping analysis. Figure 9(d) depicts the BSEI and X-ray mapping in case of conventional AlCrN coated Grade A-1 boiler steel. The coating is rich in Al and Cr with negligible presence of N and O. The X-ray mapping analysis indicates the presence of thin horizontal bands of Al and O in the coating.

4. Discussion

The nanostructured thin TiAlN and AlCrN coatings on ASTM-SA210 Grade A-1 boiler steel were obtained using a front-loading Balzer's rapid coating system (RCS) machine at Oerlikon Balzers Ltd. Gurgaon (India). The coating parameters were optimized. The coating thickness was measured along the cross-section for some randomly selected samples as has been reported in Table 5. The thin coatings were obtained and the coatings thickness was observed as 6.1 and 4.2 μm in case of nanostructured TiAlN and AlCrN coatings, respectively. The surface appearance of nanostructured TiAlN coating was violet grey in color and light grey in case of nanostructured AlCrN coating (Figure 2). Figure 3 depicts the optical micrographs of the nanostructured thin coatings. The coatings were dense and have uniform microstructure. It is evident from the microstructure that the coatings contain some pores and inclusions (Figure 3). The negligible porosity values for as-coated nanostructured thin TiAlN and AlCrN coatings were observed, which were less than 0.5% (Table 5).

In case of conventional thick TiAlN and AlCrN coatings, the plasma spray coatings were obtained at Anod Plasma Ltd. Kanpur (India) using a 40 kW Miller Thermal Plasma Spray apparatus. The self-disintegration of thicker coatings usually restricts the thickness of the coatings [17]. The coating parameters were optimized. In the present study, it was possible to obtain a thickness in range of 150–200 μm for the Ti-Al and Al-Cr coatings. After coating deposition process, the gas nitriding was done in lab in order to obtain hard metal nitride coatings. The coating thickness was measured along the cross-section for some randomly selected samples as has been reported in Table 6. The thick coatings were obtained, and the coatings thickness was observed as 172 and 166 μm in case of conventional thick TiAlN and AlCrN, coatings respectively.

The surface appearance of conventional TiAlN and AlCrN coatings was grey with some bluish shining (Figure 2).

(a) (b)

FIGURE 7: Fracture surfaces of a plasma sprayed gas nitride specimen after ASTM C633 tensile test. (a) Conventional TiAlN coating, (b) conventional AlCrN coating.

The surface of conventional thick coatings was rough. The measured porosity of plasma sprayed and gas nitrided coatings is given in Table 6. The measured values of porosity (2.00–4.30%) for as-sprayed conventional Ti-Al and Al-Cr coatings are almost in close agreement with the findings of Chen et al. [18], Erickson et al. [19], Hidalgo et al. [20, 21], Singh et al. [22], and Sidhu et al. [17, 23] for thermal plasma sprayed coatings, whereas after gas nitriding negligible porosity values were observed (Table 6), which were found to be less than 0.6% (Table 6). This may be due to the reason that gas nitriding process may have eliminated the microstructural inhomogeneities due to filling of pores and voids by nitrogen. Figure 3 depicts the optical micrographs of the conventional thick coatings. In case of conventional thick coatings (Figures 3(c) and 3(d)), the massive microstructure could be observed with irregularly shaped grains.

XRD diffractograms for nanostructured and conventional TiAlN and AlCrN coated ASTM-SA210 Grade A-1 boiler steel are depicted in Figure 4 on reduced scale. The phases identified by XRD analysis for nanostructured thin TiAlN coating are TiN and AlN. The phases analyzed are also in agreement with that reported by Yoo et al. [24], Falub et al. [25], and Man et al. [26]. Further, in case of nanostructured AlCrN coating, the prominent phases are CrN and AlN which are in agreement with the findings of Reiter et al. [27] and Endrino et al. [28]. The grain size (Table 1) of the nanostructured thin coatings was estimated from Scherrer formula [29] as follows:

$$D = \frac{0.9\lambda}{B \operatorname{Cos}\theta}, \quad (2)$$

where λ, B, and θ are the X-ray wavelength (1.54056°A), Bragg diffraction angle, and line width at half maximum, respectively. The calculated grain size for nanostructured thin TiAlN and AlCrN coatings was 16 and 26 nm, respectively. Oerlikon Balzers Ltd. Gurgaon (India) has provided the data regarding hardness of the as-coated nanostructured TiAlN and AlCrN coatings (Table 5). The coated layer on the steel substrate has provided higher hardness as compared to the substrate. In particular, nanostructured TiAlN coating showed higher hardness value than nanostructured AlCrN coating.

Figure 5 shows the AFM surface morphology (2D and 3D) of the nanostructured thin TiAlN and AlCrN coatings deposited on ASTM-SA210 Grade A-1 steel. The surface roughness and particle size in the coatings were also provided by AFM analysis, which are reported in Table 5. As the axis scale indicates, the overall roughness of the nanostructured TiAlN coating (Figure 5(b)) is less than that of AlCrN coating (Figure 5(d)). The surface roughness in case of nanostructured TiAlN coating was observed to be 3.75 nm and 6.10 nm in case of nanostructured AlCrN coating (Table 5). The nanostructured thin TiAlN coating had shown smaller particle size (18 nm) as compared to nanostructured thin AlCrN coating (28 nm). The particle size determined by AFM analysis is in good agreement with the results obtained from Scherrer formula (as reported previously). The coating surface was very rough in case of conventional thick TiAlN and AlCrN coatings due to the presence of melted/partially melted particles, and the roughness was found to be in the range of 10.35–15.13 μm and 11.84–15.23 μm, respectively. The centre line average (CLA) method was used to obtain the R_a values.

Further, the main phases identified for the conventional thick TiAlN coating are Al_2O_3, TiN, Ti_3Al, and AlN and small peaks of TiO_2 and Fe_2O_3 as depicted in Figure 4. The phases identified in case of conventional thick AlCrN coating are CrN, AlN, and Al_2O_3 with minor peaks of Cr_2O_3 (Figure 4). The presence of metal nitride phases indicates that the gas nitriding process has successfully produced the desired coatings. Further, the phases analyzed are also in agreement with that reported by Adachi and Nakata [30]. They have worked on Ti, Al, and Ti-Al based coatings under various conditions, that is, air and/or nitrogen environment by plasma spraying technique. Further, they have reported that the X-ray diffraction pattern for Ti-50 mass% Al mixed powder sprayed coatings detected Al, and $TiN_{0.3}$, Ti(N,O) and peaks of synthesized phase of Ti_3Al. The sprayed Ti coating had a porous structure and included cracks and pores in high density. The sprayed Al coating had a dense microstructure. And the sprayed Ti-50 mass% Al coatings had a dense laminated structure composed of Ti compounds phase and Al phase. According to them, the formula of Ti_3Al formation is given by

$$3Ti + Al = Ti_3Al, \qquad \Delta G^f = -29633.6 + 6.70801\,T/K, \quad (3)$$

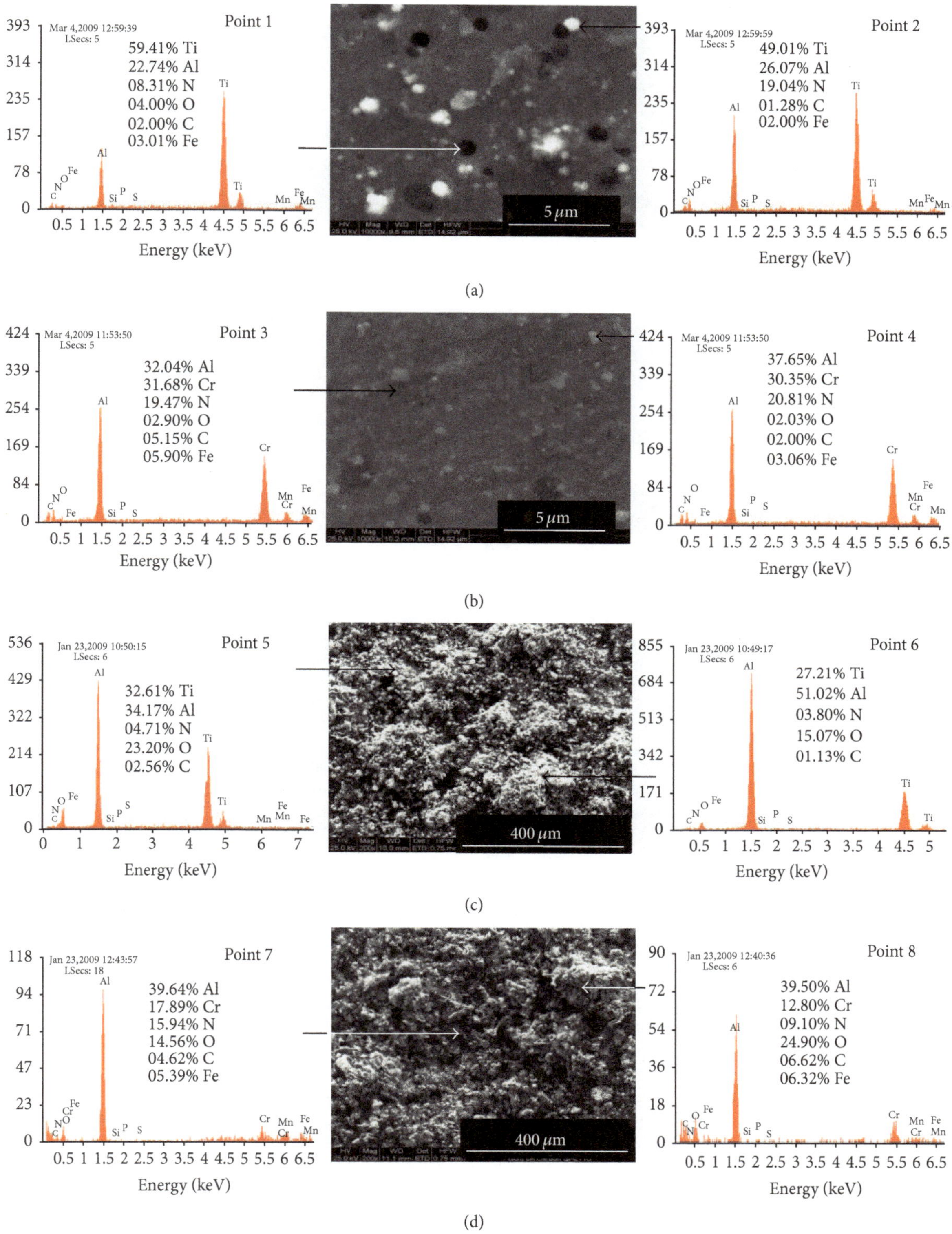

FIGURE 8: Surface-scale morphology and EDAX patterns from different spots on as-coated ASTM-SA210 Grade A-1 boiler steel. (a) Nanostructured TiAlN coating, (b) nanostructured AlCrN coating, (c) conventional TiAlN coating, and (d) conventional AlCrN coating.

where ΔG^f is the Gibbs energy of Ti_3Al formation ($Jmol^{-1}$). That is, from the view of the thermodynamics, Ti_3Al can be easily formed from Ti with Al. The pressures of O_2 and N_2 in the atmosphere are much higher than the equilibrium pressure O_2 and N_2 to form Ti_3Al from TiO with Al and TiN with Al [30]. So, Ti_3Al cannot be formed thermodynamically from TiO with Al or TiN with Al under the condition of spraying in the air atmosphere. This phenomenon indicates

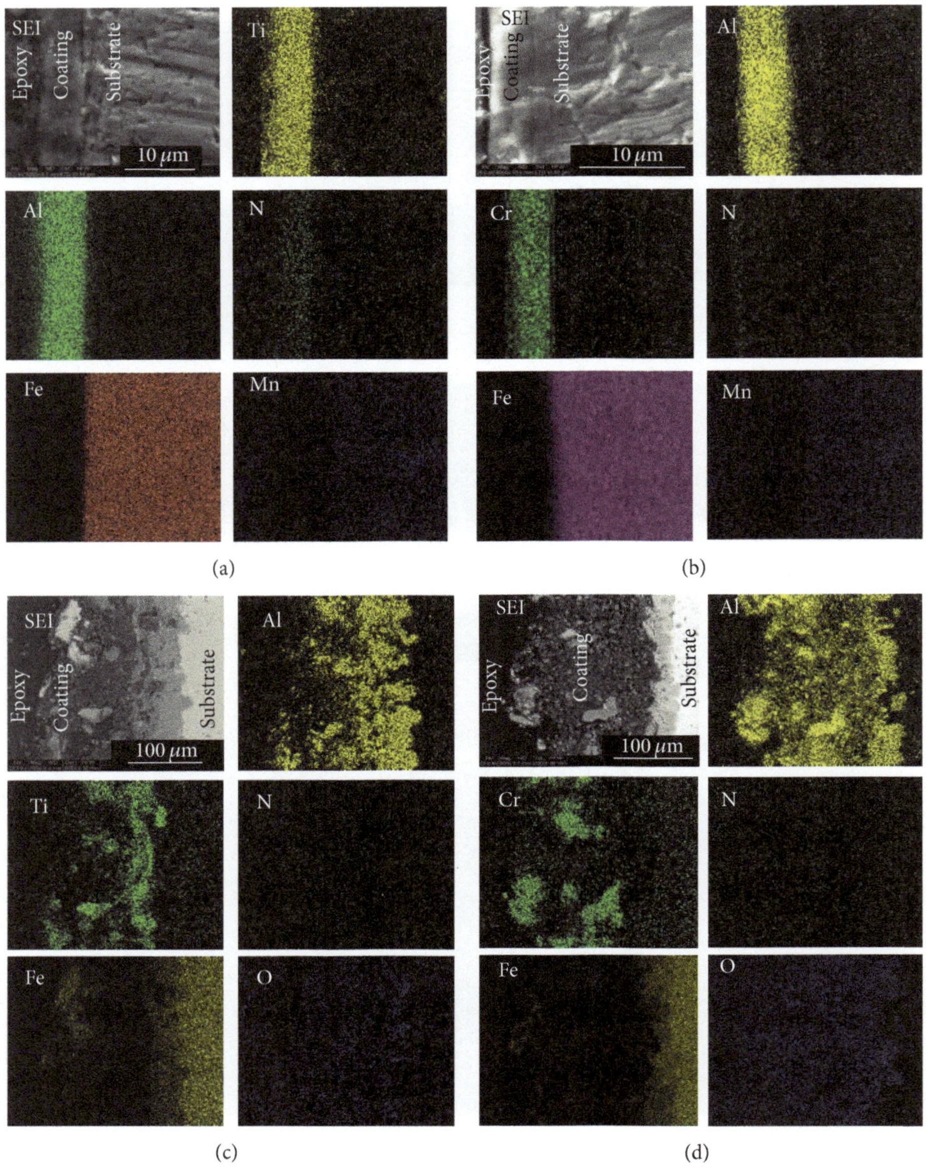

FIGURE 9: (a) Composition image (SEI) and X-ray mapping of the cross-section of as-coated nanostructured TiAlN coating on ASTM-SA210 Grade A-1 boiler steel. (b) Composition image (SEI) and X-ray mapping of the cross-section of as-coated nanostructured AlCrN coating on ASTM-SA210 Grade A-1 boiler. (c) Composition image (SEI) and X-ray mapping of the cross-section of as-coated conventional TiAlN coating on ASTM-SA210 Grade A-1 boiler steel. (d) Composition image (SEI) and X-ray mapping of the cross-section of as-coated conventional AlCrN coating on ASTM-SA210 Grade A-1 boiler steel.

that Ti_3Al would be synthesized from Ti and Al particles dynamically during flight [30]. The reaction would proceed at the interface of the Ti particle contacted with the Al particle. The formulas of compounds formation can be described as follows:

$$\begin{aligned} \mathrm{Ti} + \frac{1}{2}\mathrm{N}_2 + \frac{1}{2}\mathrm{O}_2 &= \mathrm{Ti}\,(\mathrm{N},\mathrm{O}), \\ \mathrm{Ti} + \mathrm{N}_2 &= \mathrm{TiN}_{0:3}, \\ 3\mathrm{Ti} + \mathrm{Al} &= \mathrm{Ti}_3\mathrm{Al}. \end{aligned} \tag{4}$$

Conclusively, Adachi and Nakata [30] have reported that the formation of the Ti-Al sprayed coating on the substrate would proceed as follows. At first, the compounds of Ti oxide and nitride in the fused particles would be solidified almost as soon as they adhere to the substrate, as the melting points of TiO and TiN are 2023 K and 3223 K, respectively, which are much higher than that of Al, 933 K. Then, cracks and pores between the Ti compound layers would be filled with the melting Al particles, and the coating finally became solidified as a whole.

Hardness is the most frequently quoted mechanical property of the coatings [31]. The observed microhardness values (Figure 6) for the conventional coatings are within the range of microhardness values reported for plasma coatings by Adachi and Nakata [30], Vuoristo et al. [32],

Chen and Hutchings [33], and Westergård et al. [34]. Microhardness plots (Figure 6) indicate some increase in the microhardness of substrate steel at the substrate coating interface. The hardening of the substrates as observed in the current study might have occurred due to the high-speed impact of the coating particles during plasma spray deposition. The effect has also been reported by Hidalgo et al. [20, 21, 35, 36] and Sidhu et al. [17, 23]. The observed nonuniformity in the hardness values along the thickness of the coatings may be due to the microstructural changes along the cross-section of the coatings [37]. Moreover, the microhardness and other properties of the thermal spray coatings are anisotropic because of their typical splat structure and directional solidification [31].

The bond strength of the conventional thick TiAlN and AlCrN coatings was measured on three specimens as per ASTM standard C633-01. The coatings failed at the substrate-coating interface while they remain attached to the adhesive (Figure 7). Average bond strength of 68.74 MPa and 54.69 MPa was observed in case of conventional TiAlN and AlCrN coatings, respectively (Table 6). The bond strength in case of conventional TiAlN coatings (68.74 MPa) is in good agreement (slightly more) with the results reported by Adachi and Nakata [30] regarding their work based on plasma sprayed Ti-Al coatings (60 MPa).

FE-SEM micrographs along with EDS spectrum that reveal the surface morphology of the nanostructured and conventional TiAlN and AlCrN coated ASTM-SA210 Grade A-1 boiler steel are shown in Figure 8. Surface EDAX analysis is supported by XRD (Figure 4) and X-ray mapping analysis (Figure 9) in all coatings. The presence of Ti, Al, and N with negligible amount of Fe (in nanostructured TiAlN coating), Al, Cr, and N along with small amount of Fe and oxygen (in nanostructured TiAlN coating) the higher concentration of Ti and Al in the coating along with some amount of N and oxygen (in conventional TiAlN coating) and Al and Cr along with N and oxygen (in conventional AlCrN coating); is revealed by surface EDAX analysis (Figure 8). Also, the XRD analysis (Figure 4) indicates the presence of the oxide phases in the coatings, that is, Al_2O_3, TiO_2, and Fe_2O_3 in case of conventional TiAlN coating and the presence of Al_2O_3 and Cr_2O_3 in case of conventional AlCrN coatings. According to Bluni and Mardar [38] the oxides may have formed due to the in-flight oxidation during spraying process and/or preexisting in the feed material. The latter reason for the oxides formation in the structure of coatings under study looks to be more relevant as the chances of inflight oxidation are meager in case of the shrouded plasma spraying.

Deshpande et al. [39] proposed that, during in-flight oxidation, a layer of oxide is formed on the molten particle due to chemical reactions between the surface of the liquid phase and oxygen or due to diffusion of oxygen into the liquid. The turbulent mixing of the liquid part of the powder particle during its flight destroys the surface layer of oxides and causes the oxides to be distributed more uniformly through the bulk volume of the particle. However, when temperature of the particle starts dropping during later part of the flight, these oxides tend to solidify and a thin oxide shell would be formed around the droplet.

The oxidation time during thermal spray coating is short typically less than 0.01 s and can occur in either the solid or molten state. The oxidation of coatings is not always harmful; it is equally important to control and understand the different aspects of oxidation of coatings; therefore, it is important to find an optimum level for oxidation of coatings [40–42]. The diffusion of Fe from the substrate to the coating (in conventional coatings) was identified by XRD and X-ray-mapping analysis. This type of diffusion of the base metal component into the coating has also been observed by Wang et al. [43]. They have reported the penetration of Co from the base alloy to the coating after heat treatment of coatings for 2 hrs at 1050°C.

5. Conclusions

The nanostructured thin TiAlN and AlCrN coatings (by physical vapour deposition process at Oerlikon Balzers Ltd. Gurgaon, India) and conventional thick TiAlN and AlCrN coatings (by plasma spraying process at Anod Plasma Ltd. Kanpur, India; followed by gas nitriding process) were successfully deposited on ASTM-SA210 Grade A-1 boiler steel. The coatings were characterized for their microstructural features and hardness in present work. The following observations were made based on the present study.

(1) The nanostructured thin TiAlN and AlCrN coatings exhibited negligible porosity values for as-coated, which were less than 0.5%. The conventional TiAlN and AlCrN coatings showed higher porosity values (2.00–4.30%) for as-sprayed conventional Ti-Al and Al-Cr coatings which after gas nitriding were found to be less than 0.6%.

(2) The phases identified by XRD analysis for nanostructured thin TiAlN coating are TiN and AlN. Further, in case of nanostructured AlCrN coating, the prominent phases are CrN and AlN. The main phases identified (by XRD analysis) for the conventional thick TiAlN coating were Al_2O_3, TiN, Ti_3Al, ands AlN and small peaks of TiO_2 and Fe_2O_3. The phases identified in case of conventional thick AlCrN coating were CrN, AlN, and Al_2O_3 with minor peaks of Cr_2O_3. The presence of metal nitride phases indicates that the gas nitriding process has successfully produced the desired coatings.

(3) The grain size (calculated by Scherrer formula from XRD plot) for nanostructured thin TiAlN and AlCrN coatings was 16 and 26 nm, respectively. The particle size determined by AFM analysis is in good agreement with the results obtained from Scherrer formula. From AFM analysis, the particle size observed for nanostructured thin TiAlN and AlCrN coatings was 18 nm and 28 nm respectively.

(4) The surface roughness in case of nanostructured TiAlN coating was observed as 3.75 nm and 6.10 nm in case of nanostructured AlCrN coating, as observed by AFM analysis. The coating surface was very rough

in case of conventional thick TiAlN and AlCrN coatings due to the presence of melted/partially melted particles, and the roughness was found to be in the range of 10.35–15.13 μm and 11.84–15.23 μm, respectively. The centre line average (CLA) method was used to obtain the R_a values.

(5) A good adhesion of the conventional thick TiAlN and AlCrN coatings was evident from bond test results. Average bond strength of 68.74 MPa and 54.69 MPa Psi was observed in case of conventional TiAlN and AlCrN coatings, respectively.

References

[1] V. H. Hidalgo, J. B. Varela, A. C. Menéndez, and S. P. Martínez, "High temperature erosion wear of flame and plasma-sprayed nickel-chromium coatings under simulated coal-fired boiler atmospheres," *Wear*, vol. 247, no. 2, pp. 214–222, 2001.

[2] V. H. Hidalgo, F. J. B. Varela, A. C. Menéndez, and S. P. Martínez, "A comparative study of high-temperature erosion wear of plasma-sprayed NiCrBSiFe and WC-NiCrBSiFe coatings under simulated coal-fired boiler conditions," *Tribology International*, vol. 34, no. 3, pp. 161–169, 2001.

[3] B. S. Sidhu, D. Puri, and S. Prakash, "Mechanical and metallurgical properties of plasma sprayed and laser remelted Ni-20Cr and Stellite-6 coatings," *Journal of Materials Processing Technology*, vol. 159, no. 3, pp. 347–355, 2005.

[4] J. R. Nicholls, N. J. Simms, W. Y. Chan, and H. E. Evans, "Smart overlay coatings: concept and practice," *Surface and Coatings Technology*, vol. 149, no. 2-3, pp. 236–244, 2002.

[5] V. Chawla, S. Prakash, D. Puri, and B. Singh, "State of Art: corrosion in thermal power plants and preventive measures: a review," in *Proceedings of the National Conference on Advancement & Futuristic Trends in Mechanical & Materials Engineering*, G.Z.S. College of Engineering anf Technology, Bathinda, India, October 2006.

[6] V. Chawla, S. Prakash, D. Puri, and B. Singh, "Plasma sprayed coatings for protection against hot Corrosion: a review," in *Proceedings of the of International Conference on Advances in Mechanical Engineering (AME '06)*, Baba Banda Singh Bahadaur College of Engineering, Fatehgarh Sahib, India, December 2006.

[7] J. T. DeMasi-Marcin and D. K. Gupta, "Protective coatings in the gas turbine engine," *Surface and Coatings Technology*, vol. 68-69, no. C, pp. 1–9, 1994.

[8] H. O. Pierson, "CVD/PVD coatngs," in *ASM Handbook: Corrosion*, J. R. Davis, Ed., vol. 13, pp. 456–458, ASM International, Materials Park, Ohio, USA, 1987.

[9] P. Fauchais, "Topical Review: understanding plasma spraying," *Journal of Physics D*, vol. 37, no. 9, pp. R86–R108, 2004.

[10] V. Chawla, S. Prakash, D. Puri, and B. Singh, "State of art: hot corrosion: a review," in *Proceedings of the Global Conference on Production and Industrial Engineering (CPIE '07)*, B.R. Ambedkar National Institute of Technology, Jalandhar, India, March 2007.

[11] J. L. He, K. C. Chen, C. C. Chen, A. Leyland, and A. Matthews, "Cyclic oxidation resistance of Ni-Al alloy coatings deposited on steel by a cathodic arc plasma process," *Surface and Coatings Technology*, vol. 135, no. 2-3, pp. 158–165, 2001.

[12] V. Chawla, S. Prakash, D. Puri, and B. Singh, "Corrosion behaviour of nanostructured TiAlN and AlCrN hard coatings on Superfer 800H superalloy in simulated marine environment," *Journal of Minerals and Materials Characterization and Engineering*, vol. 8, no. 9, pp. 693–700, 2009.

[13] V. Chawla, D. Puri, S. Prakash, A. Chawla, and B. Singh, "Characterization and comparison of corrosion behaviour of nanostructured TiAlN and AlCrN coatings on Superfer 800H (INCOLOY 800H) substrate," *Journal of Minerals and Materials Characterization and Engineering*, vol. 8, no. 9, pp. 715–727, 2009.

[14] V. Chawla, S. Prakash, and B. S. Sidhu, "State of the art: applications of mechanically alloyed nanomaterials: a review," *Materials and Manufacturing Processes*, vol. 22, no. 4, pp. 469–473, 2007.

[15] V. Chawla, B. S. Sidhu, D. Puri, and S. Prakash, "Performance of plasma sprayed nanostructured and conventional coatings," *Journal of the Australian Ceramic Society*, vol. 44, no. 2, pp. 56–62, 2008.

[16] V. Chawla, S. Prakash, D. Puri, and B. Singh, "State of art: plasma sprayed nanostructured coatings: a review," *Materials Forum*, vol. 32, pp. 137–143, 2008.

[17] B. S. Sidhu, D. Puri, and S. Prakash, "Characterisations of plasma sprayed and laser remelted NiCrAlY bond coats and Ni_3Al coatings on boiler tube steels," *Materials Science and Engineering A*, vol. 368, no. 1-2, pp. 149–158, 2004.

[18] H. C. Chen, Z. Y. Liu, and Y. C. Chuang, "Degradation of plasma-sprayed alumina and zirconia coatings on stainless steel during thermal cycling and hot corrosion," *Thin Solid Films*, vol. 223, no. 1, pp. 56–64, 1993.

[19] L. C. Erickson, R. Westergård, U. Wiklund, N. Axén, H. M. Hawthorne, and S. Hogmark, "Cohesion in plasma-sprayed coatings: a comparison between evaluation methods," *Wear*, vol. 214, no. 1, pp. 30–37, 1998.

[20] V. Higuera Hidalgo, F. J. Belzunce Varela, and A. Carrilez Menendez, "Characterization and high temperature behaviour of thermal sprayed coatings used in boilers," in *Proceedings of the 15th International Thermal Spray Conference*, pp. 617–621, Nice, france, May 1998.

[21] V. H. Hidalgo, F. J. B. Varela, S. P. Martinez, and S. G. Espana, "Characterization and high temperature behaviour of Cr_3C_2-NiCr plasma sprayed coatings," in *Proceedings of the United Thermal Spray Conference*, pp. 683–686, Germany, 1999.

[22] H. Singh, D. Puri, and S. Prakash, "Some studies on hot corrosion performance of plasma sprayed coatings on a Fe-based superalloy," *Surface and Coatings Technology*, vol. 192, no. 1, pp. 27–38, 2005.

[23] T. S. Sidhu, R. D. Agrawal, and S. Prakash, "Hot corrosion of some superalloys and role of high-velocity oxy-fuel spray coatings: a review," *Surface and Coatings Technology*, vol. 198, no. 1–3, pp. 441–446, 2005.

[24] Y. H. Yoo, D. P. Le, J. G. Kim, S. K. Kim, and P. V. Vinh, "Corrosion behavior of TiN, TiAlN, TiAlSiN thin films deposited on tool steel in the 3.5 wt.% NaCl solution," *Thin Solid Films*, vol. 516, no. 11, pp. 3544–3548, 2008.

[25] C. V. Falub, A. Karimi, M. Ante, and W. Kalss, "Interdependence between stress and texture in arc evaporated Ti-Al-N thin films," *Surface and Coatings Technology*, vol. 201, no. 12, pp. 5891–5898, 2007.

[26] B. Y. Man, L. Guzman, A. Miotello, and M. Adami, "Microstructure, oxidation and H2-permeation resistance of TiAlN films deposited by DC magnetron sputtering technique," *Surface and Coatings Technology*, vol. 180-181, pp. 9–14, 2004.

[27] A. E. Reiter, V. H. Derflinger, B. Hanselmann, T. Bachmann, and B. Sartory, "Investigation of the properties of $Al_{1-x}Cr_xN$ coatings prepared by cathodic arc evaporation," *Surface and Coatings Technology*, vol. 200, no. 7, pp. 2114–2122, 2005.

[28] J. L. Endrino, G. S. Fox-Rabinovich, and C. Gey, "Hard AlTiN, AlCrN PVD coatings for machining of austenitic stainless steel," *Surface and Coatings Technology*, vol. 200, no. 24, pp. 6840–6845, 2006.

[29] B. D. Cullity, *Elements of X-Ray Diffraction*, Addison-Wesley, Reading, Mass, USA, 1970.

[30] S. Adachi and K. Nakata, "Improvement of adhesive strength of Ti-Al plasma-sprayed coatings," *Surface and Coatings Technology*, vol. 201, pp. 5617–5620, 2007.

[31] R. C. Tucker Jr., "Advanced thermal spray deposition techniques," in *Handbook of Deposition Technologies For Films and Coatings*, R. F. Bunshah, Ed., chapter 11, Noyes Publications, Park Ridge, NJ, USA; William Andrew Publishing, New York, NY, USA, 1994.

[32] P. Vuoristo, K. Niemi, A. Makela, and T. Mantyla, "Abrasion and erosion wear resistance of Cr_3C_2-NiCr coatings prepared by plasma, detonation and high-velocity oxyfuel spraying," in *Proceedings of the 7th National Thermal Spray Conference*, pp. 121–126, Boston, Mass, USA, 1994.

[33] H. Chen and I. M. Hutchings, "Abrasive wear resistance of plasma-sprayed tungsten carbide-cobalt coatings," *Surface and Coatings Technology*, vol. 107, no. 2-3, pp. 106–114, 1998.

[34] R. Westergård, L. C. Erickson, N. Axén, H. M. Hawthorne, and S. Hogmark, "The erosion and abrasion characteristics of alumina coatings plasma sprayed under different spraying conditions," *Tribology International*, vol. 31, no. 5, pp. 271–279, 1998.

[35] V. Higuera Hidalgo, F. J. Belzunce Varela, and E. Fernández Rico, "Erosion wear and mechanical properties of plasma-sprayed nickel- and iron-based coatings subjected to service conditions in boilers," *Tribology International*, vol. 30, no. 9, pp. 641–649, 1997.

[36] V. Higuera Hidalgo, J. Belzunce Varela, J. Martínez De La Calle, and A. Carriles Menéndez, "Characterisation of NiCr flame and plasma sprayed coatings for use in high temperature regions of boilers," *Surface Engineering*, vol. 16, no. 2, pp. 137–142, 2000.

[37] M. H. Staia, T. Valente, C. Bartuli, D. B. Lewis, and C. P. Constable, "Part I: characterization of Cr_3C_2-25% NiCr reactive plasma sprayed coatings produced at different pressures," *Surface and Coatings Technology*, vol. 146-147, pp. 553–562, 2001.

[38] S. T. Bluni and A. R. Marder, "Effects of thermal spray coating composition and microstructure on coating response and substrate protection at high temperatures," *Corrosion*, vol. 52, no. 3, pp. 213–218, 1996.

[39] S. Deshpande, S. Sampath, and H. Zhang, "Mechanisms of oxidation and its role in microstructural evolution of metallic thermal spray coatings: case study for Ni-Al," *Surface and Coatings Technology*, vol. 200, no. 18-19, pp. 5395–5406, 2006.

[40] H. Herman, *Plasma Sprayed Coatings*, Scientific American, 1988.

[41] K. Korpiola and P. Vuoristo, "Effect of HVOF gas velocity and fuel to oxygen ratio on the wear properties of tungsten carbide coating," in *Thermal Spray: Practical Solutions For Engineering Problems*, C. C. Bernt, Ed., Cincinnati, Ohio, USA, 1996.

[42] J. E. Nerz, B. A. Kushner, and A. J. Rotolico, "Microstructural evaluation of tungsten carbide-cobalt coatings," *Journal of Thermal Spray Technology*, vol. 1, no. 2, pp. 147–152, 1992.

[43] B. Wang, R. F. Huang, G. H. Song et al., "Interdiffusion behavior of Ni-Cr-Al-Y coatings deposited by arc-ion plating," *Oxidation of Metals*, vol. 56, no. 1-2, pp. 1–13, 2001.

Corrosion Behaviour of Ti-6Al-4V Alloy with Nitride Coatings in Simulated Body Fluids at 36°C and 40°C

I. M. Pohrelyuk, O. V. Tkachuk, and R. V. Proskurnyak

Karpenko Physico-Mechanical Institute, National Academy of Sciences of Ukraine, 5 Naukova Street, Lviv 79060, Ukraine

Correspondence should be addressed to I. M. Pohrelyuk; pohreliuk@mail.ru

Academic Editors: G. Bereket, C.-H. Hsu, and R. Wang

Nitride coatings were formed on Ti-6Al-4V alloy by thermodiffusion treatment. The corrosion-electrochemical behaviour of Ti-6Al-4V alloy with nitride coatings I and II was investigated in physiological solutions (0.9% NaCl and Tyrode's) at temperatures of 36°C and 40°C. It is determined that nitride coating I provides Ti-6Al-4V alloy the higher corrosion resistance in Tyrode's solution at both temperatures of solution while nitride coating II in isotonic 0.9% NaCl.

1. Introduction

Titanium and its alloys are widely used in dentistry and orthopedics [1–4]. They provide high biomechanical properties and chemical stability in biological systems than other materials such as stainless steel and cobalt-chromium alloys. Ti-6Al-4V alloy has better physical and mechanical properties than titanium, but has a low corrosion resistance [2, 5, 6].

Surface treatment is widely used to enhance abrasion, corrosion resistance, and surface hardness [2, 3, 7, 8]. Nitriding is often used to harden the titanium alloy surface [3, 9–11]. Good biocompatibility of titanium nitride is shown in the blood and bone tissue [11]. In addition, the titanium nitride is characterized by good mechanical, tribological, and anticorrosion properties. That is why the titanium nitride is the most suitable to protect the surface of medical implants.

One of the main criteria to use material as implant is its corrosion resistance in physiological solution that simulates the environment of the human body [12]. In particular, isotonic 0.9% NaCl solution simulates blood plasma and Tyrode's solution biological tissue.

The temperature increase is a protective reaction to the penetration of infection or inflammation. When the physiological solution temperature changes, the transition from normal to inflammatory state of the human body can occur.

Purpose is to examine the corrosion-electrochemical behaviour of Ti-6Al-4V alloy with nitride coatings in physiological solutions at 36°C and 40°C.

2. Experimental

Samples of $(\alpha + \beta)$ Ti-6Al-4V alloy (wt%: C-0.1; H-0.015; N-0.04; O-0.15; Fe-0.25; Al-6.0; V-4.0; Ti-bal.) with dimensions $10 \times 15 \times 1$ mm were investigated. Before chemical-thermal treatment, the samples were polished by diamond paste to a surface roughness of $R_a = 0.4\,\mu$m. Then, they were washed in alcohol and dried.

Thermodiffusion saturation by nitrogen was carried out at $T = 850$°C for $\tau = 12$ h. The samples were heated to the nitriding temperature in a vacuum of 10^{-3} Pa. Heating rate was 0.040°C/s. After the isothermal exposure, the samples were cooled in nitrogen with the furnace at an average cooling rate of 0.028°C/s. After cooling to 500°C, the system was vacuumized. Nitride coatings I and II on Ti-6Al-4V alloy were formed at two different nitrogen partial pressures: (I) $p_{N_2} = 1$ Pa, gas flow rate $I = 7 \times 10^{-3}$ Pa/s (dynamic conditions) and (II) $p_{N_2} = 10^5$ Pa (static conditions).

Nitrogen gas of the technical purity was used. It contained oxygen not more than 0.4 vol% and water vapour up to 0.07 g/m^3. The oxygen and moisture in the nitrogen atmosphere were removed by filtering the gas through a silica gel filter and heating the titanium chips at 50°C above the nitriding temperature.

The content of each nitride in nitride zone depends on the nitrogen partial pressure during nitriding [13]. It was formed two types of nitride coating: I and II (Figure 1), by changing the gas dynamic conditions in the reaction chamber. These coatings are different because of their ratio values of nitride layer components: nitrides δ-TiN and ε-Ti_2N. Coating I is characterized by thin nitride film which contains ε-Ti_2N mainly and coating II by thick nitride film which contains δ-TiN mainly. Nitride zone is delimited from a matrix alloy by diffusion layer Ti(N). The characteristic surface topography is formed in coating II through the formation and growth of δ-TiN layer. It worsens the quality of nitrided surface.

Phase composition of the surface layers after nitriding was determined by X-ray phase analysis (CuK_α radiation). The tube focusing system was made using Bragg-Brentano method. Voltage on the anode of X-ray tube was of 40 kV and current density was of 30 mA. The scan step was 0.02° and rate was 10°/min. The diffraction patterns were refined using Sietronix and Powder Cell 2.4 programs, by means of which one conducted Fourier analysis of diffractograms and determined the places of diffraction maximums of reflection identified from data of card index of JCPDS-ASTM phases.

Surface microstructure of the surface layers was carried out by scanning electron microscope EVO-40XVP.

Chemical composition of the surface layers was determined by energy dispersive X-ray spectrometer INCA ENERGY 350 and X-ray fluorescence spectrometer EXPERT 02L.

The profilograms of samples surface were received after nitriding. Quantitative characteristics of the surface microgeometry (height parameters of R_a, R_z, and R_{max} and footpace parameters of S, and S_m) were determined by special programs.

The electrochemical tests of Ti-6Al-4V alloy with nitride coatings were carried out in isotonic 0.9% NaCl solution and Tyrode's solution (g/L: NaCl-8.00; $CaCl_2$-0.20; KCl-0.20; $NaHCO_3$-0.10; $MgCl_2$-$6H_2O$-0.10; NaH_2PO_4-0.10) using IPC-pro potentiostat. Experiments were performed at solution temperatures of 36°C and 40°C. By changing the temperature of the solution, we simulated the transition from normal to inflammatory state of the human body. Three-electrode glass cell with a platinum counter electrode and a saturated Ag/AgCl reference electrode was used. Surface of the working electrode of Ti-6Al-4V alloy was coated with epoxy resin, leaving area for exposure to the electrolyte of 1 cm^2. The polarization curves for Ti-6Al-4V alloy with nitride coatings were recorded in the potential range −1.0–2.5 V versus Ag/AgCl at a scan rate of 2 mV/s.

3. Results and Discussion

Nitride zone is formed during nitriding of Ti-6Al-4V alloy. According to the results of X-ray phase analysis, it contains titanium nitrides TiN and Ti_2N (Figure 2). Nitride coating I formed during nitriding at $p_{N_2} = 1$ Pa has 4% of TiN. However, nitride coating II formed during nitriding at $p_{N_2} = 10^5$ Pa has 67% of TiN. So, nitride coating I mainly contains ε-Ti_2N and nitride coating II δ-TiN. Nitride zone of the coating II is twice thicker than coating I (7 and 3 μm, resp.).

The quality of nitrided surface depends on the formed coating. Quantitative characteristics of the surface microgeometry indicate that the surface quality of coating II is worse than coating I (Figure 3).

We will consider the corrosion-electrochemical behaviour of Ti-6Al-4V alloy with nitride coatings I and II in physiological solutions.

3.1. Nitride Coating I

At 36°C. Two peaks of increase of anode current density at 0.07 and 0.3 V are observed on the anodic branch of the polarization curve of nitride coating I in 0.9% NaCl after the region of active dissolution (Figure 4(a), curve 1). The reason for this behaviour is the dissolution of nitride film with formation titanium oxynitride and titanium oxide of nonstoichiometric composition on the alloy surface [14–16] as follows:

$$2TiN + 2yH_2O \longrightarrow 2TiN_xO_y + (1-x)N_2 + 4yH^+ + 4ye^- \quad (1)$$

$$2TiN_xO_y + 2H_2O \longrightarrow 2Ti^{3+} + (y+1)O_2 + xN_2 + 4H^+ + 10e^- \quad (2)$$

$$Ti^{3+} + xH_2O \longrightarrow TiO_x + 2xH^+ + (2x-3)e^- \quad (3)$$

They slow the dissolution.

At the same time, the electrochemical behaviour of the nitride coating in Tyrode's solution differs. The passive region is observed in the range of potentials −0.25–0.30 V (Figure 4(a), curve 2). Perhaps ions of calcium, magnesium, and phosphorus take part in the formation of a surface film. The film inhibits the dissolution of nitride coating [17, 18]. It confirms by the results of X-ray fluorescence analysis. They fix a small mass share of these elements on the nitrided samples surface after polarization in this physiological solution (wt%: K-0.12, Ca-0.10). Passive current density is 0.045 A/m^2. Despite the fact that the corrosion potential shifts to negative potential range, corrosion current density of nitride coating I decreases in Tyrode's solution (see Table 1). Moreover, the current density of anodic dissolution decreases too.

These results indicate higher corrosion resistance of nitride coating I in a solution that simulates biological tissue (Tyrode's), compared with a solution that simulates blood plasma (0.9% NaCl).

At 40°C. Processes on the anodic branch of the polarization curve of the nitride coating I in 0.9% NaCl are similar at lower temperature (Figure 4(b), curve 1). However, this coating provides higher corrosion resistance for Ti-6Al-4V alloy. It confirms by lower corrosion current density (see Table 1). It is caused by the formation of titanium oxide films. They slow the anodic dissolution [19, 20].

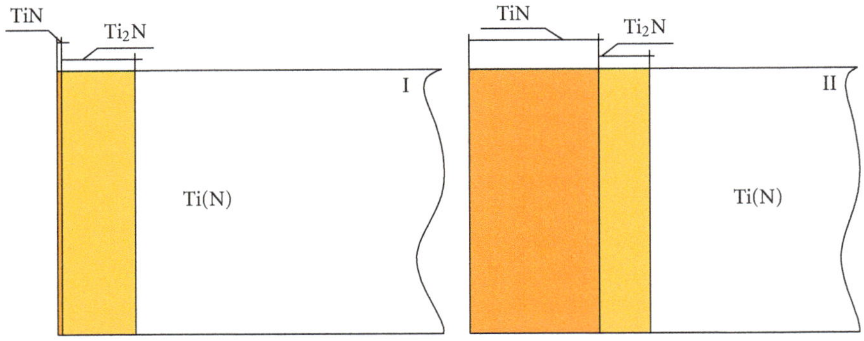

FIGURE 1: Schematic presentation of the surface layers of Ti-6Al-4V alloy after nitriding at regimes I and II. TiN, Ti_2N: titanium nitrides; Ti(N): solid solution of nitrogen in α-titanium.

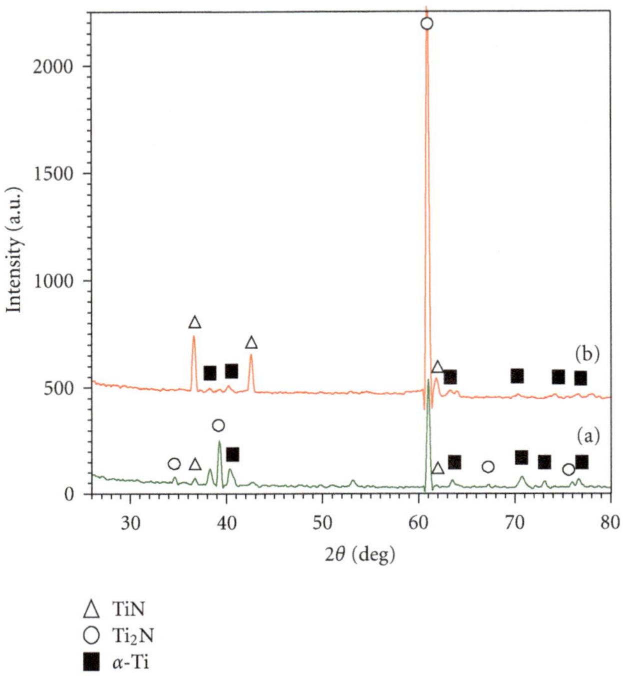

FIGURE 2: X-ray diffraction patterns of Ti-6Al-4V alloy after different regimes of nitriding. a: I; b: II.

TABLE 1: Corrosion parameters of Ti-6Al-4V alloy with nitride coatings I and II in physiological solutions.

T, °C	Corrosion parameters	0.9% NaCl		Tyrode's solution	
		I	II	I	II
36	E_κ, V	−0.200	−0.285	−0.450	−0.600
	i_κ, A/m^2	0.0060	0.0015	0.0040	0.0035
40	E_κ, V	−0.300	−0.170	−0.400	−0.300
	i_κ, A/m^2	0.0037	0.0014	0.0030	0.0015

The passive region is observed in the potential range of −0.25–0.25 V in Tyrode's solution (Figure 4(b), curve 2). The passive current density is 0.02 A/m^2. It should be noted that the mechanism of the passing of anodic processes is similar to those observed for 36°C. Microanalysis does not fix chlorine on the surface but only traces of calcium, magnesium, and phosphorus.

Corrosion current density in Tyrode's solution compared with 0.9% NaCl solution at 40°C shows higher corrosion resistance of nitride coating I in these conditions (see Table 1).

Thus, better corrosion resistance of nitride coating I is fixed in the solution that simulates the biological tissue (Tyrode's) during simulating the temperature variation from normal to inflammatory processes in the body.

3.2. Nitride Coating II

At 36°C. Two peaks of the current density increase are observed on the anodic branch of the polarization curve of nitride coating II in 0.9% NaCl at potentials 0.1–0.4 V. It occurred due to dissolution of nitride film with formation of nonstoichiometric composition of reaction (3) on the titanium oxide surface. Its further dissolution at potentials 0.5–2.1 V (Figure 5(a), curve 1) occurs with the formation of oxychloride on the titanium surface as follows:

$$TiO_x + 2Cl^- \longrightarrow TiO_xCl_2 + 2e^- \quad (4)$$

It is confirmed by the presence of oxygen and chlorine in the surface spectra of X-ray fluorescence analysis (6.59/8.74 and 4.19/2.51 wt/at%, resp.).

It was observed the passive region (passive current density is 0.003 A/m^2) in Tyrode's solution, unlike 0.9% NaCl, in the range of potentials −0.33–0.35 V (Figure 5(a), curve 2). The existence of the passive region, as in coating I, is caused by the influence of physiological solution. Chlorine ions affect the increase of the current density. They cause the formation of oxychloride by reaction (4) on the titanium surface. Oxygen and chlorine are present in the surface spectra of X-ray fluorescence analysis. From the results of the analysis (Figure 6, Table 2), the intensification of interaction of chlorine ions with the surface should be associated with the roughness increase, especially height parameters of microrelief.

It should be noted that the current density of the anodic dissolution of coating II is lower but corrosion current density is higher than 0.9% NaCl.

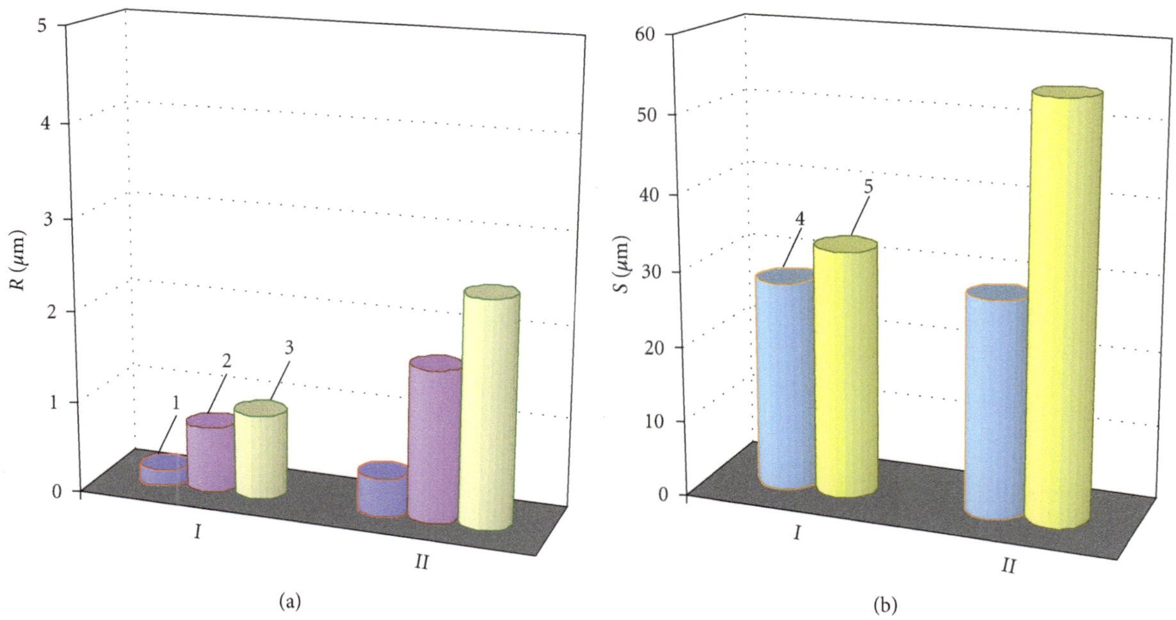

FIGURE 3: Quantitative characteristics of the surface microgeometry of Ti-6Al-4V alloy with nitride coatings I and II. (a) 1—R_a; 2—R_z, 3—R_{max}; (b) 4—S, 5—S_m.

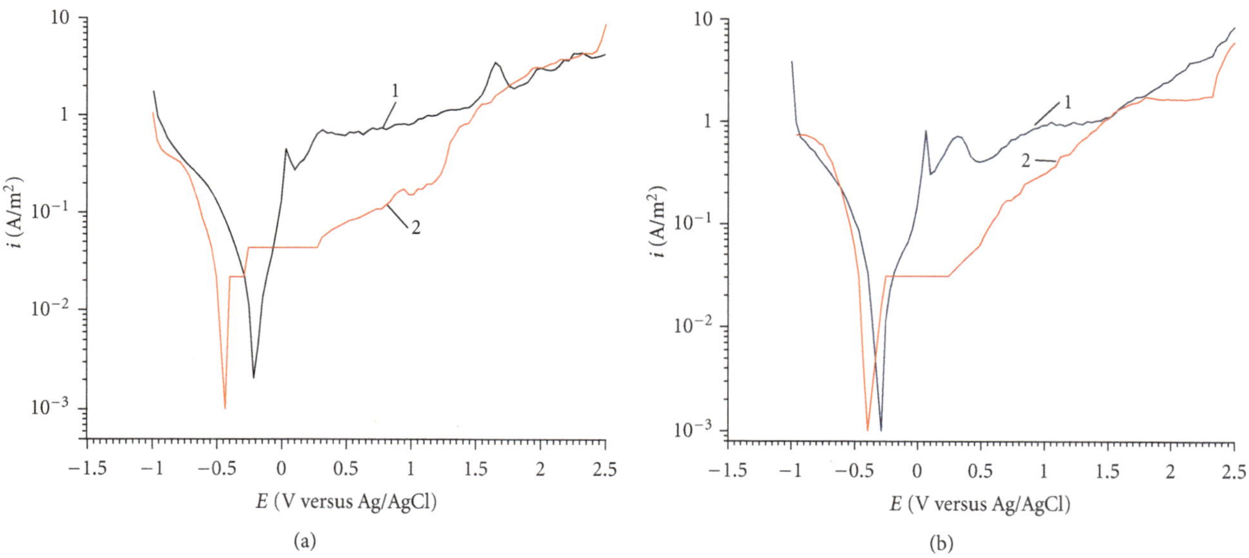

FIGURE 4: Potentiodynamic polarization curves of Ti-6Al-4V alloy with nitride coating I in 0.9% NaCl (1) and Tyrode's solution (2) at 36°C (a) and 40°C (b).

Corrosion-electrochemical behaviour of nitride coatings I and II is similar, although the protective effect of coating II is higher; corrosion current density and current density of anodic dissolution are lower in 0.9% NaCl and Tyrode's solution.

At 40°C. Processes which occur in the anode region of potentials for nitride coating II in both physiological solutions are similar. Thus, the peaks at 0.05 and 0.25 V (Figure 5(b), curves 1 and 2) are observed on the anodic branch of the polarization curve of the coating II in 0.9% NaCl and Tyrode's solution. It is caused by the formation of titanium oxynitride, oxide, and oxychloride of nonstoichiometric composition [14–16].

Although the processes for coating II in 0.9% NaCl at 40°C in the anode region are similar as processes that are observed at 36°C, processes of dissolution become slow because of the protective film formation with the solution temperature increase. It is confirmed by the deceleration of increase of current density at increase of potential.

Polarization curve of Ti-6Al-4V alloy with nitride coating II in Tyrode's solution at the temperature 40°C shows the lack of passive area. Its formation was associated with the

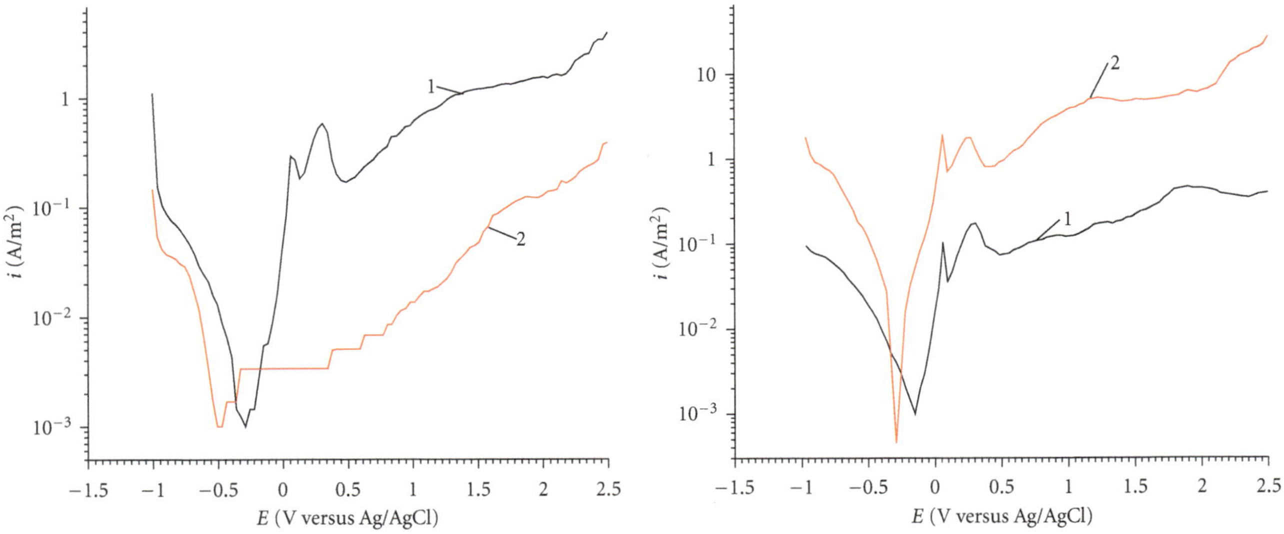

(a) (b)

FIGURE 5: Potentiodynamic polarization curves of Ti-6Al-4V alloy with nitride coating II in 0.9% NaCl (1) and Tyrode's solution (2) at 36°C (a) and 40°C (b).

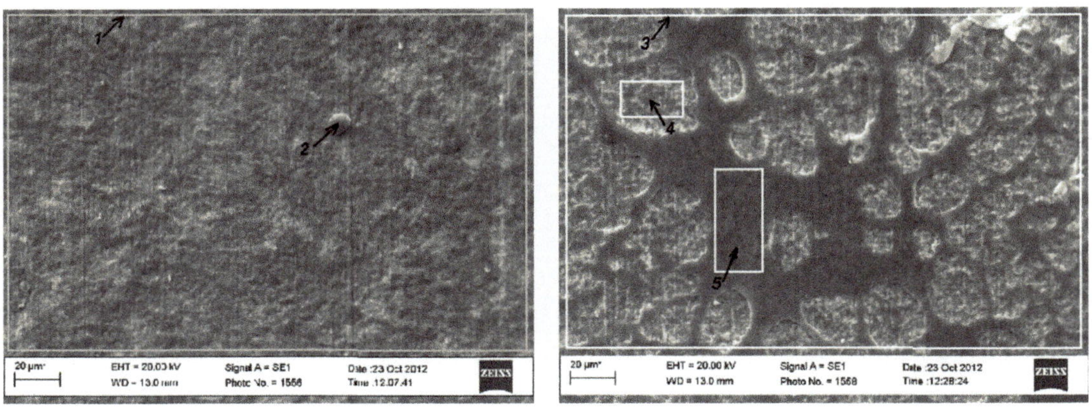

(a) (b)

FIGURE 6: The surface of Ti-6Al-4V alloy with nitride coating II after polarization in Tyrode's solution at 36°C (a) and 40°C (b).

TABLE 2: Energy dispersive X-ray analysis from surface of Ti-6Al-4V alloy with nitride coating II after polarization in Tyrode's solution.

Element	Areas of energy dispersive X-ray analysis (see Figure 6)				
	No. 1	No. 2	No. 3	No. 4	No. 5
C K	—	40.77/69.59	10.59/26.89	—	23.04/48.47
N K	13.93/35.49	—	—	9.16/25.54	—
O K	—	5.06/6.48	11.83/22.56	—	9.88/15.61
Al K	0.30/0.40	0.13/0.10	0.30/0.34	0.34/0.49	0.26/0.24
Cl K	0.72/0.72	4.65/2.69	4.44/3.82	0.61/0.68	2.43/1.73
P K	0.05/0.06	—	—	—	—
Ca K	0.03/0.03	—	—	—	—
Ti K	84.98/63.31	49.40/21.14	72.84/46.39	89.89/73.9	64.38/33.95

*Numerator: wt%; denominator: at%.

composition of physiological solution (Figure 5(b), curve 2). Perhaps the corrosion-electrochemical behaviour of the coating II in these conditions is associated with a change of the dominant processes of oxidation-reduction reaction by increasing the surface roughness. This assumption is confirmed by the results of energy dispersive X-ray and X-ray fluorescence analyses. They demonstrate the high content of chlorine on the surface of the samples after anodic polarization, and its concentration increases at high-altitude maximums of surface microrelief (Figure 6, Table 2).

Stable electrochemical properties of the nitride coating II in 0.9% NaCl confirmed the higher functionality (corrosion protection) in this solution compared with Tyrode's solution.

In general, the nitride coating II formed during nitriding at $p_{N_2} = 10^5$ Pa provides Ti-6Al-4V alloy the higher corrosion resistance in physiological solutions. It occurs by means of stronger shielding effect (thickness of nitride layer), the higher strength of chemical bonds in TiN phase, and more of its content in the nitride coating.

4. Conclusions

Nitride coating formed on Ti-6Al-4V alloy during nitriding at $p_{N_2} = 1$ Pa regardless of temperature of physiological solution shows higher corrosion resistance in Tyrode's solution, simulating biological tissue, while nitride coating formed during nitriding at $p_{N_2} = 10^5$ Pa in isotonic 0.9% NaCl, simulating blood plasma.

Increased activity of chlorine ions of physiological solution is associated with microrelief of nitride coating.

Raising the temperature of physiological solution from 36°C to 40°C, simulating the transition from normal to inflammatory state of the human body affects the electrochemical behaviour of nitride coating and, especially in solution, simulates biological tissue (Tyrode's).

Acknowledgments

This work is supported by the State Fund for Fundamental Researchers and the State Agency on Science, Innovations and Informatization of Ukraine (Project no. F41.2/011).

References

[1] S. Krischok, C. Blank, M. Engel et al., "Influence of ion implantation on titanium surfaces for medical applications," *Surface Science*, vol. 601, no. 18, pp. 3856–3860, 2007.

[2] F. Variola, J. H. Yi, L. Richert, J. D. Wuest, F. Rosei, and A. Nanci, "Tailoring the surface properties of Ti6Al4V by controlled chemical oxidation," *Biomaterials*, vol. 29, no. 10, pp. 1285–1298, 2008.

[3] F. Yildiz, A. F. Yetim, A. Alsaran, and I. Efeoglu, "Wear and corrosion behaviour of various surface treated medical grade titanium alloy in bio-simulated environment," *Wear*, vol. 267, no. 5–8, pp. 695–701, 2009.

[4] T. Chang-Bin, L. Dao-Xin, W. Zhan, and G. Yang, "Electro-spark alloying using graphite electrode on titanium alloy surface for biomedical applications," *Applied Surface Science*, vol. 257, no. 15, pp. 6364–6371, 2011.

[5] M. Metikos-Hukovic, A. Kwokal, and J. Piljac, "The influence of niobium and vanadium on passivityof titanium-based implants in physiological solution," *Biomaterials*, vol. 24, pp. 3765–3775, 2003.

[6] I. Cvijovic-Alagic, Z. Cvijovic, S. Mitrovic, V. Panic, and M. Rakin, "Wear and corrosion behaviour of Ti-13Nb-13Zr and Ti-6Al-4V alloys in simulated physiological solution," *Corrosion Science*, vol. 53, pp. 796–808, 2011.

[7] D. V. Shtansky, N. A. Gloushankova, A. N. Sheveiko et al., "Design, characterization and testing of Ti-based multicomponent coatings for load-bearing medical applications," *Biomaterials*, vol. 26, no. 16, pp. 2909–2924, 2005.

[8] L. Wang, J. F. Su, and X. Nie, "Corrosion and tribological properties and impact fatigue behaviors of TiN- and DLC-coated stainless steels in a simulated body fluid environment," *Surface and Coatings Technology*, vol. 205, no. 5, pp. 1599–1605, 2010.

[9] P. Budzynski, A. A. Youssef, and J. Sielanko, "Surface modification of Ti-6Al-4V alloy by nitrogen ion implantation," *Wear*, vol. 261, pp. 1271–1276, 2006.

[10] T. M. Manhabosco, S. M. Tamborim, C. B. dos Santos, and I. L. Müller, "Tribological, electrochemical and tribo-electrochemical characterization of bare and nitrided Ti6Al4V in simulated body fluid solution," *Corrosion Science*, vol. 53, no. 5, pp. 1786–1793, 2011.

[11] B. Subramanian, C. V. Muraleedharan, R. Ananthakumar, and M. Jayachandran, "A comparative study of titanium nitride (TiN), titanium oxy nitride (TiON) and titanium aluminum nitride (TiAlN), as surface coatings for bio implants," *Surface and Coatings Technology*, vol. 205, no. 21-22, pp. 5014–5020, 2011.

[12] S. Barril, S. Mischler, and D. Landolt, "Electrochemical effects on the fretting corrosion behaviour of Ti6Al4V in 0.9% sodium chloride solution," *Wear*, vol. 259, no. 1–6, pp. 282–291, 2005.

[13] I. M. Pohrelyuk, O. V. Tkachuk, and R. V. Proskurnyak, "Corrosion resistance of the Ti-6Al-4V titanium alloy with nitride coatings in 0.9 % NaCl," *Journal of the Minerals, Metals and Materials Society*, vol. 63, no. 6, pp. 35–40, 2011.

[14] V. A. Lavrenko, V. A. Shvets, N. V. Boshitskaya, and G. N. Makarenko, "Comparative study of the chemical resistance of titanium nitride and stainless steel in media of the oral cavity," *Powder Metallurgy and Metal Ceramics*, vol. 40, no. 11-12, pp. 630–636, 2001.

[15] T. H. Fang and K. T. Wu, "Local oxidation characteristics on titanium nitride film by electrochemical nanolithography with carbon nanotube tip," *Electrochemistry Communications*, vol. 8, no. 1, pp. 173–178, 2006.

[16] R. S. Razavi, M. Salehi, M. Ramazani, and H. C. Man, "Corrosion behaviour of laser gas nitrided Ti-6Al-4V in HCl solution," *Corrosion Science*, vol. 51, no. 10, pp. 2324–2329, 2009.

[17] H. Takadama, H. M. Kim, T. Kokubo, and T. Nakamura, "XPS study of the process of apatite formation on bioactive Ti-6Al-4V alloy in simulated body fluid," *Science and Technology of Advanced Materials*, vol. 2, no. 2, pp. 389–396, 2001.

[18] C. Vasilescu, S. I. Drob, E. I. Neacsu, and J. C. M. Rosca, "Surface analysis and corrosion resistance of a new titanium base alloy in simulated body fluids," *Corrosion Science*, vol. 65, pp. 431–440, 2012.

[19] J. Zhang, L. Duan, D. Jiang, Q. Lin, and M. Iwasa, "Dispersion of TiN in aqueous media," *Journal of Colloid and Interface Science*, vol. 286, no. 1, pp. 209–215, 2005.

[20] V. A. Lavrenko, A. D. Panasyuk, M. Desmaison-Brut, V. A. Shvets, and J. Desmaison, "Kinetics and mechanism of electrolytic corrosion of titanium-based ceramics in 3% NaCl solution," *Journal of the European Ceramic Society*, vol. 25, no. 10, pp. 1813–1818, 2005.

11

The Effect of *Achyranthes aspera* Extracts on Mild Steel Corrosion in Industrial Water Medium

Chikkur B. Pradeep Kumar and Kikkeri N. Mohana

Department of Studies in Chemistry, University of Mysore, Manasagangotri, Mysore 570006, India

Correspondence should be addressed to Kikkeri N. Mohana; drknmohana@gmail.com

Academic Editors: O. Benali, C. V. D'Alkaine, A. Y. El-Etre, A. Hermann, R. Salghi, and E. Stupnisek-Lisac

Achyranthes aspera (AA) extracts were studied as corrosion inhibitor for mild steel (MS) in industrial water medium using mass loss and electrochemical techniques. The results of the study revealed that AA extracts inhibit MS corrosion through adsorption process following Langmuir adsorption isotherm model. The protection efficiency increased with increase in inhibitor concentration and decreased with temperature. The electrochemical impedance spectroscopy (EIS) measurements showed that the charge transfer resistance increases with increase in the concentration of AA extracts. The polarization curves obtained indicate that AA extracts act as mixed type of inhibitor. Scanning electron microscopy (SEM) was used to analyze the surface adsorbed film.

1. Introduction

The use of inhibitors is one of the most practical methods for protection against corrosion and prevention of unexpected metal dissolution. Different organic and inorganic compounds have been studied as inhibitors to protect metals from corrosive attack. The efficiency of these organic corrosion inhibitors is related to the presence of polar functional groups with S, O, and N atoms in the molecule [1–5]. Such compounds can adsorb onto the metal surface and block the active surface sites, thus reducing the corrosion rate. Nevertheless, most of these compounds are not only expensive but also toxic for living beings. So, plant extracts have attracted the attention of researchers as ecofriendly corrosion inhibitors for many decades [6–9]. Plant extracts are an incredibly rich source of natural chemical compounds that can be extracted by simple procedures with low cost and are biodegradable in nature. The actual inhibitors in the plant extracts are usually alkaloids and other organic nitrogen bases, as well as carbohydrates, proteins, and their acid hydrolysis products [10]. Alkaloids have an ability to coordinate the transition metals or their alloys via the d-orbitals of metal and empty p-orbitals of the heteroatoms in the inhibitor molecules. A number of natural compounds have been used as corrosion inhibitors for metals and their alloys in acidic, alkaline, and neutral solutions [11–15].

Achyranthes aspera (AA) is one of the most important antioxidant medicinal plants, and it is an excellent source of several flavonoids, tannins, reducing sugars, alkaloids, and saponins [16]. Besides these phytoconstituents, several glycosides [17], fatty acids [18], ecdysterone [19], and triterpenoids [20] were isolated in various parts of plant. In the light of phytoconstituents reported, the present study was intended to ascertain the ability of methanolic extracts of AA to protect the mild steel in industrial water medium. Effects of inhibitor concentration, temperature, and immersion time on the corrosion rate were investigated and discussed. Gravimetric, potentiodynamic polarization, and electrochemical impedance spectroscopy (EIS) techniques were employed to study the mechanism of corrosion inhibition. Adsorption behaviour and thermodynamic parameters were obtained to establish the mechanism of adsorption. The morphology of the inhibited mild steel was analyzed using scanning electron microscopic images.

2. Experimental

2.1. Preparation of Plant Extracts. Healthy, disease free, mature leaves of *Achyranthes aspera* (AA) collected from

Western Ghats, Karnataka, India, were used for the preparation of methanolic extracts. Thoroughly washed mature leaves were shade dried and then powdered with the help of a blender. The powder was extracted with methanol using a Soxhelt extractor. The extracts were concentrated using rotary flash evaporator and preserved at ±5°C in airtight bottle until further use.

2.2. Materials. The specimens used for corrosion tests were mild steel (MS) coupons which have the following composition (wt %): 0.051 C : 0.023 Si : 0.005 P : 0.103 Al : 0.179 Mn : 0.023 S and the remainder iron. The test solution was industrial water collected from heat exchangers and reboilers of the chemical industries in and around Mysore city, India. The chemical composition of the industrial water (ppm) obtained from ionic chromatograph was 7500 Cl^-; 64 Ca^{2+}; 3440 SO_4^{2-}; 23 Mg^{2+}; 140 Na^+; 0.28 PO_4^{3-}. Prior to gravimetric and electrochemical measurements, the surface of the specimens was polished under running tap water using emery paper (SiC, grade 220–600), rinsed with distilled water, dried on a clean tissue paper, immersed in benzene for 5 s, dried and immersed in acetone for 5 s, and dried with clean tissue paper. Finally, the specimens were kept in desiccators until use. At the end of the gravimetric experiment, the specimens were carefully washed with acetone and benzene, dried, and then weighed. For polarization and electrochemical impedance studies, the MS specimen was embedded in epoxy resin to expose a geometrical surface area of 1 cm^2 to the electrolyte.

2.3. Methods

2.3.1. Gravimetric Measurements. Gravimetric experiments were carried out in a glass cell, and the solution volume was 100 cm^3. The temperature of the environment was maintained by thermostatically controlled water bath (Weiber, India) with an accuracy of ±0.2°C under aerated condition. The mild steel specimens used were rectangular with a dimension of 1 cm × 1 cm × 0.1 cm. The initial weight of the specimen was recorded using an analytical balance (precision ±0.1 mg). After the corrosion test in industrial water with and without inhibitor, the specimens were carefully washed in double distilled water, dried, and then weighed. The weight loss of the specimen was determined after an immersion period of 10 to 50 h at the temperature range of 30 to 60°C. Triplicate experiments were performed in each case, and the average mass loss was reported. The corrosion rate (CR) and inhibition efficiency (% IE) are calculated using

$$\mathrm{CR} = \frac{\Delta W}{St},$$
$$\mathrm{IE}\ \% = \frac{(\mathrm{CR})_a - (\mathrm{CR})_p}{(\mathrm{CR})_a} \times 100, \quad (1)$$

where ΔW is the weight loss, S is the surface area of the specimen (cm^2), t is the immersion time (h), and $(\mathrm{CR})_a$, $(\mathrm{CR})_p$ are corrosion rates in the absence and presence of the inhibitor, respectively.

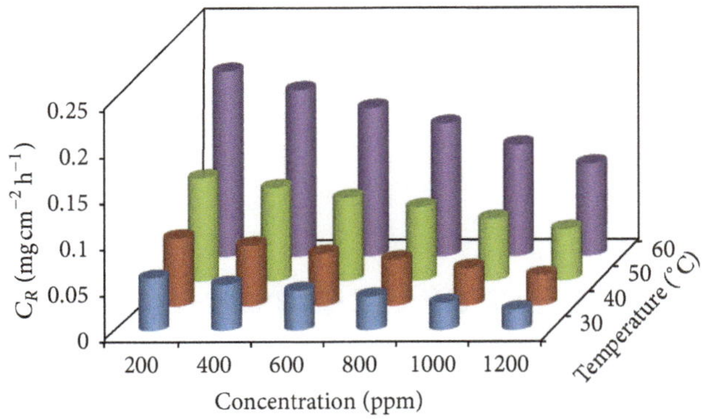

FIGURE 1: Variation of inhibition efficiency with temperature and inhibitor concentration in the absence and presence of different concentrations of AA extracts.

2.3.2. Electrochemical Measurements. Electrochemical tests were carried out with a CH-analyzer model CH1660D. The cell arrangement used was a conventional three-electrode cell with platinum counter electrode, saturated calomel electrode as reference electrode, and test material (mild steel) as working electrode. Anodic and cathodic polarization curves were recorded in the potential range from −350 to −800 mV with a scan rate of 0.4 mV s^{-1}. The AC impedance measurements were performed in the frequency range from 10 kHz to 0.05 Hz with signal amplitude of ±10 mV.

2.3.3. Morphological Investigation. The surface morphology of the mild steel samples in the absence and presence of AA extracts (1200 ppm in industrial water) at 30°C was investigated by scanning electron microscopy (SEM) technique (Model JSM-5800).

3. Results and Discussion

3.1. Mass Loss Studies. Weight loss of MS was determined in industrial water medium in the absence and presence of different concentrations of AA extracts. The inhibition efficiency of MS with and without the addition of different concentrations of AA extracts was determined after 40 h at different temperature, and the results are depicted in Table 1. It has been observed that 1200 ppm of AA extracts serve as an optimum concentration that exhibit higher inhibition efficiency. The corrosion rate of MS decreased on increasing the inhibitor concentration up to 1200 ppm. This behavior could be attributed to the increase in adsorption of the inhibitor at the metal/solution interface [21] and reduces the surface area that is available for direct attack on the metal surface. Above 1200 ppm, the corrosion rate approximately remained constant. The constant rate obtained could be attributed to the competitive adsorption effect between inhibitor molecules and the metal surface which is already covered with initial layers of molecules [22]. The % IE at different concentrations of AA extracts decreases with increasing temperature from 303 K to 333 K (Figure 1). This is due to increased rate of dissolution process of MS and partial desorption of the inhibitor from the metal surface with temperature [23].

TABLE 1: C_R and IE (%) obtained from weight loss measurements of MS in industrial water medium containing various concentrations of AA extracts at different temperatures.

C (ppm)	Temperature							
	30°C		40°C		50°C		60°C	
	C_R (mg cm^{-2} h^{-1})	IE %	C_R (mg cm^{-2} h^{-1})	IE %	C_R (mg cm^{-2} h^{-1})	IE %	C_R (mg cm^{-2} h^{-1})	IE %
0	0.1204	—	01428	—	0.2048	—	0.3326	—
200	0.0573	52.40	0.0736	48.48	0.1122	44.04	0.2001	39.84
400	0.0501	58.38	0.0656	54.08	0.1011	50.63	0.1795	46.09
600	0.0433	64.03	0.0568	60.24	0.0901	56.00	0.1601	51.87
800	0.0365	69.68	0.0499	65.07	0.0795	61.18	0.1426	57.13
1000	0.0298	75.24	0.0406	71.58	0.0668	67.38	0.1201	63.89
1200	0.0223	81.47	0.0328	77.04	0.0552	73.04	0.0995	70.08

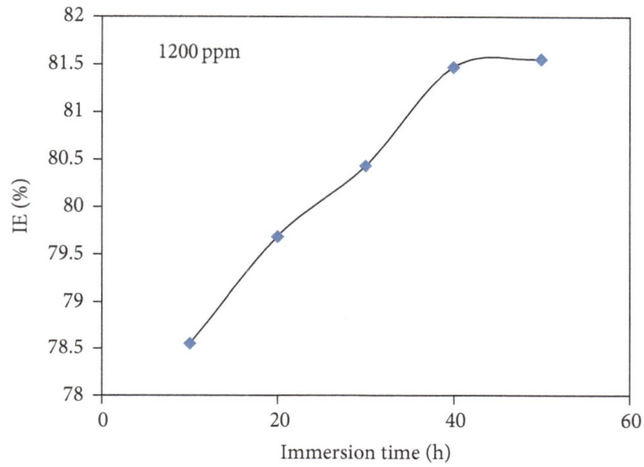

FIGURE 2: Effect of immersion time on IE % of AA extracts.

The inhibitive effect of AA extracts is ascribed to the presence of several organic compounds in the extracts. AA extracts are rich in organic compounds of high molecular weight with heteroatoms and π centers in their molecular structures. These include several flavonoids, tannins, reducing sugars, alkaloids, saponines, glycosides, fatty acids, ecdysterone, betaine, and triterpenoids. The synergistic and antagonistic effects may play an important role in the inhibition efficiency of AA extracts [24].

3.2. Effect of Immersion Time. In order to assess the stability of the inhibitor on MS surface, weight loss measurements were performed in industrial water medium in the absence and presence of AA extract for 10–50 h of immersion time at 303 K. From Figure 2, it can be observed that the inhibition efficiency of AA extracts increased with immersion time and stabilized after 40 hr. This increase in inhibition of AA extracts with immersion time reflects strong adsorption on the mild steel surface resulting in a more protective layer.

3.3. Effect of Temperature. Analysis of temperature dependence of inhibition efficiency and comparison of corrosion activation energies in the absence and presence of AA extracts give some insight into the possible mechanism of inhibitor adsorption. The weight loss measurements were carried out as a function of temperature (30 to 60°C) and concentration (200 to 1200 ppm) at 10 to 50 h immersion time.

The percentage inhibition efficiencies (IE%) and corrosion rates (CR) are summarized in Table 1. It was found that increase in temperature leads to an increase in the corrosion rate in the absence and presence of AA extracts in industrial water medium. The dependence of corrosion rate on temperature can be expressed by the following Arrhenius equation:

$$\mathrm{CR} = A \exp\left(-\frac{E_a^*}{RT}\right), \quad (2)$$

where E_a^* is the apparent activation energy, T is the absolute temperature, A is the Arrhenius preexponential constant, and R is the universal gas constant. The apparent activation energy and preexponential factor for different concentrations of the inhibitor were calculated from the plots of logarithm of CR versus $1/T$ (Figure 3), and the results are shown in Table 2. It is clear that the activation energy increases in the presence of AA extracts and consequently corrosion rate decreases. An alternative formulation of Arrhenius equation is [25]

$$\mathrm{CR} = \frac{RT}{Nh} \exp\frac{\Delta S^*}{R} \exp\left(-\frac{\Delta H^*}{RT}\right), \quad (3)$$

where h is Planck's constant, N is Avogadro's number, R is the universal gas constant, ΔH^* is the enthalpy of activation, and ΔS^* is the entropy of activation. Using (3), plots of $\log(C_R/T)$ versus $1/T$ gave straight lines with a slope of $(-\Delta H^*/2.303R)$ and an intercept of $[\log(R/Nh) + \Delta S^*/2.303R]$ (Figure 4) from which the values of ΔH^* and ΔS^* are calculated and tabulated in Table 2. Inspection of Table 2 reveals that the calculated values of ΔS^* for AA extracts in industrial water medium are large and negative. The entropy of activation values are less negative for inhibited solutions than that for the uninhibited solutions. This suggests that an increase in randomness occurred while moving from reactants to the activated complex [26]. The positive sign of ΔH^* reflects the endothermic nature of the MS dissolution process, suggesting that the dissolution of MS is slow in the presence of inhibitor

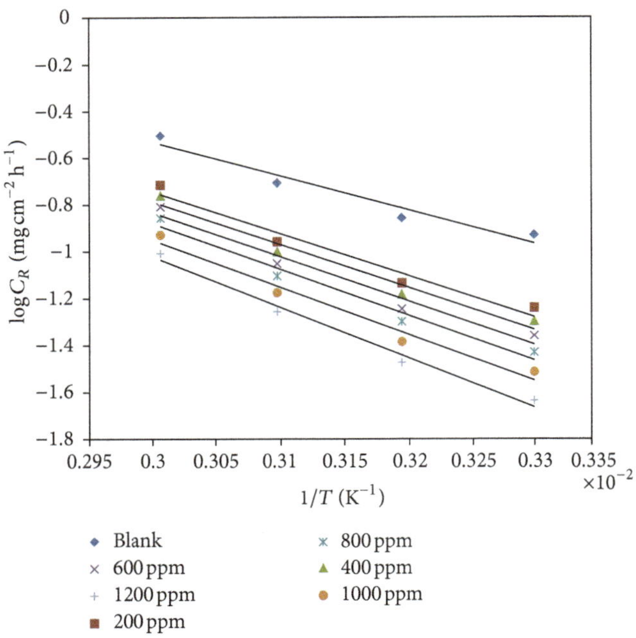

FIGURE 3: Arrhenius plots for MS in industrial water medium in the absence and presence of different concentrations of AA extracts.

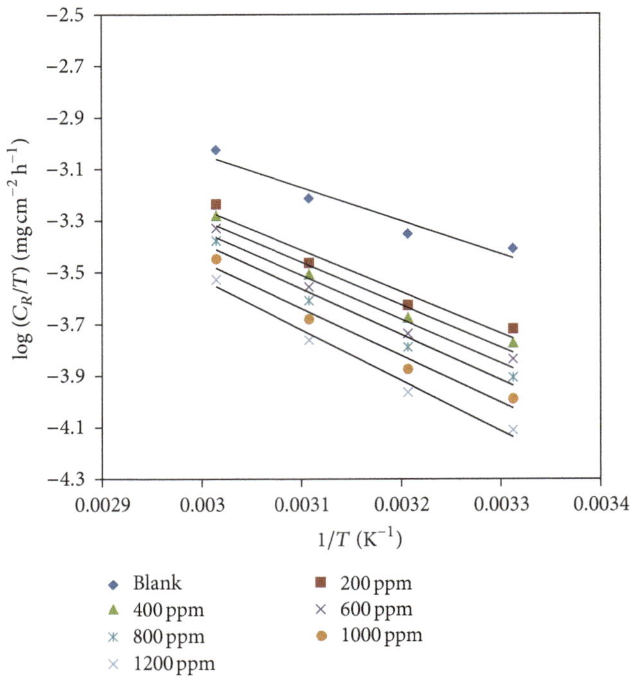

FIGURE 4: Alternative Arrhenius plots for MS in industrial water medium in the absence and presence of different concentrations of AA extracts.

[27]. The values of ΔH^* could also be obtained from the following equation:

$$\Delta H^* = E_a^* - RT. \quad (4)$$

The values of ΔH^* obtained here are in very good agreement with those obtained from (3), confirming the endothermic process of mild steel dissolution in industrial water medium.

3.4. Adsorption Isotherm. The surface coverage (θ) of corrosion inhibitor on MS is the most useful parameter for the study of the adsorption behaviour of the inhibitor. A direct relationship between inhibition efficiency and the degree of surface coverage can be assumed for different concentrations of the inhibitor. Data obtained from mass loss measurements were tested graphically for fitting various isotherms including Langmuir, Frumkin, and Temkin. However, the best fit is obtained from Langmuir isotherm (Figure 5). This isotherm assumes that the adsorbed molecules occupy only one site, and there are no interactions with other adsorbed species [28]. According to this isotherm Θ is related to inhibitor concentration as

$$\frac{C}{\theta} = \frac{1}{K_{\text{ads}}} + C, \quad (5)$$

where K_{ads} is the equilibrium constant for adsorption, C is the concentration of the inhibitor, and θ is the surface coverage. The strong correlation ($R^2 > 0.97$) suggests that the adsorption of inhibitor on the MS surface obeyed this isotherm. The K_{ads} is related to the free energy of adsorption (ΔG_{ads}) using the following equation:

$$K_{\text{ads}} = \frac{1}{55.5} \exp\left(\frac{-\Delta G_{\text{ads}}}{RT}\right), \quad (6)$$

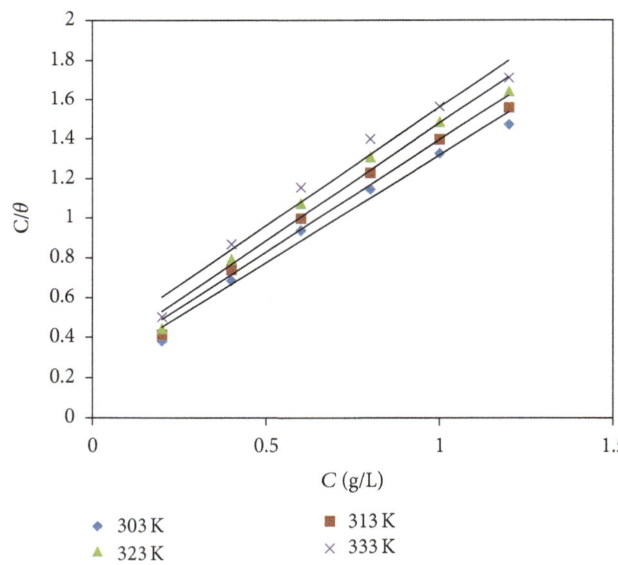

FIGURE 5: Langmuir adsorption isotherm of AA extracts on MS in industrial water medium at different temperatures.

where R is the universal gas constant, T is the absolute temperature, and 55.5 is the concentration of water in solution (mol L^{-1}). The values free energy of adsorption for AA extracts are presented in Table 3. The negative values of ΔG_{ads} means that the adsorption of AA on MS surface is a spontaneous process and furthermore shows the strong interaction of the inhibitor onto the MS surface [29].

TABLE 2: Activation parameters for MS in industrial water medium in the absence and presence of different concentrations of AA extracts.

C (ppm)	E_a^* (kJ mol^{-1})	k (mg cm^{-2} h^{-1})	ΔH^* (kJ mol^{-1})	$\Delta H^* = E_a^* - RT$ (kJ mol^{-1})	ΔS^* (J mol^{-1} K^{-1})
0	28.41	9977.06	25.77	25.89	−178.35
200	35.34	66221.65	32.16	32.82	−163.43
400	35.47	65162.84	32.91	32.96	−162.03
600	36.66	93972.33	33.94	34.14	−159.85
800	38.84	181970.1	35.40	36.33	−156.44
1000	39.80	215774.4	36.43	37.28	−154.72
1200	40.11	179060.6	39.23	37.59	−147.77

TABLE 3: Thermodynamic adsorption parameters for adsorption of AA extracts on MS in industrial water medium at different temperatures from Langmuir adsorption isotherm.

Temperature (K)	R^2	K_{ads} (L g^{-1})	ΔG_{ads} (kJ mol^{-1})	ΔH_{ads} (kJ mol^{-1})	ΔS_{ads} (J mol^{-1} K^{-1})	$\Delta G_{ads} = \Delta H_{ads} - T\Delta S_{ads}$ (kJ mol^{-1})
303	0.983	4310.34	−31.20			−30.26
313	0.981	3816.79	−31.91	−11.98[a] −12.11[b]	63.58[a] 63[b]	−30.90
323	0.977	3448.27	−32.66			−31.53
333	0.970	2762.43	−33.06			−32.17

[a]Values obtained from (7). [b]Values obtained from (8).

Generally, ΔG_{ads} values of the order of −20 kJ mol^{-1} or higher are associated with an electrostatic interaction between charged molecules and charged metal surface (physisorption), and those of −40 kJ mol^{-1} or lower indicate charge sharing or a transfer from the inhibitor molecules to the metal surface to form a coordinate covalent bond (chemisorption) [30]. The values for ΔG_{ads} for AA extracts on the MS surface are in the ranges of −31.20 to −33.06; indicating that the adsorption of AA extracts on MS surface is neither typical physisorption nor typical chemisorption, but it is a complex mixed—type adsorption; that is, it involves both physisorption and chemisorption [31]. The enthalpy and entropy of adsorption (ΔH_{ads} and ΔS_{ads}) can be calculated using

$$\ln K_{ads} = \ln \frac{1}{55.5} - \frac{\Delta H_{ads}}{RT} + \frac{\Delta S_{ads}}{R}. \quad (7)$$

From (7), the values of ΔH_{ads} and ΔS_{ads} were evaluated from the slope and intercept of the plot of $\ln K_{ads}$ versus $1/T$ (Figure 6). All the calculated thermodynamic parameters are listed in Table 3. It has been found that the values ΔH_{ads} are negative, suggesting that the adsorption of the inhibitor is an exothermic process [32]. Adsorption entropy is positive and large indicating that the driving force for the adsorption of adsorbate is the increase in entropy during the process of adsorption rather than the decrease in enthalpy [33].

The values of ΔH_{ads} and ΔS_{ads} can also be calculated by using following equation:

$$\Delta G_{ads} = \Delta H_{ads} - T\Delta S_{ads}. \quad (8)$$

Using (8), the plot of ΔG_{ads} versus T gives a straight line (Figure 7) with a slope of $-\Delta S_{ads}$ and intercept of ΔH_{ads}. The values obtained are well correlated with those obtained from (7).

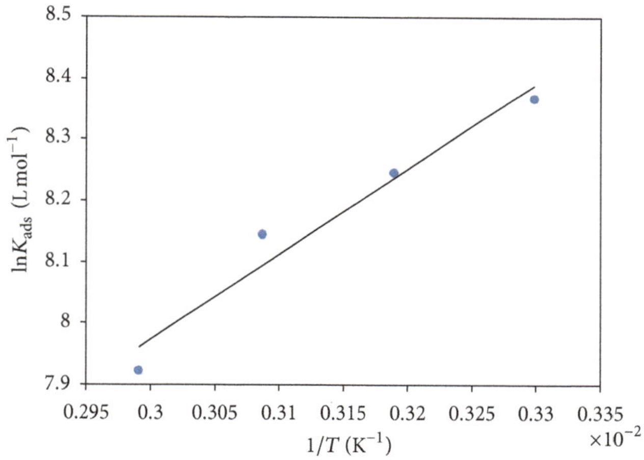

FIGURE 6: Plot of $\ln K_{ads}$ versus $1/T$.

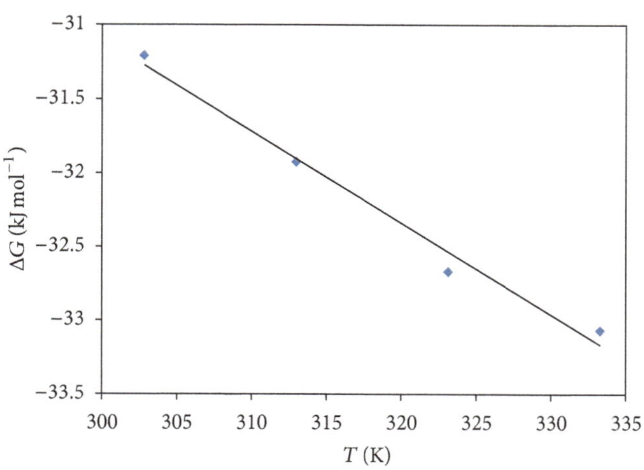

FIGURE 7: Plot of Gibbs free energy versus absolute temperature.

TABLE 4: E_{corr}, I_{corr}, R_{ct}, and IE % obtained from polarization and impedance measurements for mild steel in industrial water medium containing various concentrations of AA extracts at 30°C.

C (ppm)	EIS		Polarization				
	R_{ct} (Ω cm^2)	IE %	E_{corr} (mV)	I_{corr} (mA cm^{-2})	b_a (mA dec^{-1})	b_c (mA dec^{-1})	IE %
0	111.7		−0.785	0.00476	6.000	1.034	
200	226.7	50.72	−0.503	0.00250	10.585	2.301	47.57
400	232.7	51.99	−0.574	0.00270	12.216	3.715	52.29
600	290.6	61.56	−0.706	0.00199	10.174	1.380	58.29
800	328.7	66.01	−0.684	0.00156	9.978	1.763	67.23
1000	412.0	72.88	−0.653	0.00128	7.320	1.481	73.14
1200	559.7	80.04	−0.610	0.00089	8.949	2.423	81.23

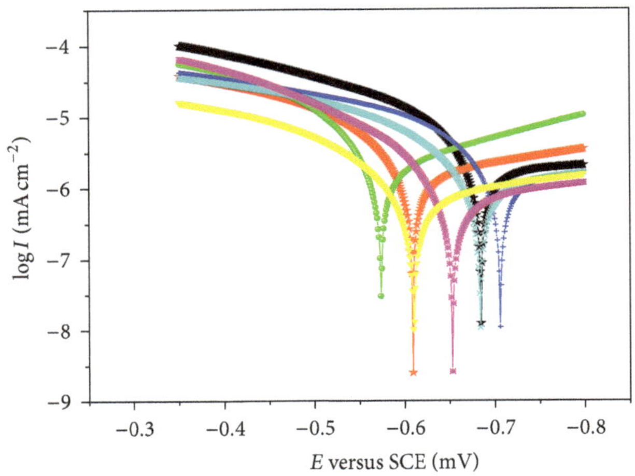

FIGURE 8: Potentiodynamic polarization curves for MS in industrial water medium containing different concentrations of compound AA extracts.

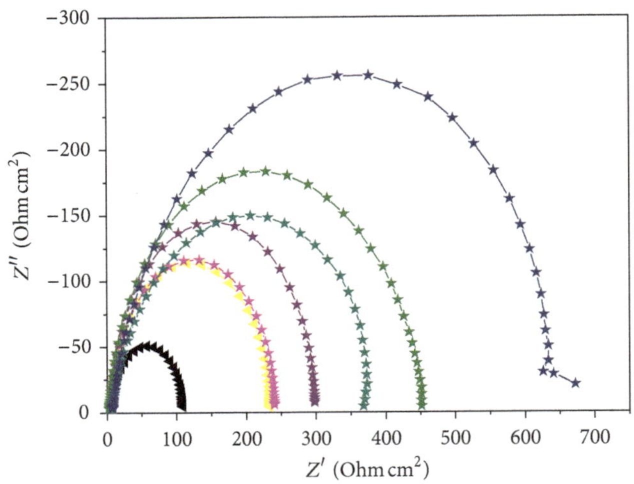

FIGURE 9: Nyquist plots for MS in industrial water medium containing different concentrations of AA extracts.

3.5. Potentiodynamic Polarization Studies. Anodic and cathodic polarization curves for MS in industrial water medium without and with different concentrations of AA extracts are shown in Figure 8. It is observed that both cathodic and anodic curves show a lower current density in the presence of AA extracts than those recorded in the industrial water medium alone. The values of corrosion potential (E_{corr}), corrosion current density (I_{corr}), anodic Tafel slope (b_a) and cathodic Tafel slope (b_c), and percentage inhibition efficiency (IE %) are presented in Table 4. The % IE was calculated from polarization measurements according to following equation:

$$\text{IE}\,(\%) = \frac{(I_{corr})_a - (I_{corr})_p}{(I_{corr})_a} \times 100, \quad (9)$$

where $(I_{corr})_a$ and $(I_{corr})_p$ are the corrosion current densities (mA cm^{-2}) in the absence and presence of the inhibitor, respectively. From Figure 8, it is evident that the addition of AA extracts shifts the corrosion potential (E_{corr}) values significantly in cathodic direction. The presence of AA extracts causes a significant decrease in the corrosion rate that is predominantly shifts the cathodic curves to lower values of current densities. On the other hand, the anodic reaction is slightly affected by AA extracts. But it alters both the tafel slope values indicating that AA extracts act as a mixed type of inhibitor [34]. The % IE increases with the inhibitor concentration due to the increase in the blocked fraction of the electrode surface by adsorption. % IE of AA extracts reaches to as high as 81.23 which indicates that AA extracts are a good inhibitor for MS in industrial water medium.

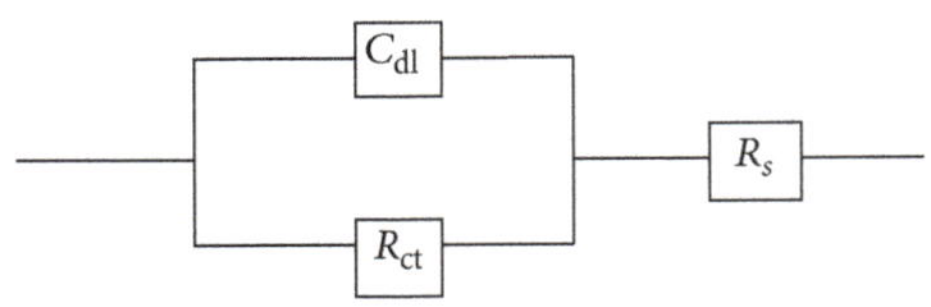

FIGURE 10: Equivalent circuit diagram.

3.6. Electrochemical Impedance Spectroscopy (EIS). Figure 9 shows the impedance response of MS in industrial water

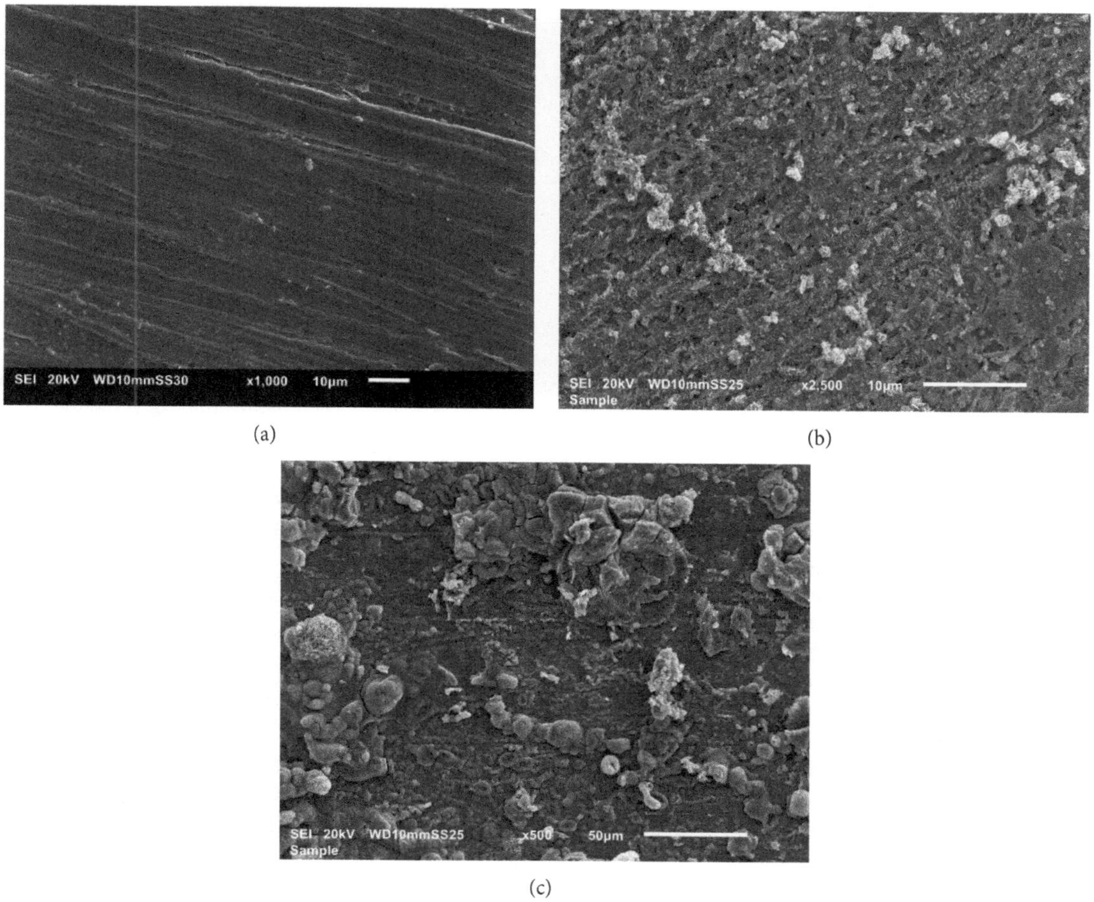

FIGURE 11: SEM images of MS in industrial water medium after 40 h of immersion at 30°C (a) before immersion (polished), (b) without inhibitor (blank), and (c) with 1200 ppm of AA extracts.

medium without and with different concentration of the inhibitor. Nyquist impedance plots were analyzed by fitting the experimental data to a simple circuit model (Figure 10). All the obtained plots show only one capacitive loop, and the diameter of the semicircle increases with increasing the inhibitor concentration suggesting that the inhibitive film formed was strengthened by the addition of inhibitor [35]. The presence of the inhibitor increases the impedance but does not change other aspects of the behavior. These results support the results of polarization measurements that the inhibitor does not alter the electrochemical reactions responsible for corrosion. The experimental results of EIS measurements for the corrosion of MS in industrial water medium in the absence and presence of AA extracts are given in Table 4. Inspection of the Table 4 shows that AA extracts enhances the charge transfer resistance which is attributed to a decrease in local dielectric constant and/or an increase in the thickness of electrical double layer [36]. % IE obtained from this technique is also comparable with those obtained from polarization measurements.

3.7. Scanning Electron Microscopy (SEM). To establish whether inhibition of MS surface is due to the formation of inhibitor protective film on the metal surface, SEM images were taken. SEM photographs obtained from MS surface after immersion in industrial water medium in the absence and presence of optimum concentration of inhibitors are shown in Figures 11(a)–11(c). Figure 11(a) shows the clear smooth surface morphology of MS examined. It can be seen from Figure 11(b) that the MS surface is strongly damaged in the absence of the inhibitor (blank). Figure 11(c) showed the protected MS surface after the addition of AA extracts, whereby the surface damage has diminished in comparison to the blank material. This is attributed to the formation of a good protective film on the MS surface.

4. Conclusion

(1) The inhibition efficiency of AA extracts increases with the inhibitor concentration and reaches a maximum at 1200 ppm of AA extracts.

(2) IE % decreases with the increasing temperature in industrial water medium and remains constant at higher temperature.

(3) The adsorption of AA extracts in industrial water medium on MS surface obeys Langmuir adsorption isotherm. The adsorption process involved is both physisorption and chemisorption.

(4) Polarization curves indicate that AA extracts act as mixed type of inhibitor.

(5) The adsorbed film containing the investigated compound was identified by SEM analysis.

Acknowledgments

The authors would like to thank University of Mysore, Mysore, MHRD and UGC, New Delhi, India, for providing financial assistance through Institution of Excellence (IOE) scheme. The authors are also thankful to Dr. Muralidhar and Mr. Yogeesh K. K., Department of Chemistry, K. S. Institute of Technology, Bangalore, India, for their help to carry out electrochemical work.

References

[1] E. Fouda, A. A. Sarawy, and E. E. Katori, "Thiazole derivatives as corrosion inhibitors for C-steel in sulphuric acid solution," *European Journal of Chemistry*, vol. 1, no. 4, pp. 312–318, 2010.

[2] A. S. Fouda, M. Diab, A. El-Sonbati, and S. A. Hassan, "Benzothiazole derivatives as corrosion inhibitors for carbon steel in 1 M phosphoric acid (H_3PO_4) solutions," *African Journal of Pure and Applied Chemistry*, vol. 7, no. 2, pp. 67–78, 2013.

[3] O. O. Xomet, N. V. Likhanova, N. Nava et al., "Thiadiazoles as corrosion inhibitors for carbon steel in H_2SO_4 solutions," *International Journal of Electrochemical Science*, vol. 8, pp. 735–752, 2013.

[4] R. Bostan, S. Varvara, L. Gaina, and L. M. Mures, "Evaluation of some phenothiazine derivatives as corrosion inhibitors for bronze in weakly acidic solution," *Corrosion Science*, vol. 63, pp. 275–286, 2012.

[5] Z. Chen, L. Huang G, Zhang, Y. Qiu, and X. Guo, "Benzotriazole as a volatile corrosion inhibitor during the early stage of copper corrosion under adsorbed thin electrolyte layers," *Corrosion Science*, vol. 65, pp. 214–222, 2012.

[6] O. K. Abiola, J. O. E. Otaigbe, and O. J. Kio, "Gossipium hirsutum L. extracts as green corrosion inhibitor for aluminum in NaOH solution," *Corrosion Science*, vol. 51, no. 8, pp. 1879–1881, 2009.

[7] S. Deng and X. Li, "Inhibition by *Jasminum nudiflorum* Lindl. leaves extract of the corrosion of aluminium in HCl solution," *Corrosion Science*, vol. 64, pp. 253–262, 2012.

[8] S. Deng and X. Li, "Inhibition by Ginkgo leaves extract of the corrosion of steel in HCl and H_2SO_4 solutions," *Corrosion Science*, vol. 55, pp. 407–415, 2012.

[9] X. Li and S. Deng, "Inhibition effect of Dendrocalamus brandisii leaves extract on aluminum in HCl, H_3PO_4 solutions," *Corrosion Science*, vol. 65, pp. 299–308, 2012.

[10] B. A. Abd-El-Nabey, A. M. A. Gaber, M. E. S. Ali, E. Khamis, and El- Housseiny, "Cannabis plant extract as inhibitor for the corrosion of nickel in 0. 5 M H_2SO_4," *International Journal Electrochemical Science*, vol. 7, no. 12, pp. 11811–11826, 2012.

[11] S. S. Shivakumar and K. N. Mohana, "Centella asiatica extracts as green corrosion inhibitor for mild steel in 0. 5 M sulphuric acid medium," *Advances in Applied Science Research*, vol. 3, no. 5, pp. 3097–3106, 2012.

[12] S. S. Shivakumar and K. N. Mohana, "Studies on the inhibitive performance of Ccnnamomum zeylanicum extracts on the corrosion of mild Steel in hydrochloric acid and sulphuric acid medium," *Journal of Materials and Environmental Science*, vol. 4, pp. 448–459, 2013.

[13] S. S. Shivakumar and K. N. Mohana, "Ziziphus mauritiana leaves extracts as corrosion inhibitor for mild steel in H_2SO_4 and HCl solutions," *European Journal of Chemistry*, vol. 3, no. 4, pp. 426–432, 2012.

[14] I. B. Obot and N. O. Obi-Egbedi, "Ipomoea involcrata as an ecofriendly inhibitor for aluminium in alkaline medium," *Portugaliae Electrochimica Acta*, vol. 27, no. 4, pp. 517–524, 2009.

[15] A. Y. El-Etre, M. Abdallah, and Z. E. El-Tantawy, "Corrosion inhibition of some metals using lawsonia extract," *Corrosion Science*, vol. 47, no. 2, pp. 385–395, 2005.

[16] M. A. Alam, N. Slahin, R. Uddin et al., "Analgesic and neuropharmacological investigations of the aerial part of *Achyranthes aspera* Linn," *Stamford Journal of Pharmaceutical Sciences*, vol. 1, no. 1, pp. 44–50, 2008.

[17] R. D. Rameshwar and N. Akito, "Three oleanoliic acid glycosides from medicinally important seeds of *Achyranthes aspera*," *Natural Product Communications*, vol. 2, pp. 727–730, 2007.

[18] H. N. Khastgir, S. K. S. Gupta, and P. S. Gupta, "The sapogenin from seeds of *Achyranthes aspera* Linn," *Journal of the Indian Chemical Society*, vol. 35, pp. 693–694, 1958.

[19] A. Banerji and M. S. Chadha, "Insect moulting hormone from *Achyranthes aspera*," *Phytochemistry*, vol. 9, no. 7, p. 1671, 1970.

[20] G. Michl, D. Abebe, F. Bucar et al., "New triterpenoid saponins from *Achyranthes aspera* Linn," *Helvetica Chimica Acta*, vol. 83, no. 2, pp. 359–363, 2000.

[21] M. H. Hussin and M. J. Kassim, "The corrosion inhibition and adsorption behavior of Uncaria gambir extract on mild steel in 1 M HCl," *Materials Chemistry and Physics*, vol. 125, no. 3, pp. 461–468, 2011.

[22] S. S. Shivakumar, K. N. Mohana, and D. M. Gurudatt, "Inhibition performance and adsorption behavior of spinacia oleracea leaves extracts on mild steel corrosion in hydrochloric acid Medium," *Chemical Science Transactions*, vol. 2, no. 1, pp. 163–175, 2013.

[23] M. Schorr and J. Yahalom, "The significance of the energy of activation for the dissolution reaction of metal in acids," *Corrosion Science*, vol. 12, no. 11, pp. 867–868, 1972.

[24] O. K. Abiola and A. O. James, "The effects of Aloe vera extract on corrosion and kinetics of corrosion process of zinc in HCl solution," *Corrosion Science*, vol. 52, no. 2, pp. 661–664, 2010.

[25] H. Cang, Z. Fei, J. Shao, W. Shi, and Q. Xu, "Corrosion inhibition of mild steel by *aloes* extract in HCl solution medium," *International Journal of Electrochemical Science*, vol. 8, pp. 720–734, 2013.

[26] M. Bouklah, B. Hammouti, M. Lagrenée, and F. Bentiss, "Thermodynamic properties of 2,5-bis(4-methoxyphenyl)-1,3,4-oxadiazole as a corrosion inhibitor for mild steel in normal sulfuric acid medium," *Corrosion Science*, vol. 48, no. 9, pp. 2831–2842, 2006.

[27] G. N. Mu, X. Li, and F. Li, "Synergistic inhibition between o-phenanthroline and chloride ion on cold rolled steel corrosion in phosphoric acid," *Materials Chemistry and Physics*, vol. 86, no. 1, pp. 59–68, 2004.

[28] S. A. Ali, M. T. Saeed, and S. U. Rahman, "The isoxazolidines: a new class of corrosion inhibitors of mild steel in acidic medium," *Corrosion Science*, vol. 45, no. 2, pp. 253–266, 2003.

[29] M. Scendo, "The effect of purine on the corrosion of copper in chloride solutions," *Corrosion Science*, vol. 49, no. 2, pp. 373–390, 2007.

[30] M. Hosseini, S. F. L. Mertens, and M. R. Arshadi, "Synergism and antagonism in mild steel corrosion inhibition by sodium dodecylbenzenesulphonate and hexamethylenetetramine," *Corrosion Science*, vol. 45, no. 7, pp. 1473–1489, 2003.

[31] I. Ahamad, R. Prasad, and M. A. Quraishi, "Inhibition of mild steel corrosion in acid solution by Pheniramine drug: Experimental and theoretical study," *Corrosion Science*, vol. 52, no. 9, pp. 3033–3041, 2010.

[32] X. Li and G. Mu, "Tween-40 as corrosion inhibitor for cold rolled steel in sulphuric acid: weight loss study, electrochemical characterization, and AFM," *Applied Surface Science*, vol. 252, no. 5, pp. 1254–1265, 2005.

[33] O. Benali, L. Larabi, M. Traisnel, L. Gengembre, and Y. Harek, "Electrochemical, theoretical and XPS studies of 2-mercapto-1-methylimidazole adsorption on carbon steel in 1 M $HClO_4$," *Applied Surface Science*, vol. 253, no. 14, pp. 6130–6139, 2007.

[34] P. B. Raja and M. G. Sethuraman, "Inhibitive effect of black pepper extract on the sulphuric acid corrosion of mild steel," *Materials Letters*, vol. 62, no. 17-18, pp. 2977–2979, 2008.

[35] H. Amar, T. Braisaz, D. Villemin, and B. Moreau, "Thiomorpholin-4-ylmethyl-phosphonic acid and morpholin-4-methyl-phosphonic acid as corrosion inhibitors for carbon steel in natural seawater," *Materials Chemistry and Physics*, vol. 110, no. 1, pp. 1–6, 2008.

[36] M. A. Quraishi, A. Singh, V. K. Singh, D. K. Yadav, and A. K. Singh, "Green approach to corrosion inhibition of mild steel in hydrochloric acid and sulphuric acid solutions by the extract of Murraya koenigii leaves," *Materials Chemistry and Physics*, vol. 122, no. 1, pp. 114–122, 2010.

12

Cuminum cyminum Extracts as Eco-Friendly Corrosion Inhibitor for Mild Steel in Seawater

V. Sribharathy[1] and Susai Rajendran[2]

[1] *PG and Research Department, GTN Arts College, Dindigul 624005, Tamil Nadu, India*
[2] *Department of Chemistry, RVS School of Engineering and Technology, Dindigul 624005, Tamil Nadu, India*

Correspondence should be addressed to V. Sribharathy; bharathy_chem@yahoo.co.in

Academic Editors: C. Gervasi, I. Obot, and Q. Qu

The inhibitive effect of Jeera *(Cuminum cyminum)* plant extracts on the corrosion of mild steel in an aqueous solution of seawater was investigated using potentiodynamic polarization and electrochemical impedance spectroscopy (EIS) techniques. The stability of the inhibition efficiency of Jeera extracts was examined by weight-loss method. Potentiodynamic polarization curves indicated that the Jeera extract behaves as an anodic type inhibitor. EIS measurements showed that the dissolution process occurs under activation control. The corrosion rates of steel and the inhibition efficiencies of the extract obtained from impedance and polarization measurements were in good agreement. Inhibition was found to increase with an increasing concentration of the plant extract. The results obtained show that the Jeera extract could serve as an effective inhibitor for the corrosion of mild steel in seawater.

1. Introduction

Recently, plant extracts have again become important as an environmentally acceptable, readily available, and renewable source for a wide range of needed inhibitors. Plant extracts are viewed as an incredibly rich source of naturally synthesized chemical compounds that can be extracted by simple procedures with low cost. However, synergistic (and antagonistic) effects are often expected with these mixtures of inhibitors that may affect their inhibition efficiency. Several investigations have been reported using such economic plant extracts. El Hosary et al. [1] studied the corrosion inhibition of aluminium and zinc in 2 N HCl using naturally occurring *Hibiscus sabdariffa* (Karkade) extract. The inhibition of corrosion of steel, aluminium, and copper in HCl, H_2SO_4, and citric acid by molasses was also studied [2], and 83% and 13% inhibition efficiencies were obtained for HCl and H_2SO_4 solutions, respectively, containing 0.75% molasses. Loto reported the inhibitive action of *Vernonia amygdalina* (bitter leaf) on the corrosion of mild steel in 0.5 M HCl at 28°C [3]. Avwiri and Igho studied the inhibitive action of *V. amygdalina* on the corrosion of aluminium alloys in HCl and HNO_3 at concentrations of 0.2 and 0.4 g/L at 29°C [4]. They showed that the solution extract of the leaves serves as an excellent inhibitor. The inhibition effect of *Zanthoxylum alatum* plant extract on the corrosion of mild steel in 20%, 50%, and 88% aqueous orthophosphoric acid has been investigated by weight loss and electrochemical impedance spectroscopy (EIS). Plant extract was found to reduce the corrosion of steel more effectively in 88% than in 20% phosphoric acid [5]. An inhibition efficiency of 75.11% was observed with the extract of the leaves of *Nypa fruticans* Wurmb [6] for the corrosion of mild steel in hydrochloric acid solutions. El-Etre et al. examined some naturally occurring substances as corrosion inhibitors for different metals in various environments [7–11].

The application of extracts of henna, thyme, bgugaine, and inriine was investigated for their anticorrosion activity [12–15]. The effect of addition of bgugaine on steel corrosion in HCl is patented [16]. Saleh and El-Hosaray studied the peel of pomegranate [17] and beetroot [18, 19] as corrosion inhibitor for mild steel in acid media. Sanghvi et al. have investigated the anticorrosion activity of *Emblica officinalis*, *Terminalia chebula*, *Terminalia bolivia* [20], *Sapindus trifoliatus*, and *Acacia concinna* [21]. Corrosion inhibition has also been studied for the extracts of *Swertia angustifolia* [22], *Eucalyptus leaves* [23], *Eugenia jambolana* [24], *Pongamia*

Table 1: Water analysis.

Parameters	Result
Total dissolved salts (mg/L)	8000 ppm
pH	8.1
Calcium as Ca (mg/L)	160 ppm
Total hardness ($CaCO_3$ equivalent)	100 ppm
Magnesium as Mg (mg/L)	100 ppm
Sodium as Na (mg/L)	0
Chloride as Cl (mg/L)	17600 ppm
Sulphate as SO_4 (mg/L)	3254 ppm

glabra, *Annona squamosa* [25], *Acacia arabica* [26], *Carica papaya* [27], *Azadirachta indica* [28], and *Vernonia amygdalina* [29] for steel in acid media. The anticorrosion effect of *Andrographis paniculata* [30] and tea wastes [31] has been reported from our laboratories. Kliskic et al. analyzed aqueous extract of *Rosmarinus officinalis* [32] as corrosion inhibitor for aluminium alloy corrosion in chloride solution. Guar gum was analyzed for its anticorrosion activity by Abdallah [33]. Martinez and Stern have studied the inhibitory mechanism of low carbon steel corrosion of mimosa tannin in H_2SO_4 media [34]. Oguzie investigated the efficiency of *Telfairia occidentalis* extract as corrosion inhibitor in both HCl and H_2SO_4 media [35]. The extracts of chamomile, halfabar, black cumin, and kidney bean were analyzed for their inhibitive action of corrosion of steel in acid media by Abdel-Gaber et al. [36]. El-Hosary et al. [37] studied the corrosion inhibition of aluminium and zinc in HCl using *Hibiscus sabdariffa* extract. Rajendran et al. [38, 39] studied the corrosion inhibition of aluminium in rain water containing garlic extract. Aqueous extracts of rhizome powder [40], beet root extract [41], and *Hibiscus rosa-sinensis* have been used as corrosion inhibitors.

The present work is to examine the aqueous Jeera extract as an inhibitor for corrosion of carbon steel in solution-containing seawater. Weight loss measurements, potentiodynamic polarization, fluorescence, and UV-spectroscopy are used to study surface film formed on the metal surface.

2. Experimental

2.1. Preparation of Plant Extract of Jeera Extract. An aqueous plant extract was prepared by grinding 10 g of (*Cuminum cyminum*) extract, filtering, and making up to 100 using double distilled water.

2.2. Preparation of Specimens. Carbon steel specimens (0.0267% S, 0.06% P, 0.4% Mn, 0.1% C, and the rest iron) of dimensions 1.0 cm × 4.0 cm × 0.2 cm were polished to a mirror finish and degreased with trichloroethylene.

2.3. Weight Loss Method. Relevant data of seawater used in the study are given in Table 1.

Carbon steel specimens in triplicate were immersed in 100 mL of the solutions containing various concentrations of the inhibitor for one day. The weight of the specimens before and after immersion was determined using Shimadzu balance, model AY 62. The corrosion products were cleansed with Clarke's solution [42]. The inhibition efficiency (IE) was then calculated using

$$\text{I.E} = 100\left[1 - \left(\frac{W_2}{W_1}\right)\right]\%, \quad (1)$$

where W_1 = corrosion rate in the absence of the inhibitor, and W_2 = corrosion rate in the presence of the inhibitor.

2.4. Surface Examination. The carbon steel specimens were immersed in various test solutions for a period of one day, taken out, and dried. The nature of the film formed on the surface of metal specimens was analyzed by FTIR spectroscopic study.

2.4.1. FTIR Spectra. FTIR spectra were recorded in a Perkin-Elmer 1600 spectrophotometer. The film was carefully removed and mixed thoroughly with KBr made in to pellets, and FTIR spectra were recorded.

The fluorescence spectra of the film formed on the CS samples recorded with a Hitachi F-4500 fluorescence spectrophotometer.

2.5. Potentiodynamic Polarization. A three-electrode cell consisting of mild steel as working electrode (WE), a platinum wire counter electrode (CE), and a saturated reference electrode was used for measurements. All the potential values reported here were VS SCE. The working electrode was mechanically polished on various grades of emery sheet, rinsed with double distilled water, and degreased with trichloroethylene. Potentiodynamic polarization curves were recorded using an H and CH electrochemical work station impedance analyzer model CHI 660A provided with *iR* compensation option. Polarization curve measurements were carried out at scan rate of 0.01 V s^{-1}. The exposed area (1 cm^2) was mechanically polished with a series of emery sheets of variable grades. The samples were washed thoroughly with double distilled water before insertion in the cell. During the polarization study, the scan rate was 0.01 V s^{-1}, hold time at *Ef* was 0 s, and quiet time was 2 s.

2.6. Ac Impedance Measurements. The instrument used for polarization was used for AC impedance study also. The cell set-up was the same as that had been used for polarization measurements. The real part and imaginary part of the cell impedance were measured in ohms at various frequencies. The values of charge transfer resistance, R_{ct}, and the double layer capacitance, C_{dl}, were calculated. The equivalent electrical circuit diagram is shown in Figure 7.

3. Result and Discussion

3.1. Analysis of Results from Weight Loss Method. Table 2 shows the values of corrosion rates and inhibition efficiencies obtained from weight loss measurements of different concentrations of Jeera extract. 4 mL of the Jeera extract offered 93% corrosion inhibition efficiency to carbon steel immersed in 100 ml solution containing seawater. When the

Table 2: Corrosion rates (milligram per square decimeter per day) (Mdd) of carbon steel immersed in seawater in the presence and absence of inhibitors and the inhibition efficiencies (Ie) obtained by mass loss method. Inhibitor: Jeera extracts (*Cuminum cyminum*). Period of immersion: 1 day.

Jeera extract	CR/mdd	Inhibition efficiency IE%
0	21.52	—
2	18.49	45
4	2.35	93
6	4.37	87
8	14.46	51

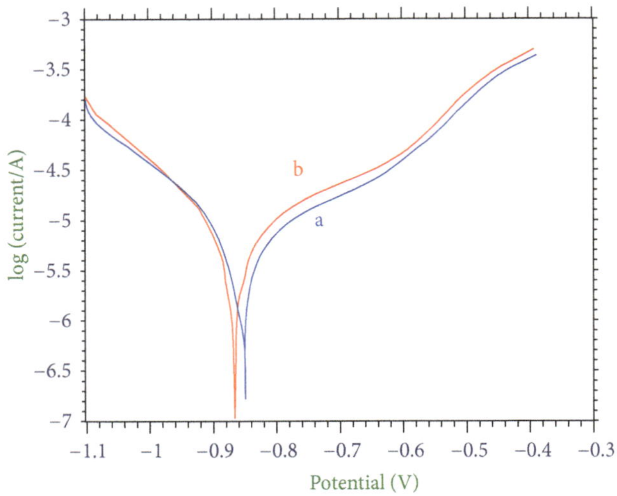

Figure 1: Polarization curves of carbon steel immersed in various test solutions: (a) seawater, (b) seawater + Jeera extract 4 mL.

concentration of Jeera extract was increased, the inhibition efficiency decreased, and the corrosion rate increased, and this is due to the fact that when higher concentrations of Jeera extract are added, the protective film (Fe^{2+}-*Cuminum cyminum* complex) formed on the metal surface goes in to the solution and thus destroying the protective film. It may be considered that the protective film formed may go into transpassive state, where the film is broken [43].

3.2. Analysis of Polarization Curves. A polarization study has been used to detect the formation of protective film on the metal surface [44]. When a protective film formed on the metal surface, the linear polarization resistance (LPR) increases, and the corrosion current (I_{corr}) decreases. The potentiodynamic polarization curves of carbon steel immersed in various test solution are shown in Figure 1. The corrosion parameters, namely, corrosion potential (E_{corr}) tafel slopes (b_c = cathodic; b_a = anodic), linear polarization resistance (LPR), and the corrosion current (I_{corr}) are given in Table 3 when carbon steel is immersed in seawater. The corrosion potential of seawater is −926 mV VS SCE. The formulation consisting of 4 mL of Jeera extract in presence of seawater solution shifts the corrosion potential −883 mV VS SCE. This suggested that anodic reaction is controlled predominantly. The (LPR) value increases from 51.67×10^{-2} to 63.67×10^{-2} ohm cm^2. This suggests that a protective film is formed

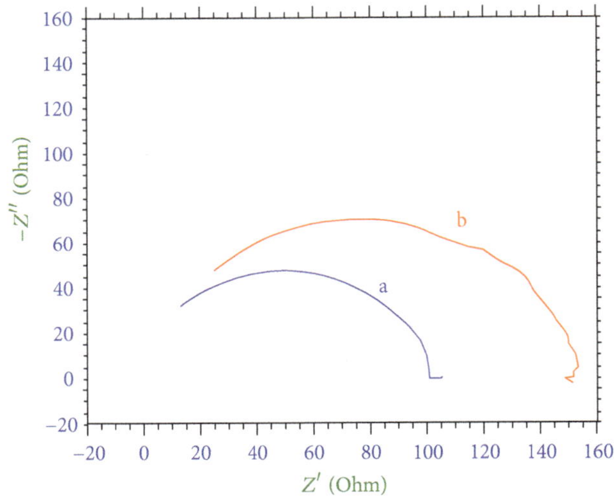

Figure 2: AC impedance spectra (Nyquist plots) of carbon steel immersed in various test solutions: (a) seawater, (b) seawater + *Phyllanthus amarus* 4 mL.

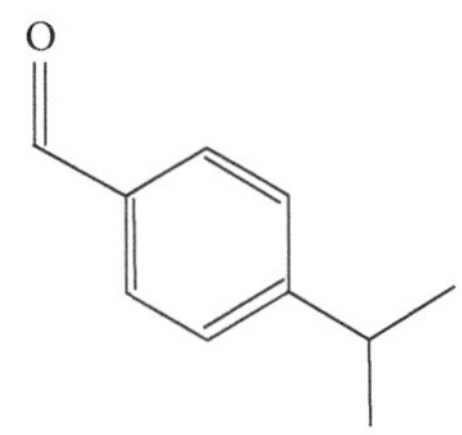

Scheme 1: Cuminaldehyde.

on the metal surface. Further the corrosion current decreases from 7.963×10^{-6} to 6.295×10^{-6} Acm^2 [45, 46].

3.3. Analysis of Ac Impedance Spectra. AC impedance spectra have been used to detect the formation of the film formed on the metal surface. If the protective film is formed, the charges transfer resistance increases, and double layer capacitance value decreases [40]. The AC impedance spectra of carbon steel immersed in various solutions are shown in Figure 2 (Nyquist) and Figure 3 (impedance-Bode plots). The AC impedance parameter, namely, charge transfer resistance (R_{ct}) and double layer capacitance (C_{dl}) are given in Table 4.

When carbon steel is immersed in aqueous solution containing seawater, the R_{ct} value is 88 ohm cm^2, and C_{dl} value is 1.027×10^{-7} F/cm^2. When 4 ml of (*Cuminum cyminum*) extract is added, the R_{ct} value increases from 88 ohm cm^2 to 116 ohm cm^2, and C_{dl} value decreases from 1.027×10^{-7} F/cm^2 to 0.7817×10^{-8}. This suggest that a protective film is formed on the metal surface of the metal. Further there is increase in impedance log (Z/ohm), value from 2.02 to 2.12 (derived from Bode plot shown in Figure 3).

3.4. Fourier Transfer-Infrared Spectra. The main constituent of Jeera extract is cuminaldehyde [47]. The structure of cuminaldehyde is shown in Scheme 1. It contain carboxylic group, cumin structure.

The *Cuminum cyminum* extract was evaporated to dryness to a solid mass. Its FTIR spectrum is shown in Figure 4(a). −C=C stretching frequency appeared at 2141 cm^{-1}.

TABLE 3: Corrosion parameters of carbon steel immersed in seawater in the absence and presence of inhibitors.

System	E_{corr} mV versus SCE	b_c mV/decade	b_a mV/decade	LPR Ohm cm^2	I_{corr} A/cm^2
Seawater	−926	152	236	65.32×10^{-2}	6.164×10^{-6}
Seawater + (*Cuminum cyminum*) extract (4 mL)	−883	153	247	51.67×10^{-2}	0.7963×10^{-7}

TABLE 4: AC impedance parameters of carbon steel immersed in various solutions. Inhibitors: Jeera extract.

System	R_{ct} ohm cm^2	C_{dl} F/cm^2	Impedance value log (z/ohm)
Seawater	88	1.027×10^{-7}	2.02
CC (4 mL) + Seawater	128	7.0703×10^{-8}	2.17

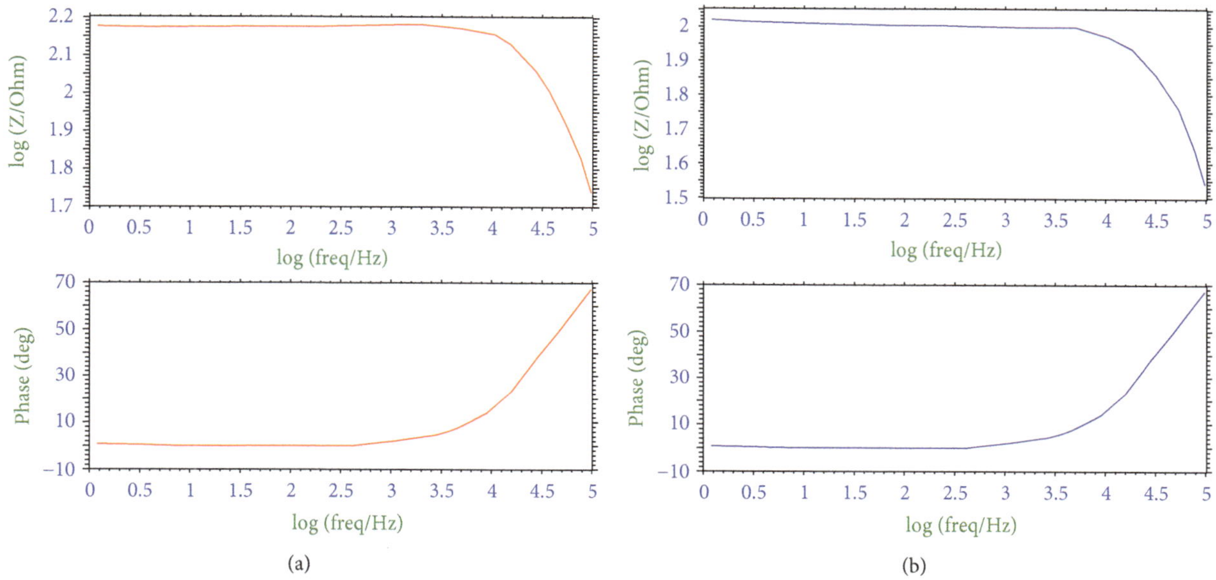

FIGURE 3: Bode plot of carbon steel immersed in various solutions (Bode plots). (a) Seawater, (b) Seawater + 4 ml Jeera extract.

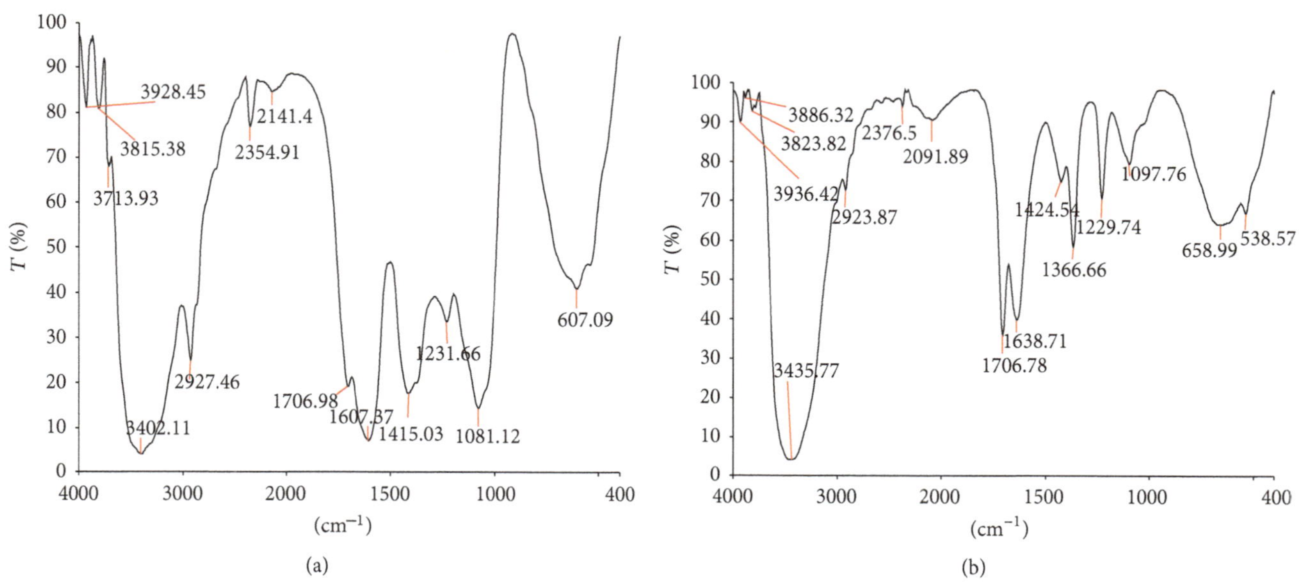

FIGURE 4: FTIR spectra. (a) Aqueous solution of Jeera extract, (b) Aqueous solution of seawater + Jeera extract 4 mL.

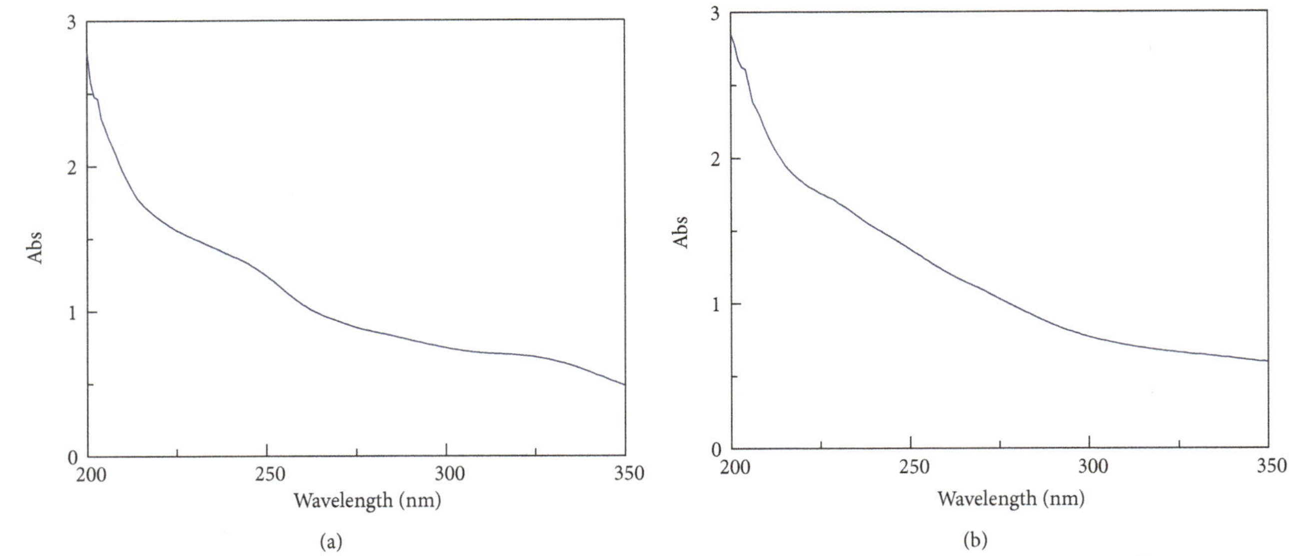

FIGURE 5: UV-spectra. (a) Aqueous solution of Jeera extract, (b) Aqueous solution of Jeera extract + Fe^{2+}.

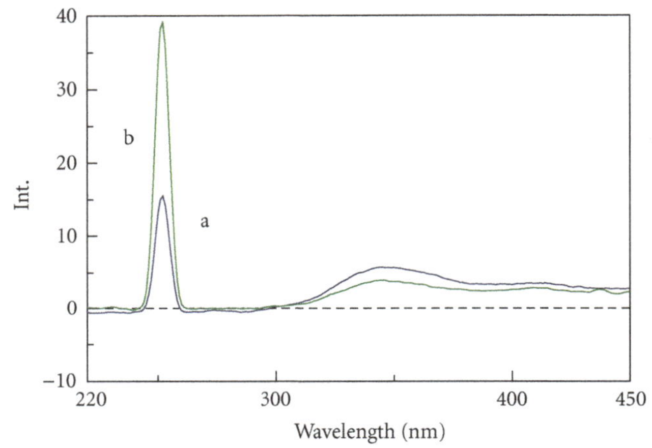

FIGURE 6: Fluorescence spectra. (a) Aqueous solution of Jeera extract, (b) film formed on the metal surface after immersion Jeera extract + 4 ml *Cuminum cyminum*.

The aromatic -CH stretching frequency appeared at 2927 cm^{-1}. -C=O stretching frequency appeared at 1607 cm^{-1} [48, 49].

The FTIR spectrum of the protective film formed on the surface of the metal after immersion in the aqueous solution containing seawater and Seawater containing 4 mL of Jeera extract is shown in Figure 4(b). -OH stretching frequency appeared at 3402 cm^{-1} to 3435 cm^{-1}. -C=C stretching frequency appeared at 2141 cm^{-1} to 2091 cm^{-1}. The aromatic -CH stretching frequency appeared at 2927 cm^{-1} to 2923 cm^{-1}. -C=O stretching frequency appeared at 1607 cm^{-1} to 1638 cm^{-1}.

3.5. Fluorescence Spectra. The UV-visible adsorption spectrum of aqueous solution of Cuminum cyminum and Fe^{2+} is shown in Figure 5. Peaks appear at 228 nm, 268 nm. When the

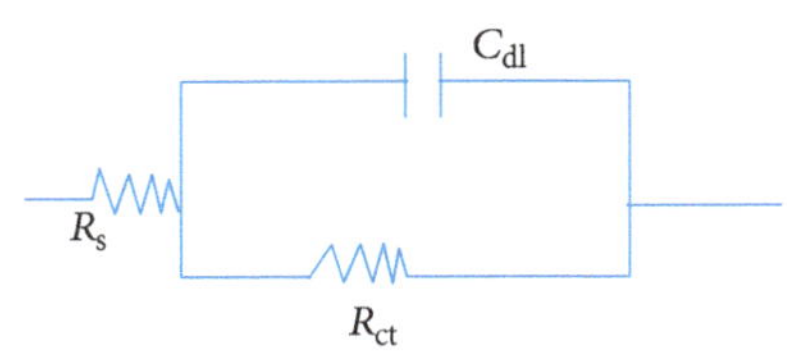

FIGURE 7: Equivalent electrical circuit diagram. R_s: solution resistance; R_{ct}: charge transfer resistance; C_{dl}: double layer capacitance.

Fe^{2+} sis added to the aqueous solution of *Cuminum cyminum*, peak appears at 225 nm and 265 mm [46].

The fluorescence spectrum (λ_{ex} = 225 nm) of an aqueous solution of Fe^{2+}-*Cuminum cyminum* is shown in Figure 6(a). A peak appears at 252 nm. This due to that Fe^{2+}-active principle complex is formed in solution [40].

The fluorescence spectrum (λ_{ex} = 225 nm) of the film formed on the surface of the metal after immersion in the solution containing seawater and 4 ml of *Cuminum cyminum* extract is shown in Figure 6(b). The peaks appeared at 252 nm confirming the presence of Fe^{2+}-*Cuminum cyminum* extract complex formed on the metal surface.

4. Conclusion

The corrosion inhibition by aqueous solution of Jeera extract in the presence and absence of inhibitor was studied by weight-loss study and electrochemical measurements. The results show that inhibitor has the ability of reducing the corrosion rate of carbon steel in aqueous solution containing Jeera extract, and it acts as an anodic inhibitor. This effectiveness is confirmed by electrochemical impedance spectra and potential polarization analysis.

Acknowledgment

The authors are thankful to DRDO, India, for help and encouragement and to Mrs R. Nagalakshmi for useful suggestions.

References

[1] A. A. El Hosary, R. M. Saleh, and A. M. Shams El Din, "Corrosion inhibition by naturally occurringsubstances-I. The effect of Hibiscus subdariffa (karkade) extract on the dissolution of Al and Zn," *Corrosion Science*, vol. 12, no. 12, pp. 897–904, 1972.

[2] A. A. El Hosary and R. M. Saleh, *Progress in Understanding and Prevention of Corrosion*, vol. 2, The Institute of Materials, London, UK, 1993.

[3] C. A. Loto, "The effect of Vernonia amygdalina (bitter leaf) solution extract on the corrosion inhibition of mild steel," *Nigerian Corrosion Journal*, vol. 19, pp. 20–28, 1998.

[4] G. O. Avwiri and F. O. Igho, "Inhibitive action of Vernonia amygdalina on the corrosion of aluminium alloys in acidic media," *Materials Letters*, vol. 57, no. 22-23, pp. 3705–3711, 2003.

[5] G. Gunasekaran and L. R. Chauhan, "Eco friendly inhibitor for corrosion inhibition of mild steel in phosphoric acid medium," *Electrochimica Acta*, vol. 49, no. 25, pp. 4387–4395, 2004.

[6] K. O. Orubite and N. C. Oforka, "Inhibition of the corrosion of mild steel in hydrochloric acid solutions by the extracts of leaves of Nypa fruticans Wurmb," *Materials Letters*, vol. 58, no. 11, pp. 1768–1772, 2004.

[7] A. Y. El-Etre, "Natural honey as corrosion inhibitor for metals and alloys. I. Copper in neutral aqueous solution," *Corrosion Science*, vol. 40, no. 11, pp. 1845–1850, 1998.

[8] A. Y. El-Etre and M. Abdallah, "Natural honey as corrosion inhibitor for metals and alloys. II. C-steel in high saline water," *Corrosion Science*, vol. 42, no. 4, pp. 738–738, 2000.

[9] A. Y. El-Etre, "Inhibition of acid corrosion of aluminum using vanillin," *Corrosion Science*, vol. 43, no. 6, pp. 1031–1039, 2001.

[10] A. Y. El-Etre, "Inhibition of aluminum corrosion using *Opuntia* extract," *Corrosion Science*, vol. 45, no. 11, pp. 2485–2495, 2003.

[11] A. Y. El-Etre, M. Abdallah, and Z. E. El-Tantawy, "Corrosion inhibition of some metals using lawsonia extract," *Corrosion Science*, vol. 47, no. 2, pp. 385–395, 2005.

[12] A. Chetouani and B. Hammouti, "Corrosion inhibition of iron in hydrochloric acid solutions by naturally henna," *Bulletin of Electrochemistry*, vol. 19, no. 1, pp. 23–25, 2003.

[13] Chetouani, "Etude Numerique de problemes non linear at application aux problemes de dynamique de population," *[Ph.D. thesis]*, University Oujda, Oujda, Morocco, 2003.

[14] B. Hammouti, S. Kertit, and M. Mellhaoui, "Bgugaine: a natural pyrollidine alkaloid product as corrosion inhibitor of iron in HCl medium," *Bulletin of Electrochemistry*, vol. 11, p. 553, 1995.

[15] B. Hammouti, S. Kertit, and A. Melhaoui, "Electrochemical behaviour of bgugaine as a corrosion inhibitor of iron in 1 M HCl," *Bulletin of Electrochemistry*, vol. 13, no. 3, pp. 97–98, 1997.

[16] S. Kertit, B. Hammouti, and M. Mellhaoui, "Bgugaine comme inhibitor de corrosion du fer en milieu acide chlorhydrique," Moroccan Patent no. 23910, 1995.

[17] R. M. Saleh and A. A. El-Hosaray, "Corrosion inhibition by naturally occuring substance-I the effect of Hibiscus subdariffa (Karkade) extract on the dissolution of Al and Zn," in *Proceedings of the 13th Seminar on Electrochemistry*, CECRI, Karaikudi, India, 1972.

[18] A. El-Hosary, M. M. Gowish, and R. M. Saleh, "Evaluation of quinolines as inhibitors by electrochemical techniques," in *Proceedings of the 2nd International Symposium and Oriented Basic Electrochemistry*, vol. 7, SAEST, IIT, Madras, Technical Session, 1980, paper 6.24.

[19] J. Arockia Selvi, S. Rajendran, V. Ganga Sri, A. John Amalraj, and B. Narayanasamy, "Corrosion inhibition by beet root extract," *Portugaliae Electrochimica Acta*, vol. 27, no. 1, pp. 1–11, 2009.

[20] M. J. Sanghvi, S. K. Shukla, A. N. Mishra, M. R. Padh, and G. N. Mehta, *5th National Congress on Corrosion Control*, New Delhi, India, 1995.

[21] M. J. Sanghvi, S. K. Shukla, A. N. Mishra, M. R. Padh, and G. N. Mehta, "Formulation of 5% HCl corrosion inhibition for mild steel using weight loss and DC polarization technique," in *International Conference on Maintenance, Inspection, Corrosion, Materials Engineering and Plant Reliability Baroda (MICMEP '99)*, vol. 84–87, p. 87, Baroda, India, February 1999.

[22] S. J. Zakvi and G. N. Mehta, "Acid corrosion of mild steel and its inhibition by swertia aungustifolia-study by electrochemical techniques," *Transactions of the SAEST*, vol. 23, no. 4, pp. 407–410, 1988.

[23] P. Kar, A. Hussein, G. Varkey, and G. Singh, "Corrosion inhibition of mild steel in hydrochloric acid by acid entacts of sapindus trifolianus, acaciaconcian and Trifla, Trans," *MFAI*, vol. 51, pp. 143–147, 1996.

[24] S. A. Verma and G. N. Mehta, "Effects of acid extracts of powered seeds of Eugenia Jambolans on corrosion of mild steel in HCl-study by DC polarisation techniques," *Transactions of the SAEST*, vol. 32, no. 4, pp. 89–93, 1997.

[25] P. Sakthivel, P. V. Nirmala, S. Umamaheswari et al., "Corrosion inhibition of mild steel by extracts of Pongamia Glabra and Annona Squamosa in acidic media," *Bulletin of Electrochemistry*, vol. 15, no. 2, pp. 83–86, 1999.

[26] S. Verma and G. N. Mehta, "Effect of acid extracts of acacia arabica on acid corrosion of mild steel," *Bulletin of Electrochemistry*, vol. 15, no. 2, pp. 67–70, 1999.

[27] U. J. Ekpe, E. E. Ebenso, and U. J. Ibok, "Inhibitory action of Azadirachta indica leaves extract on the corrosion of mild steel in H_2SO_4," *West African Journal of Biological and Applied Chemistry*, vol. 37, pp. 13–30, 1994.

[28] U. J. Ekpe, E. E. Ebenso, and U. J. I. Ibok, *Journal of the West African Science Association*, vol. 37, p. 13, 1994.

[29] A. Loto, "The effect of bitterleaf extracts on corrosion of mild steel in 0.5M HCl and H_2SO_4 solutions," *Nigerian Corrosion Journal*, pp. 19–20, 1998.

[30] S. P. Ramesh, K. P. Vinod Kumar, and M. G. Sethuraman, "Extract of andrographis paniculata as corrosion inhibitor of mild steel in acid medium," *Bulletin of Electrochemistry*, vol. 17, no. 3, pp. 141–144, 2001.

[31] M. G. Sethuraman, P. Vadivel, and K. P. Vinod Kumar, "Tea wastes as corrosion inhibitor for mild steel in acid medium," *Journal of the Electrochemical Society of India*, vol. 50, p. 143, 2001.

[32] M. Kliskic, J. Radosevic, S. Gudic, and V. Katalinic, "Aqueous extract of Rosmarinus officinalis L. as inhibitor of Al-Mg alloy corrosion in chloride solution ," *Journal of Applied Electrochemistry*, vol. 30, no. 7, pp. 823–830, 2000.

[33] M. Abdallah, "Guar gum as corrosion inhibitor for carbon steel in sulfuric acid solutions," *Portugaliae Electrochimica Acta*, vol. 22, pp. 161–175, 2004.

[34] S. Martinez and I. Stern, "Inhibitory mechanism of low-carbon steel corrosion by mimosa tannin in sulphuric acid solutions ," *Journal of Applied Electrochemistry*, vol. 31, no. 9, pp. 973–978, 2001.

[35] E. Oguzie, "Inhibition of acid corrosion of mild steel by Telfaria occidentalis," *Pigment and Resin Technology*, vol. 34, no. 6, pp. 321–326, 2005.

[36] M. Abdel-Gaber, B. A. Abd-El-Nabey, I. M. Sidahmed, A. M. El-Zayaday, and M. Saa dawy, "Inhibitive action of some plant extracts on the corrosion of steel in acidic media," *Corrosion Science*, vol. 48, no. 9, pp. 2765–2779, 2006.

[37] A. El-Hosary, R. M. Saleh, and A. M. Sharns El Din, "Corrosion inhibition by naturally occurringsubstances-I. The effect of *Hibiscus subdariffa* (karkade) extract on the dissolution of Al and Zn," *Corrosion Science*, vol. 12, no. 12, pp. 897–904, 1972.

[38] S. Rajendran, S. Muthulakshmi, R. Rajeswari, and A. Vijitha, "Emerging trends in corrosion science and engineering," *Journal of the Electrochemical Society of India*, vol. 54, no. 2, p. 61.

[39] S. Rajendran and K. Anuradha, "Corrosion behaviour of aluminium in rain water containing garlic extract," *Surface Engineering*, vol. 21, no. 3, pp. 229–231, 2005.

[40] S. Rajendran, S. Shanmugapriya, T. Rajalakshmi, and A. J. Amal Raj, "Corrosion inhibition by an aqueous extract of rhizome powder," *Corrosion*, vol. 61, no. 7, pp. 685–692, 2005.

[41] S. Rajendran, J. Arokiaselvi, and V. Gangasree, "Corrosion inhibition by beet root extract," *Protugalia Electrochemica Acta*, vol. 27, no. 1, pp. 1–11, 2009.

[42] G. Wranglen, *Introduction to Corrosion and Protection of Metals*, vol. 236, Chapman & Hall, London, UK, 1985.

[43] J. Sathyabama, R. Susai, S. J. Arokia, and A. A. John, "Methyl orange as corrosion inhibitor for carbon steel in well water," *Indian Journal of Chemical Technology*, vol. 15, pp. 462–466, 2008.

[44] S. Rajendran, S. P. Sridevi, N. Anthony, A. J. Amalraj, and M. Sundaravadivelu, "Corrosion behaviour of carbon steel in polyvinyl alcohol," *Anti-Corrosion Methods and Materials*, vol. 52, no. 2, pp. 102–107, 2005.

[45] F. R. Selvarani, S. Santhamadharasi, J. W. Sahayaraj, A. J. Amalraj, and S. Rajendran, "Synergistic effect of succinic acid and Zn^{2+} in controlling corrosion of carbon steel," *Bulletin of Electrochemistry*, vol. 20, no. 12, pp. 561–565, 2004.

[46] S. Rajendran, S. M. Reenkala, N. Anthony, and R. Ramaraj, "Synergistic corrosion inhibition by the sodium dodecylsulphate-Zn^{2+} system," *Corrosion Science*, vol. 44, no. 10, pp. 2243–2252, 2002.

[47] N. S. Iacobellis, P. Lo Cantore, F. Capasso, and F. Senatore, "Antibacterial activity of *Cuminum cyminum* L. and *Carum carvi* L. essential oils," *Journal of Agricultural and Food Chemistry*, vol. 53, no. 1, pp. 57–61, 2005.

[48] S. Agnesia Kanimozhi and S. Rajendran, "Inhibitive properties of sodium tungstate-Zn^{2+} system and its synergism with HEDP," *International Journal of Electrochemical Science*, vol. 4, pp. 353–368, 2009.

[49] R. M. Silver stein, G. C. Bassler, and T. C. Morrill, *Spectroscopic Identification of Organic Compound*, vol. 95, John Wiley & Sons, New York, NY, USA, 1986.

13

Evaluation of Corrosion Growth on SS304 Based on Textural and Color Features from Image Analysis

Ramana M. Pidaparti, Brian Hinderliter, and Darshan Maskey

Department of Mechanical and Nuclear Engineering, Virginia Commonwealth University, 401 West Main Street, Richmond, VA 23284, USA

Correspondence should be addressed to Ramana M. Pidaparti; rmpidaparti@vcu.edu

Academic Editors: W. Fürbeth and R. Salghi

Corrosion surface damage in the form of pitting and microcracks is observed in many systems and affects the integrity of steel structures in nuclear, civil, and industrial engineering. In order to gain a better understanding and develop nondestructive and automatic detection/assessment of corrosion damage and its growth, an image analysis based on texture using wavelet transforms and color features was carried out. Experiments were conducted on steel 304 panels under three different electrolyte solutions, and periodic scans were used to obtain the images for analysis over time. The results obtained from the image analysis are presented to illustrate the metrics which best characterize early stage corrosion damage growth behavior. The results obtained indicate that textural features in combination with color features are more effective and may be used for correlating service/failure conditions based on corrosion morphology.

1. Introduction

Engineering components made from structural steel 304 metals are being used in many industries, commercial and domestic fields, because of their chemical durability, mechanical properties, weldability, good corrosion, and heat resistant properties. Corrosion of stainless steel in aggressive environmental conditions in particular is a fundamental concern to academia and industry due to the destructive nature of corrosion on mechanical properties of components. Ships, storage tanks, bridges, and pipelines commonly made from structural steel are not sufficiently resistant to corrosion in their operating environments which impacts corrosion costs associated with maintenance and also safety risks. Therefore, corrosion monitoring is an important issue in detecting corrosion damage and its growth before failures occur [1–3].

There are many different experimental and analysis methods used for corrosion inspection and monitoring purposes. These include mechanical measurements (weight loss), chemical analysis, and visual inspections. In visual inspections, the corrosion damage identification requires an expert who can clearly demarcate the corrosion based on experience as well as types of corrosion, with red rust as a common experience. Usually, the corrosion process produces rough surfaces, and image analysis based on textural features can be used for quantification and discriminate corrosion extent and type. Several authors investigated automatic corrosion detection using image processing techniques [4–7]. Computer image processing involves definition and development of techniques and algorithms for processing pictorial data and its pixel value. Parameters related to the image are extracted and analyzed correlating with the corrosion time. In addition to textural features, color changes of metallic surfaces are also used for the detection of corrosion [8] due to various metal oxides and other corrosion products.

Image analysis has been used to characterize corrosion morphology in aluminium alloys subjected to a variety of environmental conditions. Wang and Song [9] used wavelet packet decomposition energies of images at different subbands as features to study the atmospheric corrosion behavior of zinc samples. They acquired a relation between the selected image features and the corrosion weight loss. Tao et al. [10] analyzed the atmospheric corrosion of field exposed high strength aluminum alloys. They used wavelet packed decomposition energies at different subbands to estimate the corrosion loss of five types of aluminum alloys. It was found that the values of subimage energies decreased with increasing exposure time.

In order to qualitatively and quantitatively characterize corrosion damage growth, an investigation of image analysis was carried out. Specifically, in this paper, we explored the idea of developing textural/color features that are sensitive to corrosion images, with a low-tech tool using a commercial color scanner. An investigation of textural features from wavelet transforms and color features through image analysis was carried out in order to identify the metrics for corrosion damage growth detection under three different electrolyte solutions. Our approach is to integrate functionality reducing properties due to material compromise (textural features) with local aspects (color values) to describe corrosion damage growth with increasing time. Results obtained from the analysis are presented and discussed.

2. Experiments

Stainless steel 304 metal samples with dimensions $6'' \times 3''$ and thickness $0.020''$ were obtained from Q-Panel. Initially, the specimens were first inspected visually to make sure they are in good condition. They were washed with clean water and ground with sand paper 300 grit first and then 600 grit. Then they were washed again and dried in a desiccator for about 10 minutes. Afterwards, these specimens were immersed in three different solutions, sodium chloride (saturated), distilled water, and sodium sulphate (1% weight). The specimens were kept in these solutions completely immersed for various times. After a specified time, the specimen is taken for image analysis using a 3200 dpi resolution scanner. The specimens were scanned at 5, 10, 15, 30, 45, 60, 75, 90, 105, 120, and 135 minutes in order to quantify the corrosion damage growth then returned to the solution until the next characterization. The experimental setup and process are shown in Figure 1. Typical scanned images and subimages were used in image analysis. This setup allows the evolution of the surface corrosion, along with some pitting corrosion, to be tracked, with three color components of red, green, and blue, energy, and entropy.

3. Image Analysis Methods

Several image analysis methods are available for characterizing corrosion damage surface morphology. In this study, texture features from wavelet transform image processing as well as color attributes from optical imaging were used. These methods are described below briefly.

3.1. Textural Features from Wavelet Packet Transform. In order to extract textural features from scanned images, multiresolution wavelet transforms can be used, the details of which can be found in [11]. As an extension of the standard wavelets, wavelet packets which represent a generalization of the multiresolution analysis use the entire family of subband decompositions to generate a complete representation of images. Two-dimensional discrete wavelet packet transform (DWPT) decomposition allows us to analyze an image simultaneously at different resolution levels and orientations. In 2D DWPT method, an image is decomposed into one approximation and three detail images. The approximation and the detail images are then decomposed into a second-level approximation and detail images, and the process is repeated. The standard 2D DWPT can be implemented with a low-pass filter and a high-pass filter [11]. The 2D DWPT of a $N \times M$ discrete image A up to level $P+1$ ($P \leq \min(\log_2 N, \log_2 m)$) is recursively defined in terms of the coefficients at level p as follows:

$$\begin{aligned}
C_{4k,(i,j)}^{p+1} &= \sum_m \sum_n h(m)\, h(n)\, C_{k,(m+2i,n+2j)}^{p}, \\
C_{4k+1,(i,j)}^{p+1} &= \sum_m \sum_n h(m)\, g(n)\, C_{k,(m+2i,n+2j)}^{p}, \\
C_{4k+2,(i,j)}^{p+1} &= \sum_m \sum_n g(m)\, h(n)\, C_{k,(m+2i,n+2j)}^{p}, \\
C_{4k+3,(i,j)}^{p+1} &= \sum_m \sum_n g(m)\, g(n)\, C_{k,(m+2i,n+2j)}^{p},
\end{aligned} \tag{1}$$

where C_0^0 is the image A and k is an index of the nodes in the wavelet packet tree denoting each subband. At each step, the image C_k^p is decomposed into four-quarter-size images C_{4k}^{p+1}, C_{4k+1}^{p+1}, C_{4k+2}^{p+1}, and C_{4k+3}^{p+1}.

The Shannon entropy in different subbands is computed from the subband coefficient matrix as

$$\begin{aligned}
\text{Entropy}_p(k) &= -\sum_i \sum_j \left| C_{k,(i,j)}^{p} \right|^2 \log \left| C_{k,(i,j)}^{p} \right|^2, \\
\text{Energy}_p(k) &= \sum_i \sum_j \left| C_{k,(i,j)}^{p} \right|^2,
\end{aligned} \tag{2}$$

where $\text{Energy}_p(k)$ and $\text{Entropy}_p(k)$ are the energy and entropy of the image projected onto the subspace at node (p, k). The entropy of each subband provides a measure of the image characteristics in that subband. The energy distribution has important discriminatory properties for images and as such can be used as a feature for texture classification. From the equations above, it follows that the wavelet entropy is minimum when the image represents an ordered activity characterized by a narrow frequency distribution, whereas the entropy is high when an image contains a broad spectrum of frequency distribution. In this study, Haar mother wavelet was employed to decompose the corrosion images to a second level of decomposition, and Shannon entropies of images at each subband have been used as image features. In the corrosion damage region, both the energy and entropy features describe how the pixel values are distributed within the corroded region and indicate the degree of corrosion reflecting the texture of corrosion surface.

3.2. Color Features Analysis. Three primary color features, namely, red (R), green (G), and blue (B), which are commonly found in the spectral distribution of light, were related to

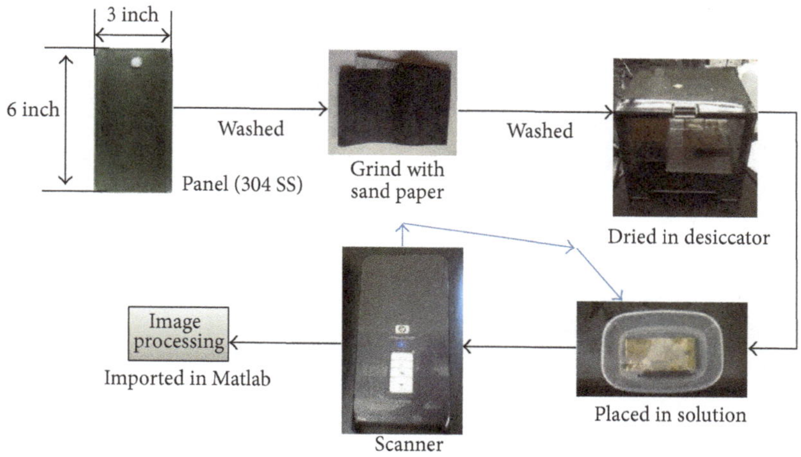

FIGURE 1: An overview of corrosion damage growth evaluation procedure.

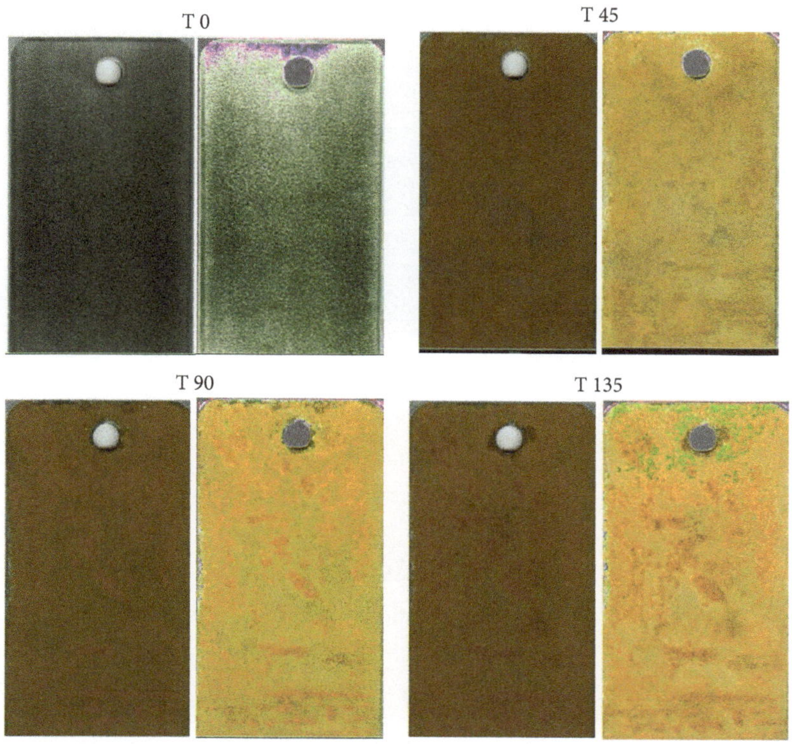

(a) Energy feature from wavelet transformation for corroded specimens

(b) Entropy feature from wavelet transformation for corroded specimens

FIGURE 2: Typical scanned images (a) and wavelet transformed images (b) at four different times for specimens corroded in NaCl solution.

corrosion damage, specifically iron oxide ions. "Imtool" in Matlab software [12] was used for obtaining the RGB values of each image along a specific profile line. RGB value of each pixel was calculated, and average values were correlated with respect to corrosion damage time. Subimages with diagonal and bisector lines along visible corrosion damage were used as the fundamental profile line to extract RGB values of each image. The R, G, and B values obtained are presented and correlated in order to quantify the corrosion damage growth. In addition, a color saturation range of 128 (8 bit color density) was selected. The first saturation layer of images at their respective times was imported in Matlab to observe the highest range of pixel for a quantification of 128.

4. Results and Discussion

Typical scanned images of specimens in NaCl solution along with wavelet transformed images of specimens at four different times are shown in Figure 2. As shown in the figure, it can be seen that it is very difficult to identify the corrosion

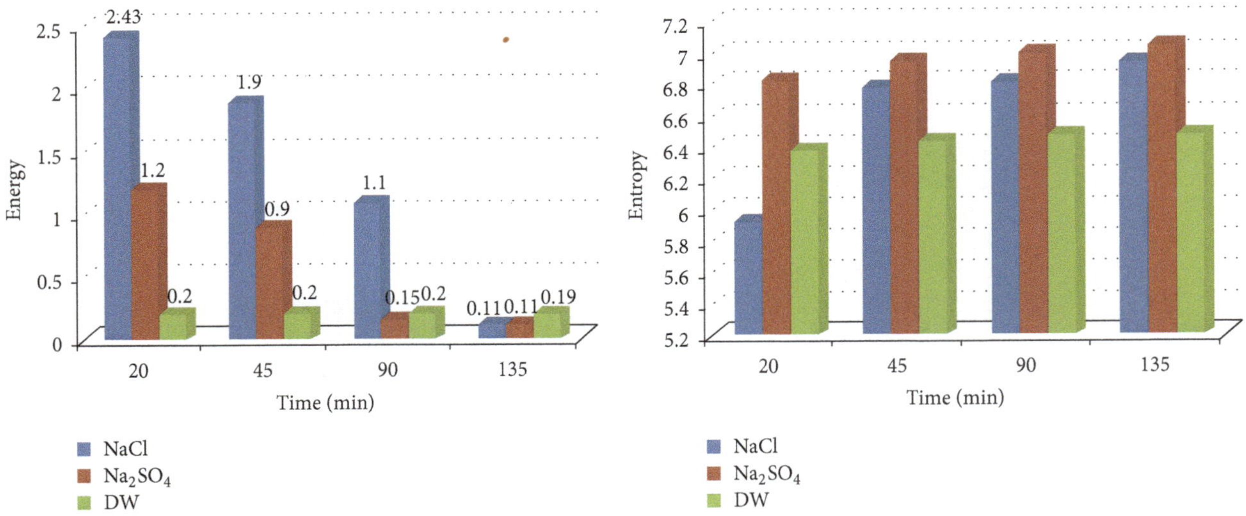

(a) Energy feature from wavelet transformation for corroded specimens (b) Entropy feature from wavelet transformation for corroded specimens

FIGURE 3: Energy (a) and Entropy (b) of corrosion damage growth calculated from wavelet image analysis under different solutions.

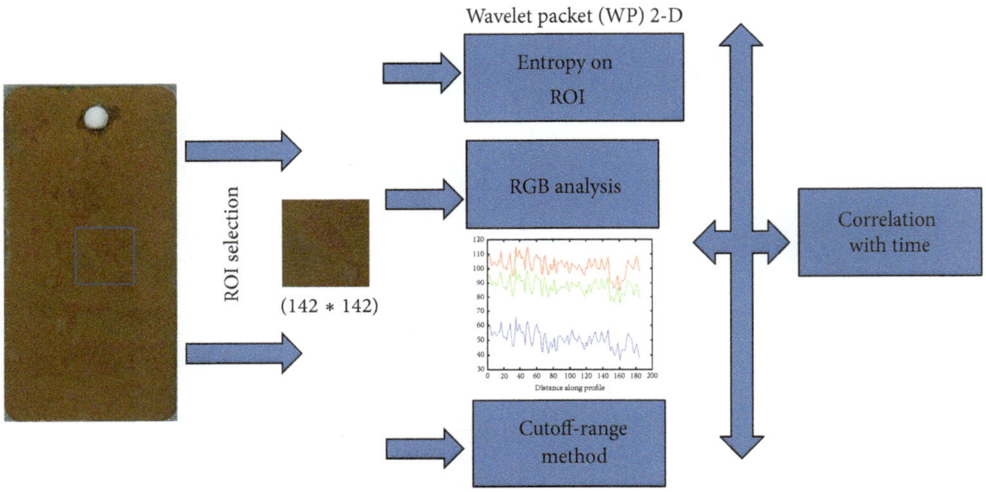

FIGURE 4: An overview of the approach for correlation of corrosion growth evaluation based on ROI.

damage regions (Figure 2(a)). To further improve the damage identification, a wavelet transformation is applied to those images and selective features, specifically if energy and entropy values were extracted. The wavelet transformed image is shown in Figure 2(b). It can be seen from Figure 2 that some of the corrosion damage features are now evident in the wavelet transformed images. The energy feature values of corroded specimens under three different solutions are presented in Figure 3(a). In general, the energy values decrease with increasing corrosion time. However, the energy values are higher for NaCl solution and lower for deionized water. In the first 90 minutes, the rate of decrease in energy value is about 55% for NaCl solution, whereas it is about 88% for sodium sulphate solution. However, the energy values are pretty much constant for distilled water. The results presented in Figure 3(a) show that a decrease in energy values illustrate the corrosion damage, which is higher under NaCl solution in comparison to the sodium sulphate solution.

The entropy feature values of corroded specimens under three different solutions are presented in Figure 3(b). In general, the entropy values increase with increasing corrosion time. However, the entropy values are lower for NaCl solution and higher for sodium sulphate solution. The entropy values are lower and constant for distilled water condition. In the first 90 minutes, the rate of increase in entropy value is about 17% for NaCl solution, whereas it is about 2% for sodium sulphate solution. However, the entropy values are pretty much constant for distilled water. The results presented in Figure 3(b) show that an increase in entropy values represents corrosion damage, which is higher under sodium sulphate solution in comparison to NaCl solution. In order to quantify local defects such as pits, an approach based on region of interest (ROI) with entropy, RGB analysis, and cutoff-range method was used as shown in Figure 4. The entropy values specific to a corroded of region of interest (ROI) under NaCl solution

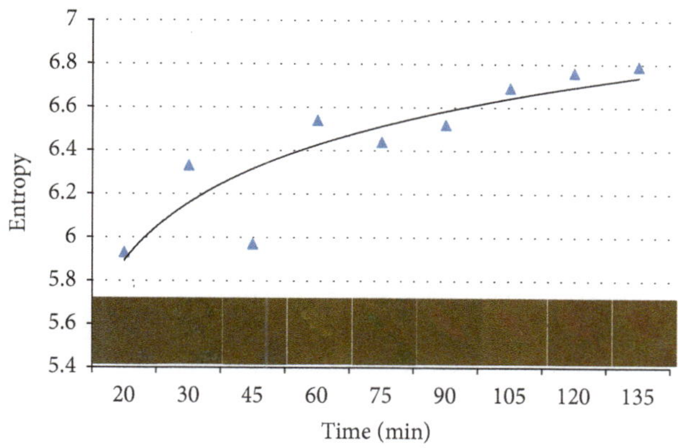

FIGURE 5: Entropy values calculated based on ROI.

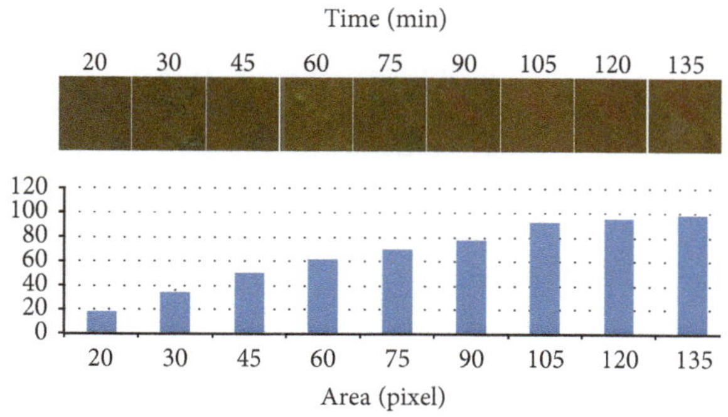

FIGURE 6: Pit initiation and growth with increasing time for a specific ROI.

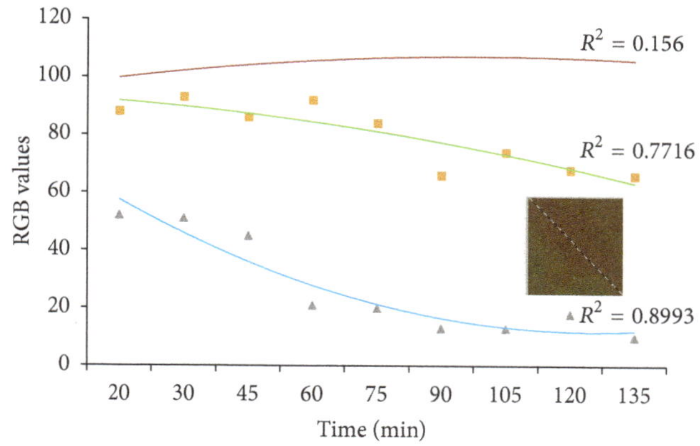

FIGURE 7: Average RGB values of corroded specimen along diagonal profile damage line.

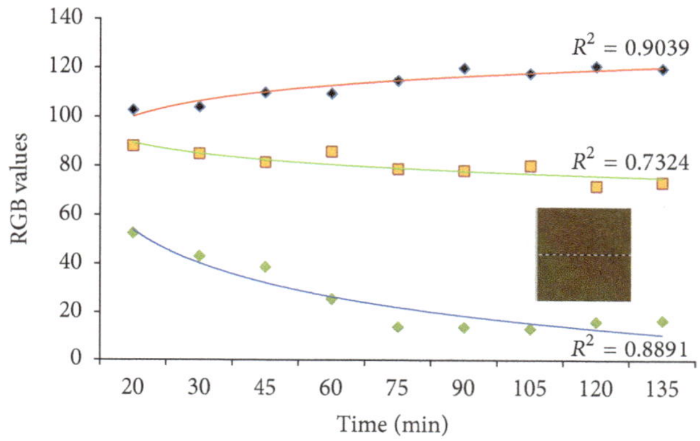

FIGURE 8: Average RGB values of corroded specimen along bisection profile damage line.

were analyzed, and the results are presented in Figure 5. It can be seen from Figure 5 that entropy values increase about 14% for about 135 minutes from the initial 20 minutes. The pit nucleation (area in pixels) with time for a specific selected pit profile is shown in Figure 6. It can be seen from Figure 6 that the pit area increases (in terms of pixels) with time and reaches a plateau. The results of energy and entropy features presented in Figures 3–6 illustrate that entropy is more sensitive and can be correlated to corrosion damage with increasing time.

The results of RGB analysis of specimens under NaCl solution at different times are shown in Figures 7 and 8 for two specific profiles, one along a diagonal line and the other a straight line profile. It can be seen from both Figures 7 and 8 that G and B color values decrease whereas the R color increases with increasing corrosion time. It is interesting to note that the R color, which relates to iron oxide ions, might give a better indication of corrosion damage with increasing time. In comparing the diagonal profile with the straight line profile, the R color values are more sensitive for the straight line profile. In order to detect corrosion damage based on RGB color values, the cutoff-range method was used based on specific pixel range. The overall cutoff-range method is shown in Figure 9. Initially, a color saturation range of 25 was selected for the cutoff. The first saturation layer of images at their respective time was analyzed using Matlab to observe the highest range of pixels. The result of applying the cutoff-range method in first saturation layer is presented in Figure 10. It can be seen from Figure 10(b) that the pixel range of 101–125 showed increasing trend whereas pixel range of 76–100 showed decreasing trend. No distinct features were seen at other pixel ranges. This result indicates that, at the pixel range of 100–125, more corrosive pixel values are increasing and the corrosion in this particular case must fall around this range. Overall, the results presented in Figure 10 suggest that selecting a suitable cutoff-range in pixel values can be used as an indication of corrosion damage growth with increasing time.

5. Conclusions

In this paper, we investigated textural features from wavelet transforms and color features through image analysis in order to identify the metrics for corrosion damage growth detection under three different electrolyte solutions. We explored the feasibility of studying textural/color features that are sensitive to corrosion images, with a low-tech tool, a commercial desktop scanner, and a system that can easily be replicated in a portable device. Our approach of integrating global textural features with local features (color values) was helpful in assessing corrosion damage growth with increasing time.

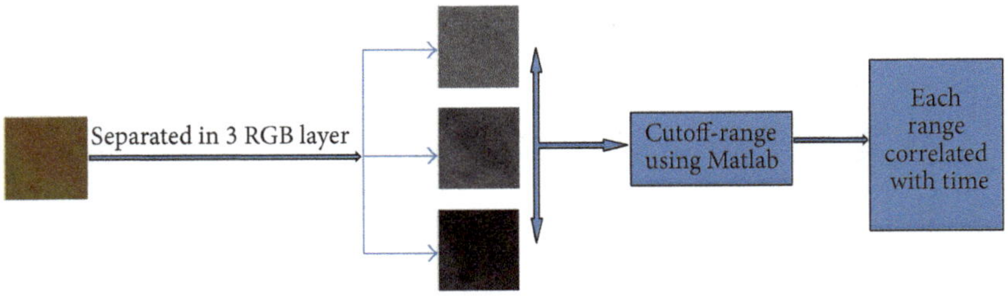

FIGURE 9: Schematic of cutoff-range approach.

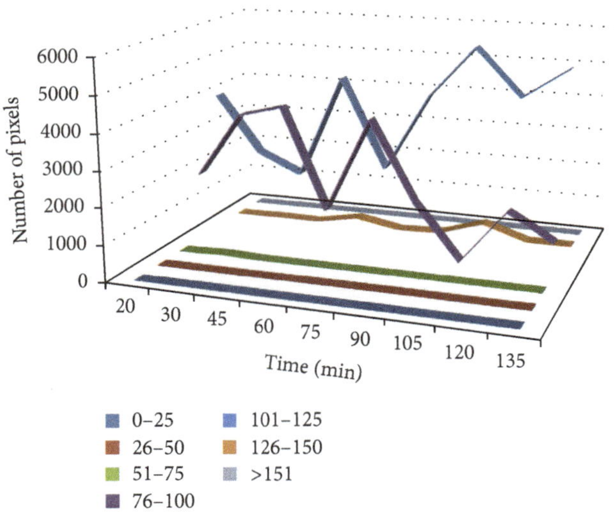

(a) RGB color values at various cutoff pixels

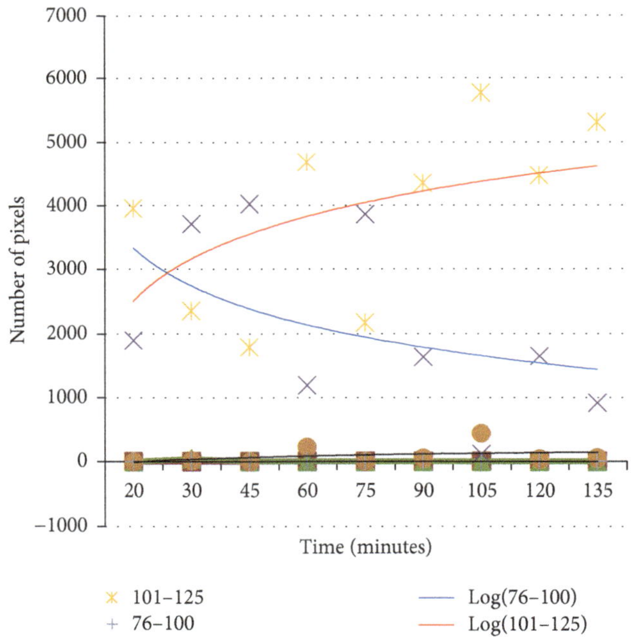

(b) Variation of pixel ranges with corrosion time

FIGURE 10: Results of cutoff-method analysis.

The energy of the images decreased, and the entropy increased with respect to corrosion time for most of the corrosive media. However, the energy and entropy values remained fairly constant in distilled water and verified the response that steel takes longer time to corrode in distilled water. It was also found that red color increased with time whereas green and blue pixel color values decreased with time. The 101–125 pixel range was found to have an increasing trend with the corrosion time. The approach presented in this paper may be used for correlating service/failure conditions based on corrosion morphology in metals. However, more work is needed for identifying multiple descriptors from image processing methods. Our future work will address this issue with controlled experiments to further validate the methods.

Acknowledgments

The first two authors thank Nuclear Regulatory Commission (NRC-38-10-929) for supporting this work. Also, the first author thanks US National Science Foundation for supporting this work through Grant DMR-0505496.

References

[1] P. R. Roberge, *Corrosion Inspection and Monitoring*, John Wiley & Sons, New York, NY, USA, 2007.

[2] E. Bardal, *Corrosion and Protection*, Springer, Berlin, Germany, 2003.

[3] M. Kutz, *Handbook of Environmental Degradation of Materials*, William Andew, New York, NY, USA, 2007.

[4] D. Itzhak, I. Dinstein, and T. Zilberberg, "Pitting corrosion evaluation by computer image processing," *Corrosion Science*, vol. 21, no. 1, pp. 17–22, 1981.

[5] M. Trujillo and M. Sadki, "Sensitivity analysis for texture models applied to rust steel classification," in *Electronic Imaging Science and Technology*, vol. 5303 of *Proceedings of SPIE*, pp. 161–169, January 2004.

[6] K. Y. Choi and S. S. Kim, "Morphological analysis and classification of types of surface corrosion damage by digital image processing," *Corrosion Science*, vol. 47, no. 1, pp. 1–15, 2005.

[7] R. M. Pidaparti, B. S. Aghazadeh, A. Whitfield, A. S. Rao, and G. P. Mercier, "Classification of corrosion defects in NiAl bronze through image analysis," *Corrosion Science*, vol. 52, no. 11, pp. 3661–3666, 2010.

[8] F. N. S. Medeiros, G. L. B. Ramalho, M. P. Bento, and L. C. L. Medeiros, "On the evaluation of texture and color features for nondestructive corrosion detection," *EURASIP Journal on Advances in Signal Processing*, vol. 2010, Article ID 817473, 7 pages, 2010.

[9] S. Wang and S. Song, "Image analysis of atmospheric corrosion exposure of zinc," *Materials Science and Engineering A*, vol. 385, no. 1-2, pp. 377–381, 2004.

[10] L. Tao, S. Song, X. Zhang, Z. Zhang, and F. Lu, "Image analysis of atmospheric corrosion of field exposure high strength aluminium alloys," *Applied Surface Science*, vol. 254, no. 21, pp. 6870–6874, 2008.

[11] S. Mallat, *A Wavelet Tour of Signal Processing*, Academic Press, New York, NY, USA, 1999.

[12] MATLAB software, Math Works, 2011.

14

Study of Synergistic Effect of Iodide on the Corrosion Antagonistic Behaviour of a Heterocyclic Phenylhydrazone in Sulphuric Acid Medium on Carbon Steel

Vinod P. Raphael, Joby Thomas Kakkassery, Shaju K. Shanmughan, and Aby Paul

Research Division, Department of Chemistry, St. Thomas' College (University of Calicut), Thrissur, Kerala 680005, India

Correspondence should be addressed to Joby Thomas Kakkassery; drjobythomask@gmail.com

Academic Editors: K. N. Allahar, G. Bereket, I. Obot, and R. Salghi

Synergistic effect of KI on the corrosion inhibition efficiency of 3-acetylpyridine phenylhydrazone (3APPH) on carbon steel (CS) in 0.5 M sulphuric acid solution has been investigated using gravimetric studies, electrochemical impedance spectroscopy (EIS), and potentiodynamic polarization studies. Gravimetric corrosion studies revealed that 3APPH showed moderate corrosion inhibition efficiency up to 8 h and beyond this period it showed corrosion accelerating behavior. This antagonistic effect of 3APPH is due to the hydrolysis of the molecule in acidic medium. A very high percentage of inhibition efficiency at 24 h was obtained on the addition of KI due to the synergistic effect of iodide ions. The adsorption of 3APPH and 3APPH + KI on the surfaces of the corroding metal obey Langmuir isotherm as obtained by impedance measurements. Polarization studies revealed that 3APPH act as a mixed type inhibitor. Thermodynamic parameters (K_{ads}, ΔG^0_{ads}) were derived from the adsorption isotherms. Surface morphology of the corroding metal was investigated by SEM analysis.

1. Introduction

Industrial processes such as pickling and descaling will escalate the corrosion rate of metals enormously and the addition of certain organic compounds which contain π-bonds and heteroatoms like N, O, and S is considered as the most useful method for reducing the corrosion of carbon steel in acidic media [1–3]. Molecules equipped with the above mentioned active probes may not act as a good inhibitors against corrosion in severe aggressive solutions such as sulphuric acid. It has been found that the addition of halide ions to the sulphuric acid enhance the corrosion inhibition capacity of the organic molecules on carbon steel synergistically. The synergistic effect of the halides has increased [4, 5] in the order $Cl^- < Br^- < I^-$. Due to large size and ease of polarizability, iodide (I^-) shows the highest synergistic effect [6–10]. The present study was conducted to examine the corrosion response of carbon steel in sulphuric acid containing the phenylhydrazone 3APPH and the synergistic behavior of 3APPH with I^- on the corrosion of carbon steel [11–13].

2. Experimental

2.1. Synthesis. The heterocyclic phenylhydrazone (3APPH) was obtained by the condensation of equimolar mixture of 3-acetylpyridine and phenylhydrazine hydrochloride in ethanol. The reaction mixture was refluxed for 2 hours, cooled to obtain yellow precipitate, filtered, washed with water, and dried. Figure 1 shows the molecular structure of phenylhydrazone 3APPH. Anal. calcd for $C_{13}H_{13}N_3$: C, 73.93; H, 6.16; N, 19.9%. Found C, 73.47; H, 6.42; N, 19.44%; m.p. = 245°C; IR (KBr): $\nu_{C=N}$ = 1590 cm^{-1} Mass: M^+ peak m/z: 211.

2.2. Solutions. The aggressive solution of 0.5 M H_2SO_4 was prepared by the dilution of A.R grade 98% of H_2SO_4 (Merck) with deionized water. Solutions of 3APPH were prepared in the range of 0.2 mM–1.0 mM concentrations in 0.5 M H_2SO_4.

2.3. Gravimetric Corrosion Studies. Carbon steel specimens of dimension 1 × 1 × 0.1 cm (composition: C, 0.58%; Mn, 0.07%; P, 0.02%; S, 0.015%; Si, 0.02%, and the rest Fe) were

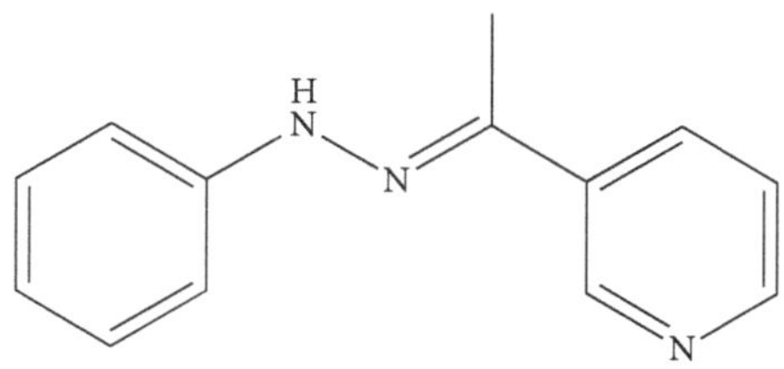

FIGURE 1: Molecular structure of 3APPH.

cut and abraded with various grades of silicon carbide papers (200, 400, 600, 800, 1000, and 1200). The exact area and thickness of each specimen were measured. Specimens were then degreased with acetone, washed with distilled water and dried. After weighing, specimens were immersed in 50 mL acid solutions at 30°C in the absence and presence of the 3APPH and 3APPH + KI (0.2 mM) at different concentrations using hooks and fishing lines. Weight loss of metal specimens was noted after 24 h. The experiments were carried out in duplicate and the average values were reported. The corrosion rate (ν) and the percentage of inhibition efficiency (η_w%) were calculated by the following equations [14–16]:

$$\nu = \frac{W}{St}, \quad (1)$$

$$\eta_w\% = \frac{\nu_0 - \nu}{\nu_0} \times 100, \quad (2)$$

where W is the average weight loss of CS specimen, S is the total area of specimen, t is the time of treatment, and ν_0 and ν are the corrosion rates of uninhibited and inhibited specimens, respectively.

2.4. AC Impedance Studies. The impedance measurements were performed in a three electrode assembly. Saturated calomel electrode (SCE) was used as the reference electrode. Platinum electrode having 1 cm^2 area was taken as the counter electrode. Metal specimens with an exposed area of 1 cm^2 were used as the working electrode. The EIS experiments were carried out on an Ivium compactstat-e electrochemical system. 0.5 M H_2SO_4 acid was taken as the electrolyte and the working area of the metal specimens were exposed to the electrolyte for 30 minutes prior to the measurement. Impedance measurements were performed at constant potential (OCP) in the frequency range from 1 KHz to 100 mHz with an amplitude of 10 mV as excitation signal. The percentage of inhibition from impedance measurements were calculated using charge transfer resistance by the following expression [17, 18]:

$$\eta_{EIS}\% = \frac{R_{ct} - R'_{ct}}{R_{ct}} \times 100, \quad (3)$$

where R_{ct} and R'_{ct} are the charge transfer resistances of working electrode with and without inhibitors, respectively.

2.5. Potentiodynamic Polarization Measurements. Electrochemical polarization studies of CS specimens in 0.5 M H_2SO_4 with and without 3APPH and 3APPH + KI were performed by recording anodic and cathodic potentiodynamic polarization curves. Polarization plots were obtained in the electrode potential range from −100 to +100 mV versus corrosion potential (E_{corr}) at a sweep rate of 1 mV/sec. To obtain corrosion current densities, Tafel polarization analyses were done by extrapolating anodic and cathodic curves. The percentage of inhibition efficiency (η_{pol}%) was evaluated from the measured I_{corr} values using the following relation [19]:

$$\eta_{pol}\% = \frac{I_{corr} - I'_{corr}}{I_{corr}} \times 100, \quad (4)$$

where I_{corr} and I'_{corr} are the corrosion current densities of the exposed area of the working electrode in the absence and presence of inhibitor.

2.6. Scanning Electron Microscopy. Surface analyses of CS specimens were performed using scanning electron microscope (model Hitachi SU6600). SEM images of CS surface were taken by treating 0.5 M H_2SO_4 solutions in the absence and presence of the 3APPH and 3APPH (0.8 mM) + KI (0.2 mM) for 24 h at 30°C.

3. Results and Discussions

3.1. Gravimetric Studies. Weight loss of CS specimens in 0.5 M H_2SO_4 at 30°C was determined at 24 h in the presence of various concentrations of 3APPH and 3APPH + KI (0.2 mM). The corrosion rates and inhibition efficiencies are listed in Table 1. It is evident from the table that the rate of corrosion in the presence of 3APPH increased with concentration at 24 h (corrosion accelerating/antagonistic behavior). A marked rise in the inhibition efficiency was noted for CS specimens in the presence of 3APPH + KI. To investigate a detailed corrosion inhibition response of the 3APPH on CS, the gravimetric measurements were performed every 2 hours. The corrosion rate of CS was found to be decreased with the concentration of 3APPH up to 8 h or η_w% increased with the concentration. Beyond this period, 3APPH tend to behave as a corrosion accelerator. From the UV spectral studies it was confirmed that the 3APPH molecules undergo slow hydrolysis in sulphuric acid medium at room temperature into the parent ketone and phenylhydrazine. The hydrolyzed products were inhibited less on the CS surface when compared to the hydrazone (3APPH), and the same was confirmed by performing the gravimetric corrosion studies of CS in sulphuric acid by taking parent amine and ketone as the inhibitor (Figure 2). To check the synergistic effect of I^- with the parent amine, parent ketone and their mixture, gravimetric corrosion analysis of the systems like phenylhydrazine (PH) + KI, 3-acetylpyridine (3AP) + KI, PH + 3AP + KI was performed independently and compared with 3APPH + KI system (Figure 2). These studies clearly showed that the highest η_w% was obtained for 3APPH + KI for all the studied concentrations. The present investigation also emphasizes that 3APPH molecules

TABLE 1: Inhibition efficiency (η_w%) from gravimetric measurements for different systems on CS corrosion at 24 h in 0.5 M H_2SO_4 at 30°C.

System/conc.	Corrosion rate (ν)	% of inhibition efficiency (η_w%)
Blank	23.53	—
Blank + 0.2 mM KI	19.40	24.0
0.2 mM 3APPH	27.83	−18.28
0.4 mM 3APPH	32.67	−38.83
0.6 mM 3APPH	44.77	−90.26
0.8 mM 3APPH	45.31	−92.54
1.0 mM 3APPH	45.48	−93.27
0.2 mM 3APPH + 0.2 mM KI	2.84	87.90
0.4 mM 3APPH + 0.2 mM KI	2.50	89.38
0.6 mM 3APPH + 0.2 mM KI	1.06	95.47
0.8 mM 3APPH + 0.2 mM KI	0.99	95.80
1.0 mM 3APPH + 0.2 mM KI	0.82	96.50

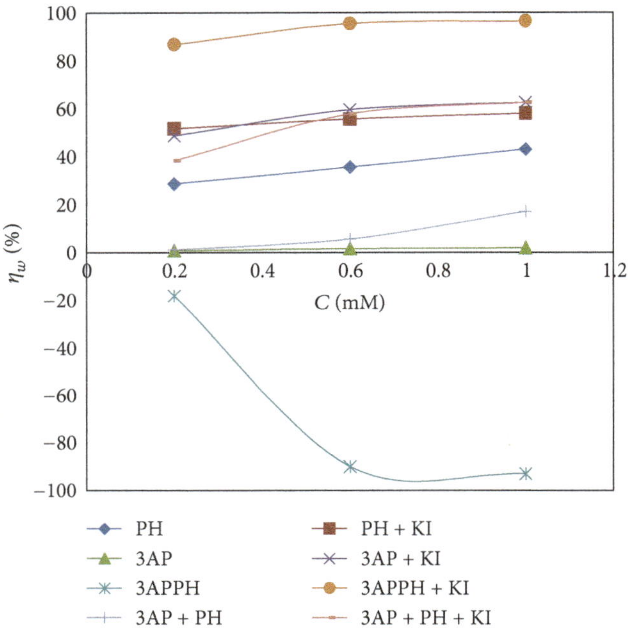

FIGURE 2: Comparison between the corrosion inhibition efficiencies of various systems.

attaching through the I^- ions on CS surface do not undergo hydrolysis by hydronium ions.

3.2. AC Impedance Studies. To emphasize the role of iodide ions on the corrosion inhibition of CS by 3APPH, impedance study was performed by exposing the corroding metal surface towards the 3APPH and 3APPH + KI solutions for 30 minutes. Figures 3 and 4 represent the Nyquist plots of CS specimens in 0.5 M H_2SO_4 in the presence of various concentrations of 3APPH and 3APPH + KI, respectively. It is evident from the plots that the impedance response of metal specimens has marked differences in the presence and absence of the KI with 3APPH. Equivalent circuit that fit into many electrochemical systems composed of a double layer capacitance, R_s and R_{ct} [20–22], can also be accepted in this

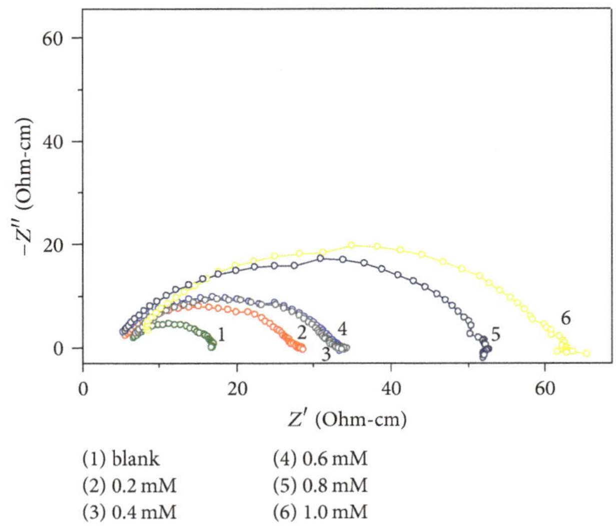

FIGURE 3: Nyquist plots for CS specimens in 0.5 M H_2SO_4 and 3APPH.

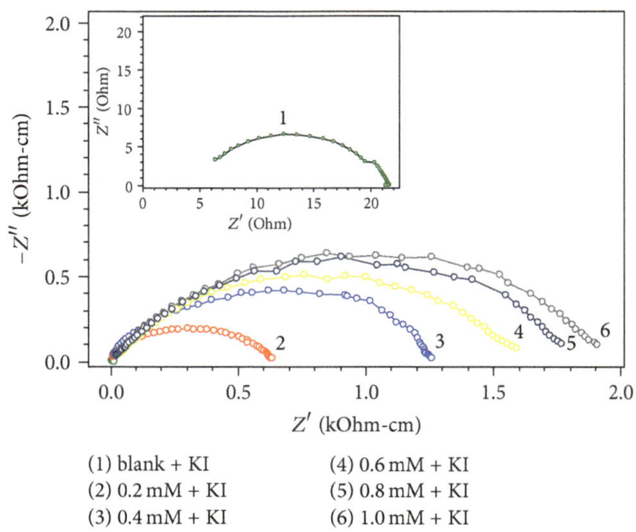

FIGURE 4: Nyquist plots for CS specimens in 0.5 M H_2SO_4 and 3APPH + KI.

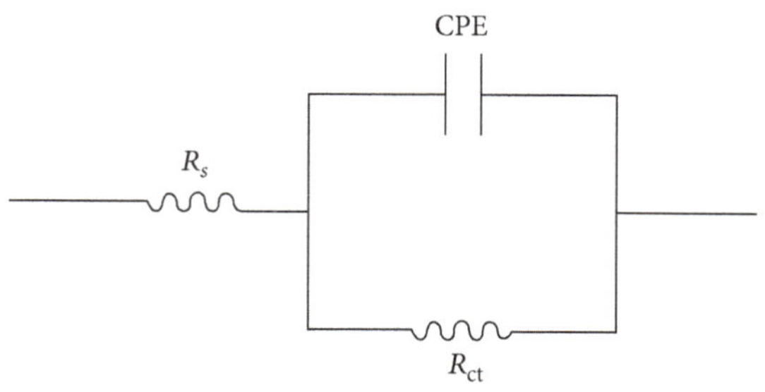

FIGURE 5: Equivalent circuit fitting for EIS.

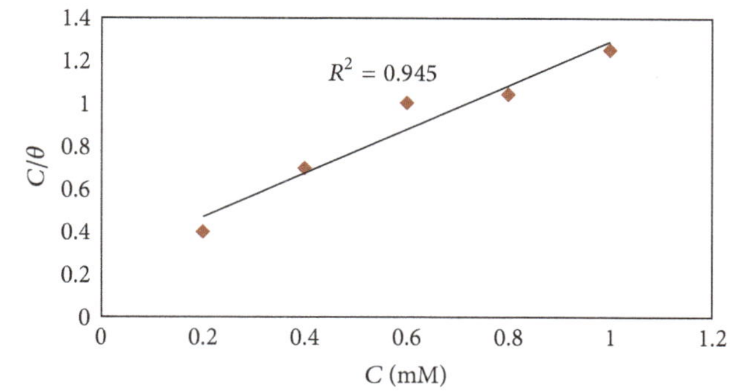

FIGURE 6: Langmuir adsorption isotherm for CS specimens in 0.5 M H_2SO_4 and 3APPH.

study. To reduce the effects due to surface irregularities of metal, constant phase element (CPE) is introduced into the circuit instead of a pure double layer capacitance which gives more accurate fit as shown in the Figure 5 [23–25].

The impedance of CPE can be expressed as

$$Z_{\mathrm{CPE}} = \frac{1}{Y_0 (j\omega)^n}, \tag{5}$$

where Y_0 is the magnitude of CPE, n is the exponent (phase shift), ω is the angular frequency, and j is the imaginary unit. CPE may be resistance, capacitance, and inductance [26] depending upon the values of n. In all measurements the value of n ranges between 0.75 and 0.99, suggesting the capacitive response of CPE. The impedance parameters and the percentage of inhibition (η_{EIS}%) are listed in Tables 2(a) and 2(b), respectively. From the data it is clear that R_{ct} values are increased with the 3APPH concentration. The capacitance values (C_{dL}) decreased with 3APPH concentration and this decrease in C_{dL} was enhanced upon addition of I^- ions to the corrosive environment. These results suggest that the 3APPH molecules function by adsorption at the metal/solution interface [27, 28] and this adsorption is reinforced by I^- ions. The addition of KI drastically increased the η_w% values. From Table 2 it is also evident that η_w% for KI in combination with 3APPH is higher than the sum of η_w% for single KI and single 3APPH, which is considered due to synergistic effect.

3.3. Adsorption Isotherm and Free Energy of Adsorption. In order to reveal the mechanism of inhibition, various adsorption isotherms such as Langmuir, Freundlich, Temkin, and Frumkin were plotted and the most suitable one was selected with the aid of correlation coefficient. Among the isotherms considered above, the most adequate one for both experiments, that is, impedance measurements in the presence and absence of KI, was Langmuir adsorption isotherm. Figures 6 and 7, respectively, portray the adsorption isotherms for the inhibition of 3APPH in the absence and presence of KI. Langmuir adsorption isotherm can be expressed as [29]

$$\frac{C}{\theta} = \frac{1}{K_{\mathrm{ads}}} + C, \tag{6}$$

where C is the concentration of the 3APPH, θ is the fractional surface coverage, and K_{ads} is the adsorption equilibrium constant. Table 3 explores the thermodynamic parameters such as adsorption equilibrium constants (K_{ads}) and free energies of adsorption $\Delta G^0_{\mathrm{ads}}$.

It is evident from Table 3 that $\Delta G^0_{\mathrm{ads}}$ is negative for both investigations, which suggest the spontaneity of the processes. For the inhibition process by 3APPH on CS without KI, $\Delta G^0_{\mathrm{ads}}$ value implies that the mechanism of adsorption involve both physisorption and chemisorption. Since $\Delta G^0_{\mathrm{ads}}$ exhibited for the 3APPH solution containing KI was more than 40 kJ, the chemisorptions [30, 31] is the sole factor responsible for the inhibition process. This investigation clearly establishes the synergistic effect of I^- ions with 3APPH which cause adsorbing firmly the 3APPH molecules without hydrolysis on the metal surface by a strong chemical interaction.

TABLE 2: Electrochemical impedance parameters of CS specimens in 0.5 M H_2SO_4 at 30°C in the absence and presence of 3APPH (a) and 3APPH + 0.2 mM KI (b).

(a)

C (mM)	R_{ct} (Ωcm^2)	C_{dL} (μF cm^{-2})	η_{EIS}%
0	9.72	133	—
0.2	19.54	134	50.25
0.4	22.77	121	57.31
0.6	24.09	119	59.65
0.8	41.39	118	76.51
1.0	48.22	116	79.84

(b)

C (mM)	R_{ct} (Ωcm^2)	C_{dL} (μF cm^{-2})	η_{EIS}%
Blank + KI	12.21	128	20.39
0.2 + KI	518.4	28.9	98.13
0.4 + KI	1071	26.2	99.09
0.6 + KI	1301	20.2	99.25
0.8 + KI	1499	21.5	99.35
1.0 + KI	1607	14.3	99.39

TABLE 3: Thermodynamic parameters from adsorption isotherms.

System	Adsorption isotherm	K_{ads}	$\Delta G^\circ_{\mathrm{ads}}$ (kJ)
3APPH	Langmuir	3.8×10^3	−30.9
3APPH + KI	Langmuir	5.0×10^5	−43.17

TABLE 4: Potentiodynamic polarization parameters of CS specimens in 0.5 M H_2SO_4 at 30°C in the absence and presence of 3APPH.

C (mM)	E_{corr} (mV/SCE)	I_{corr} (μA/cm^2)	b_a (mV/dec)	$-b_c$ (mV/dec)	η_{pol}%
Blank	−468	1041	84	122	—
0.2	−461	591	67	131	43.1
0.4	−464	432	63	123	58.5
0.6	−455	395	68	124	62.0
0.8	−465	255	54	114	75.5
1.0	−462	175	46	106	83.2

TABLE 5: Potentiodynamic polarization parameters of CS specimens in 0.5 M H_2SO_4 at 30°C in the presence of KI and 3APPH + KI.

C (mM)	E_{corr} (mV/SCE)	I_{corr} (μA/cm^2)	b_a (mV/dec)	$-b_c$ (mV/dec)	η_{pol}%
Blank + KI	−475	799	80	123	23.3
0.2 + KI	−507	18.8	85	102	98.2
0.4 + KI	−540	12.6	118	121	98.8
0.6 + KI	−526	11.8	117	105	98.9
0.8 + KI	−515	8.9	102	119	99.1
1.0 + KI	−534	8.5	127	106	99.2

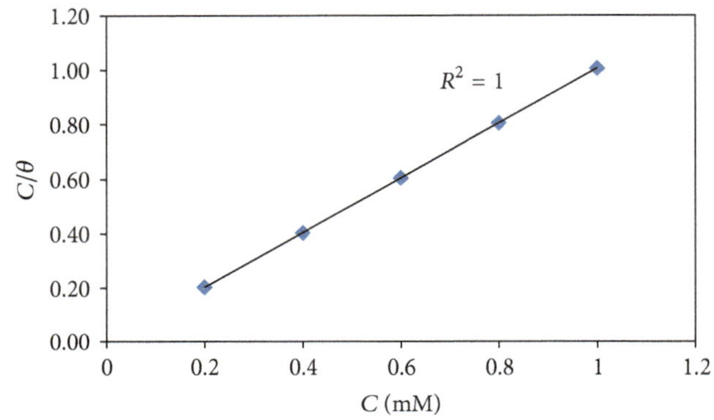

FIGURE 7: Langmuir adsorption isotherm for CS specimens in 0.5 M H_2SO_4 and 3APPH + KI.

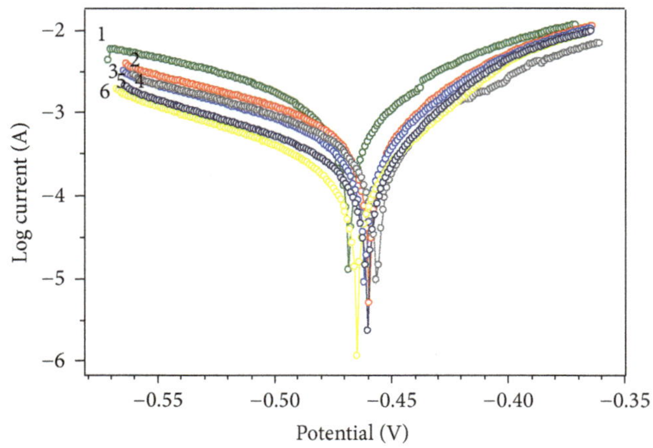

FIGURE 8: Tafel plots of CS specimens in 0.5 M H_2SO_4 at 30°C, with and without 3APPH.

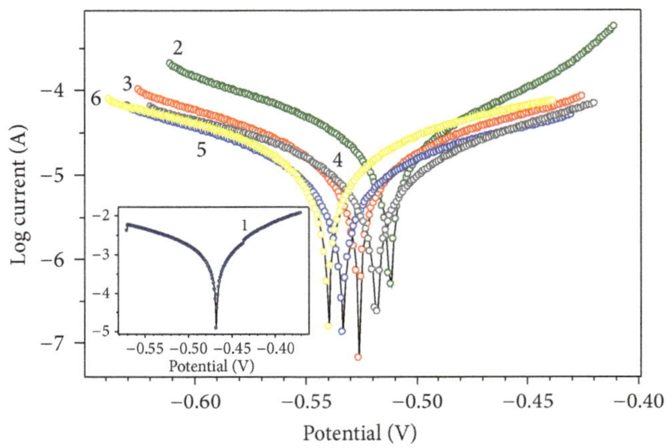

FIGURE 9: Tafel plots of CS specimens in 0.5 M H_2SO_4 at 30°C, with KI and 3APPH + KI.

3.4. Potentiodynamic Polarization Studies. Potentiodynamic polarization curves for 3APPH in 0.5 M H_2SO_4 at 30°C for CS specimens in the presence of various concentrations of 3APPH and 3APPH + KI are shown in Figures 8 and 9 respectively and the corresponding polarization parameters like corrosion current densities (I_{corr}), corrosion potential (E_{corr}), cathodic Tafel slope (b_c), anodic Tafel slope (b_a), and inhibition efficiency (η_{pol}%) are listed in Tables 4 and 5 respectively. The data show that addition of 3APPH to acid medium affects both the cathodic and anodic parts of the curves. Addition of I^- ions to 3APPH/H_2SO_4 systems results in marked decrease in the corrosion current density (I_{corr}). In other words, both anodic and cathodic reactions are drastically inhibited and the 3APPH can be viewed as mixed type inhibitor for CS specimens in 0.5 M H_2SO_4. The values

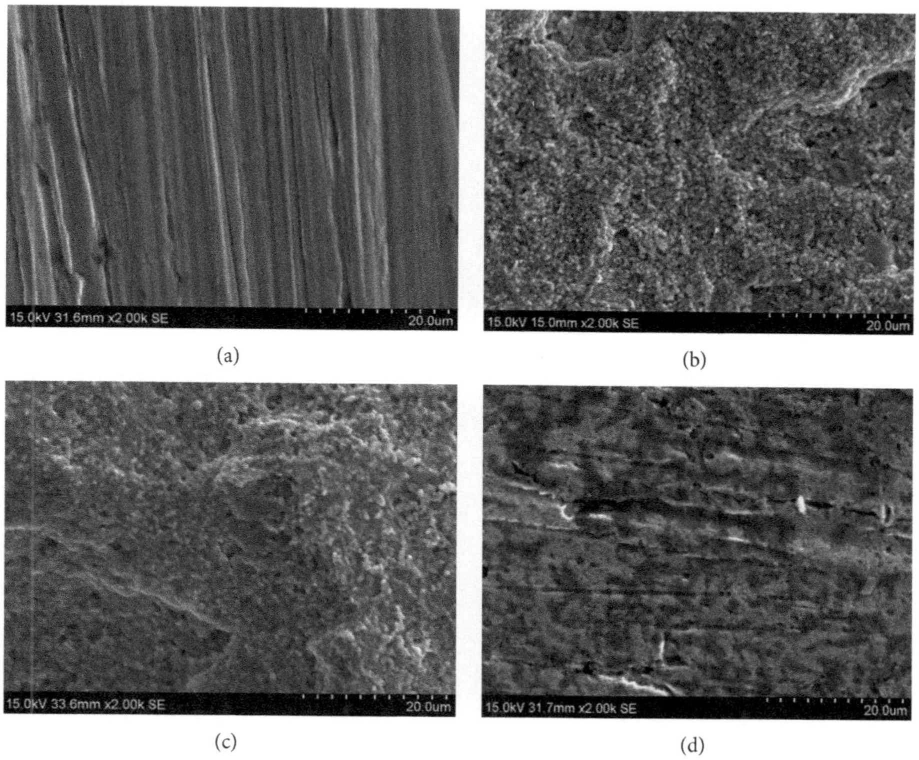

FIGURE 10: (a) SEM image of bare CS surface (b) SEM image of CS surface in 0.5 M H_2SO_4, (c) SEM image of CS surface in 0.5 M H_2SO_4 and 3APPH (0.8 mM), and (d) SEM image of CS surface in 0.5 M H_2SO_4 and 3APPH (0.8 mM) + 0.2 mM KI.

establish that the η_{pol}% of 3APPH is increased in presence of KI and these results also assure the existence of strong synergism between 3APPH and KI in the corrosion inhibition process of CS.

3.5. SEM Analysis. To examine the effect of 3APPH molecules on CS surface, SEM analyses were performed [32, 33]. SEM images of CS surfaces are given in Figures 10(a)–10(d). On comparison of the Figures 10(a) and 10(b), it is evident that the surface of CS was seriously corroded in the acidic solution. Figure 10(c) represents the surface image of the metal in the presence of the 3APPH (0.8 mM, 24 h). Comparison of the textures of Figures 10(b) and 10(c) revealed that severe damage on the surface of CS happened in the presence of the 3APPH due to the antagonistic behaviour. Figure 10(d) is the image of CS surface treated with 3APPH and KI (0.8 mM + 0.2 mM KI for 24 h). Morphology of this image is entirely different from that of other images and least damage occurred for the surface. This suggests an increased inhibition of the 3APPH molecules on the CS surface through I^- ions.

4. Conclusions

3APPH undergo slow hydrolysis in sulphuric acid medium and showed moderate/poor corrosion inhibition efficiency up to 8 h. Beyond this period, 3APPH shows corrosion antagonistic character. In the presence of KI, the 3APPH molecules are firmly attached on the CS surface and do not undergo hydrolysis and act as excellent corrosion inhibitor. Adsorption studies revealed that chemical interaction occurs between the 3APPH molecules and CS surface trough I^- ions. This causes the enhancing of the inhibition efficiency greatly. Parent ketone and parent amine exhibit poorer corrosion inhibition efficiency than the 3APPH molecules. Polarization studies reveal that 3APPH act as a mixed type inhibitor.

References

[1] O. James, N. C. Oforka, and O. K. Abiola, "Inhibition of acid corrosion of mild steel by pyridoxal and pyridoxol hydrochlorides," *International Journal of Electrochemical Science*, vol. 2, pp. 278–284, 2007.

[2] S. A. Abd El-Maksoud, "The effect of organic compounds on the electrochemical behaviour of steel in acidic media. A review," *International Journal of Electrochemical Science*, vol. 3, pp. 528–555, 2008.

[3] E. E. Ebenso, P. C. Okafor, O. E. Offiong, B. I. Ita, U. J. Ibok, and U. J. Ekpe, "Comparative investigation into the kinetics of corrosion inhibition of aluminium alloy AA 1060 in acidic medium," *Bulletin of Electrochemistry*, vol. 17, no. 10, pp. 459–464, 2001.

[4] Y. Feng, K. S. Siow, W. K. Teo, and A. K. Hsieh, "The synergistic effects of propargyl alcohol and potassium iodide on the inhibition of mild steel in 0.5 M sulfuric acid solution," *Corrosion Science*, vol. 41, no. 5, pp. 829–852, 1999.

[5] P. C. Okafor and Y. Zheng, "Synergistic inhibition behaviour of methylbenzyl quaternary imidazoline derivative and iodide

ions on mild steel in H_2SO_4 solutions," *Corrosion Science*, vol. 51, no. 4, pp. 850–859, 2009.

[6] E. E. Ebenso, H. Alemu, S. A. Umoren, and I. B. Obot, "Inhibition of mild steel corrosion in sulphuric acid using alizarin yellow GG dye and synergistic Iodide Additive," *International Journal of Electrochemical Science*, vol. 3, pp. 1325–1339, 2008.

[7] S. A. Umoren, U. M. Eduok, and E. E. Oguzie, "Corrosioninhibition ofmild steel in 1 M H_2SO_1 by polyvinyl pyrrolidone and synergistic iodide additives," *Portugaliae Electrochimica Acta*, vol. 26, no. 6, pp. 533–546, 2008.

[8] F. Bentiss, M. Bouanis, B. Mernari, M. Traisnel, and M. Lagrenee, "Effect of iodide ions on corrosion inhibition of mild steel by 3,5-bis(4-methylthiophenyl)-4H-1,2,4-triazole in sulfuric acid solution," *Journal of Applied Electrochemistry*, vol. 32, no. 6, pp. 671–678, 2002.

[9] M. A. Quraishi and J. Rawat, "Influence of iodide ions on inhibitive performance of tetraphenyl-dithia-octaaza-cyclotetradeca-hexaene (PTAT) during pickling of mild steel in hot sulfuric acid," *Materials Chemistry and Physics*, vol. 70, no. 1, pp. 95–99, 2001.

[10] C. Jeyaprabha, S. Sathiyanarayanan, S. Muralidharan, and G. Venkatachari, "Corrosion inhibition of iron in 0.5 mol L-1 H_2SO_4 by halide ions," *Journal of the Brazilian Chemical Society*, vol. 17, no. 1, pp. 61–67, 2006.

[11] Z. A. Iofa, V. V. Batrakov, and C.-N. Cho-Ngok-Ba, "Influence of anion adsorption on the action of inhibitors on the acid corrosion of iron and cobalt," *Electrochimica Acta*, vol. 9, no. 12, pp. 1645–1653, 1964.

[12] I. F. Fishtik, I. I. Vataman, and P. A. Spater, "The mechanism of ion-pair formation in the inner part of the double layer," *Journal of Electroanalytical Chemistry*, vol. 165, no. 1-2, pp. 1–8, 1984.

[13] K. S. Shaju, K. J. Thomas, V. P. Raphael, and A. Paul, "Synergistic effect of KI on corrosion inhibition of mild steel by polynuclear Schiff base in sulphuric acid," *ISRN Corrosion*, vol. 2012, Article ID 425878, 8 pages, 2012.

[14] X. Li, S. Deng, and H. Fu, "Synergistic inhibition effect of 6-benzylaminopurine and iodide ion on the corrosion of cold rolled steel in H_3PO_4 solution," *Corrosion Science*, vol. 53, no. 11, pp. 3704–3711, 2011.

[15] ASTM, "Standard recommended practice for the laboratory immersion corrosion testing of metals," Tech. Rep. ASTM G-31-72, ASTM, Philadelphia, Pa, USA, 1990.

[16] K. C. Emregül and O. Atakol, "Corrosion inhibition of iron in 1 M HCl solution with Schiff base compounds and derivatives," *Materials Chemistry and Physics*, vol. 83, no. 2-3, pp. 373–379, 2004.

[17] A. Raman and P. Labine, *Reviews on Corrosion Inhibitor Science and Technology*, NACE, Houston, Tex, USA, 1986.

[18] S. John and A. Joseph, "Electroanalytical studies of the corrosion-protection properties of 4-amino-4 H-1,2,4-triazole-3,5-dimethanol (ATD) on mild steel in 0.5 N sulfuric acid," *Research on Chemical Intermediates*, vol. 38, no. 7, pp. 1359–1373, 2012.

[19] H. Ashassi-Sorkhabi, B. Shaabani, and D. Seifzadeh, "Effect of some pyrimidinic Shciff bases on the corrosion of mild steel in hydrochloric acid solution," *Electrochimica Acta*, vol. 50, no. 16-17, pp. 3446–3452, 2005.

[20] A. Subramania, A. R. S. Priya, and V. S. Muralidharan, "Development of novel acidizing inhibitors for carbon steel corrosion in 15% boiling hydrochloric acid," *Corrosion*, vol. 64, no. 6, pp. 541–552, 2008.

[21] M. El Azhar, B. Mernari, M. Traisnel, F. Bentiss, and M. Lagrenée, "Corrosion inhibition of mild steel by the new class of inhibitors [2,5-bis(n-pyridyl)-1,3,4-thiadiazoles] in acidic media," *Corrosion Science*, vol. 43, no. 12, pp. 2229–2238, 2001.

[22] A. Yurt, A. Balaban, S. U. Kandemir, G. Bereket, and B. Erk, "Investigation on some Schiff bases as HCl corrosion inhibitors for carbon steel," *Materials Chemistry and Physics*, vol. 85, no. 2-3, pp. 420–426, 2004.

[23] J. R. Macdonald, W. B. Johnson, and J. R. Macdonald, *Theory in Impedance Spectroscopy*, John Wiley & Sons, New York, NY, USA, 1987.

[24] L. Larabi, Y. Harek, M. Traisnel, and A. Mansri, "Synergistic influence of poly(4-vinylpyridine) and potassium iodide on inhibition of corrosion of mild steel in 1M HCl," *Journal of Applied Electrochemistry*, vol. 34, no. 8, pp. 833–839, 2004.

[25] M. Lebrini, M. Lagrenée, H. Vezin, M. Traisnel, and F. Bentiss, "Experimental and theoretical study for corrosion inhibition of mild steel in normal hydrochloric acid solution by some new macrocyclic polyether compounds," *Corrosion Science*, vol. 49, no. 5, pp. 2254–2269, 2007.

[26] A. K. Singh, S. K. Shukla, M. Singh, and M. A. Quraishi, "Inhibitive effect of ceftazidime on corrosion of mild steel in hydrochloric acid solution," *Materials Chemistry and Physics*, vol. 129, no. 1-2, pp. 68–76, 2011.

[27] M. MaCafferty and N. Hackerman, "Double layer capacitance of iron and corrosion inhibition with polymethylene diamines," *Journal of the Electrochemical Society*, vol. 119, no. 2, pp. 146–154, 1972.

[28] D. A. López, S. N. Simison, and S. R. de Sánchez, "The influence of steel microstructure on CO_2 corrosion. EIS studies on the inhibition efficiency of benzimidazole," *Electrochimica Acta*, vol. 48, no. 7, pp. 845–854, 2003.

[29] M. Bouklah, B. Hammouti, M. Lagrenée, and F. Bentiss, "Thermodynamic properties of 2,5-bis(4-methoxyphenyl)-1,3,4-oxadiazole as a corrosion inhibitor for mild steel in normal sulfuric acid medium," *Corrosion Science*, vol. 48, no. 9, pp. 2831–2842, 2006.

[30] F. Bentiss, M. Lebrini, and M. Lagrenée, "Thermodynamic characterization of metal dissolution and inhibitor adsorption processes in mild steel/2,5-bis(n-thienyl)-1,3,4-thiadiazoles/hydrochloric acid system," *Corrosion Science*, vol. 47, no. 12, pp. 2915–2931, 2005.

[31] W.-H. Li, Q. He, S.-T. Zhang, C.-L. Pei, and B.-R. Hou, "Some new triazole derivatives as inhibitors for mild steel corrosion in acidic medium," *Journal of Applied Electrochemistry*, vol. 38, no. 3, pp. 289–295, 2008.

[32] F. El-Taib Heakal, A. S. Fouda, and M. S. Radwan, "Inhibitive effect of some thiadiazole derivatives on C-steel corrosion in neutral sodium chloride solution," *Materials Chemistry and Physics*, vol. 125, no. 1-2, pp. 26–36, 2011.

[33] S. John and A. Joseph, "Electro analytical, surface morphological and theoretical studies on the corrosion inhibition behavior of different 1,2,4-triazole precursors on mild steel in 1 M hydrochloric acid," *Materials Chemistry and Physics*, vol. 133, no. 2-3, pp. 1083–1091, 2012.

15

One-Step Anodization/Sol-Gel Deposition of Ce^{3+}-Doped Silica-Zirconia Self-Healing Coating on Aluminum

N. Kumar, A. Jyothirmayi, K. R. C. Soma Raju, V. Uma, and R. Subasri

International Advanced Research Centre for Powder Metallurgy and New Materials (ARCI), Balapur, Hyderabad, Andhra Pradesh 500005, India

Correspondence should be addressed to R. Subasri; subasri@arci.res.in

Academic Editors: M. Aparicio, G. Bereket, and M. Sakairi

A novel process was used for the preparation of dense, thick, and stable silica-zirconia coatings on aluminum by an *in situ* anodization along with sol-gel deposition. Anodic electrophoretic deposition was carried on aluminum using a SiO_2-ZrO_2 sol that was synthesized from an epoxy modified silane and zirconium n-propoxide along with a cerium salt ($Ce(NO_3)_3{\cdot}6H_2O$). Current density and time were varied during the deposition. The optimal parameters that yielded uniform coatings were determined. Coatings were characterized for their crystallinity, scratch hardness, and microstructure. The barrier properties of the coatings were tested using potentiodynamic polarization studies, electrochemical impedance spectroscopy, and neutral salt spray tests. Grazing angle incidence X-ray diffraction studies revealed that the coating comprised crystalline Al_2SiO_5 along with an amorphous phase. The novelty of the process was that the crystalline aluminosilicate phase was formed even at room temperature and could be deposited on aluminum by a simultaneous anodization of aluminum and sol-gel deposition. The coated substrates withstood more than 400 hours of salt spray tests. Polarization measurements reveal that the composite layer of aluminosilicate along with the Ce^{3+}-doped silica-zirconia sol enhances the corrosion properties by forming a passive layer, which acts as a good barrier against corrosion.

1. Introduction

Aluminum, that is widely used as a structural material due to its high strength to weight ratio and low cost, is highly susceptible to corrosion attack in chloride containing environment [1, 2]. Chromate conversion coatings have been the most widely used self-healing, anticorrosion treatments for aluminum and its alloys. Due to the increasing demand for the development of an environmentally-friendly, effective, inexpensive, and technologically simple method for corrosion protection, anodization of aluminum, and conversion coatings based on vanadates, permanganates, tungstates, and rare-earth salts have recently attracted a lot of attention as a method for corrosion protection [3]. Anodization is an electrochemical oxidation process employed to increase the thickness of the native oxide layer on metals like Al, Mg, Ti, and so forth [4–15]. The anodized layer is porous and, hence, there are reports in which a sol-gel coating is deposited on top of the anodized layer to render good barrier properties [16, 17]. There are also reports where, initially, a purely inorganic sol-gel coating is deposited on the aluminum substrate, followed by heat treatment at high temperatures (300–500°C) for densification, and then the substrate is subsequently anodized [18, 19] in an effort to obtain the cumulative benefits of both processes, namely, anodization and sol-gel coating. However, a low processing temperature for coating densification is one of the most important requisites, since some metals like aluminum undergo structural changes with temperature which degrade the mechanical properties of the substrate and promote corrosion [20]. The sol-gel organic-inorganic hybrid coating is a promising alternative that offers several advantages such as low temperature densification, good adhesion with the substrate, cost-effectiveness, being eco-friendly, and simple application procedures, which are easily adaptable by the user-industry. The sol-gel hybrid matrices can also be used for encapsulating self-healing materials [21, 22].

The present work describes a novel process where aluminum substrate is dipped in a silica-zirconia hybrid sol containing Ce^{3+} which acts as a self-healing material and anodic electrodeposition is carried out. Under these conditions, aluminum can be simultaneously anodized forming, *in situ*, a crystalline aluminosilicate layer on the substrate surface, along with the deposition of sol-gel coating on the crystalline aluminosilicate layer. The pH of the sol, current density, deposition time, and curing temperature were optimized to form a defect-free, dense, and corrosion resistant oxide layer.

2. Experimental

2.1. Materials and Synthesis. 3-Glycidoxypropyltrimethoxysilane (GPTMS, Gelest Inc., USA, purity > 97%), zirconium n-propoxide (GELEST Inc., USA), methacrylic acid (ABCR GmbH, Germany, purity > 97%), and high purity solvent 2-butoxyethanol (LR grade) (Sd Fine-Chem. Ltd., India) were used as the starting materials without further purification. Cerium nitrate ($Ce(NO_3)_3{\cdot}6H_2O$) 99.9% (LOBA CHEMIE Pvt., India) was used as the source of Ce^{3+}. Aluminum specimens, obtained from Q-Lab Florida, USA, with dimensions 40 mm × 30 mm × 1 mm, were used as anodes after thorough degreasing with acetone. The chemical composition of the aluminum substrate in wt% was Al-98; Mn-1; Si-0.6; Zn-0.1; Cu-0.05–0.3. The sol was synthesized in two parts and then mixed together to obtain the nanocomposite sol. In the first step, zirconium-n-propoxide was complexed with methacrylic acid in a mole ratio of 1 : 1 under vigorous stirring to reduce the hydrolysis rate of Zr-n-propoxide. In the second step, GPTMS and water mixture was stirred with 0.1 N HCl for prehydrolysis of GPTMS and after ensuring hydrolysis of GPTMS, both parts were mixed and stirred continuously for 5 h. The mole ratio of GPTMS : Zr-n-propoxide was 3 : 1. The sol (henceforth abbreviated as GZ) was then appropriately diluted using 2-butoxy-ethanol and cerium nitrate was added so that final concentration of Ce^{3+} in the sol was 0.01 M.

2.2. Anodization/Deposition. The coating was deposited electrochemically using the inorganic-organic hybrid sol (GZ sol), along with Ce^{3+} as the source of self-healing material. A programmable current source (Keithley 224) was used for supplying the current during the deposition process. The process of anodization/deposition was monitored with current source as well as withdrawal speed. The pH of the sol was varied in three steps, namely, 4, 8, and 10. The current density was varied between 1.2 and 4 mA/cm^2 in three steps, namely, 1.2 mA/cm^2, 2.4 mA/cm^2, and 4 mA/cm^2, and the time of immersion of substrates in the sol was also varied for 1, 5, and 10 minutes. The optimum conditions were found to be the deposition of a sol maintained at pH 8–8.5 using a current density 4 mA/cm^2 for a deposition time of 10 minutes. After the layer formation on the substrate, it was removed from the sol tank using a dip coater with 1 mm/s withdrawal speed. The coatings were cured in air at 130°C for 1 h. Some of the coatings were also cured in air at 300°C for 1 h. For the sake of comparison, normal dip coating of the sol was also carried out on some of the substrates using 1 mm/s withdrawal speed and coatings were cured at 130°C for 1 h.

2.3. Characterization. The coating thickness, morphology, and composition were analyzed by SEM/EDAX, using a scanning electron microscope (Hitachi model S/3400N) and the crystalline nature was ascertained using an X-ray diffractometer (Bruker D8 AXS Advance X-ray diffractometer), using both normal incidence (NI) and grazing angle incidence (GI) of 1 degree. The adhesion of the coatings was tested according to ASTM D3359-02. Scratch test is a simple and rapid method to characterize the coatings, but results obtained are influenced by various factors such as coating thickness, mechanical properties of the substrate, interfacial bond strength, and test conditions such as scratch speed, load, and indenter tip radius. Scratch test was carried out using a microscratch tester (Revtest, CSM make). A Rockwell diamond indenter with a tip radius of 200 μm and a progressive loading method from a minimum of 0.9 N at the point of contact to a maximum of 10 N normal load at the end of scratch length were employed for evaluating the coated substrates. Scratch length, loading rate, and scratching speed were fixed at 6 mm, 7.58 N/min, and 5 mm/min, respectively. In the present investigation, the scratch tester tip was brought in contact with the coated surface and the sample was moved at a constant speed, while the tip normal load was progressively increased to the set maximum value. The output was measured in terms of acoustic emission, penetration depth, and tangential frictional force. The scratch tester has a tip, which is placed with a controlled scratch load F_z on the surface to be tested. When the sample is scratched, the tip is stationary and sample moves. The resulting frictional force F_x can be monitored, while the scratch track is generated.

Electrochemical tests were carried out using an electrochemical interface (Solartron SI 1287) with an impedance analyzer (Solartron SI 1260). The corrosion test cell had the classic configuration of three electrodes, platinum electrode as counter electrode, a saturated calomel electrode as a reference electrode, and the uncoated/coated aluminum substrate as the working electrode. Polarization studies of the coated/uncoated substrates were carried out at 25°C in a N_2 purged 3.5% NaCl solution with an exposure time of 1 h and 24 h. Potentiodynamic scans were recorded by applying potentials from −1.6 V to 0.0 V with a scan rate of 1 mV/s. The electrochemical impedance scan was carried out using an AC signal of 10 mV amplitude applied over a bandwidth from 100 kHz to 0.03 Hz. The corrosion behavior was also tested by neutral salt spray tests, using 5% NaCl solution according to ASTM B117.

3. Results and Discussion

3.1. XRD Analysis. Structural properties were investigated with both normal incidence XRD and GIXRD, the latter being a surface sensitive technique. The grazing incidence method was employed due to the large penetration depth of

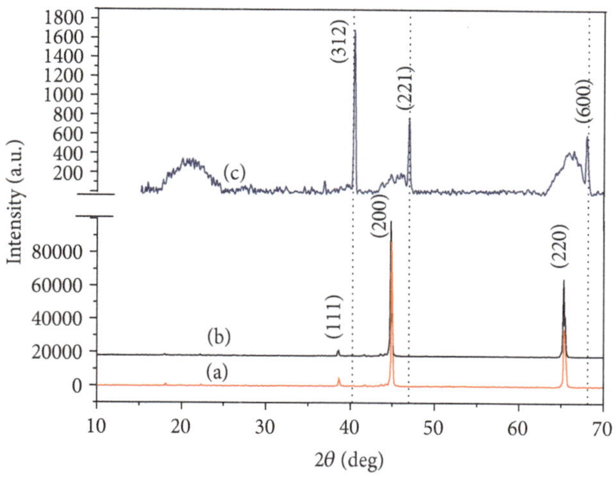

FIGURE 1: Comparison of XRD patterns of (a) bare aluminum, (b) coated aluminum cured at 130°C, both acquired at normal incidence and (c) that of coated aluminum, cured at 130°C, acquired at GI of 1 degree.

X-rays into the bulk during normal incidence, which tend to add diffraction peaks from the underlying crystalline matrix of the substrate to that of the coating on the surface, thereby masking the information due to the coating. Figure 1 presents the comparison of the normal incidence XRD patterns of bare aluminum (Figure 1(a)) and that of the coated aluminum (Figure 1(b)), which was cured at 130°C. However, it can be clearly seen that both patterns look similar. Due to high penetration depth of X-rays, the information from the coating has been masked due to the highly crystalline nature of the metallic substrate. Figure 1(c) shows the XRD pattern obtained from the GIXRD analysis. Here, it can be clearly seen that there is an initial hump, which is an indication of presence of an amorphous phase along with sharp peaks that indicate presence of a crystalline material. Since the peaks do not coincide with those of the aluminum substrate (JCPDS 04-0787), it can be inferred that these peaks are due to a different phase in the coating, which correspond to an aluminosilicate with the formula Al_2SiO_5 (JCPDS 44-0027). The XRD analysis shows that the formed material during the *in situ* anodization/sol-gel deposition process is crystalline on an as-deposited sample, even at room temperature. The amorphous phase in the coating could be silica and/or zirconia from the GZ sol.

3.2. SEM/EDAX Analysis. An image of cross section of the coated aluminum substrate is shown in Figure 2(a). The coating is seen to be 15–20 microns thick and the EDAX spectrum of the coating is shown in Figure 2(b). It can be seen from Figure 2(a) that the coating is highly dense in nature. The SEM images of the surface of the coated sample as shown in Figure 3 confirm that the coating is dense with negligible surface porosity. The coatings cured at 130°C (as shown in Figure 3(a)) do not show any defects. However, when the coatings are heat-treated at 300°C, the magnified image of the surface (inset shown in Figure 3(b)) confirms that cracks had

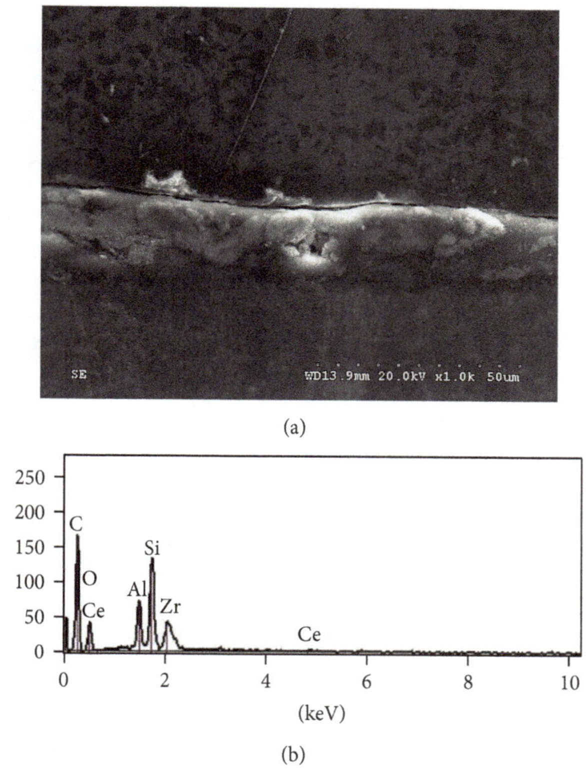

FIGURE 2: (a) SEM image of cross section of coated Al. (b) EDAX spectrum of the indicated area of (a).

appeared in the coating. Hence, it was concluded that a low temperature of 130°C for curing was sufficient to remove the residual solvent, densify, and generate a defect-free coating.

3.3. Scratch and Adhesion Test. The onset load for coating cracking is commonly referred to as the critical load and is associated with the rise in values of coefficient of friction and acoustic emission. In the present case, the frictional load and the acoustic emission were not clearly distinguishable as the coatings started to crack at about 2 N normal loads. Hence, morphology of the scratch was considered to rank the coatings for simple dip coated and electrolytically deposited coatings that were abbreviated as SD and ED, respectively.

From the scratch test results, the simple dip coated specimen started showing visible cracks and delamination even at the bare minimum load of 0.9 N applied during point of contact, whereas there were no cracks in the electrolytically coated specimens till the normal load increased to 2 N. The loads mentioned here are the average values of three scratch tests carried on each sample at different locations. These are the critical loads for the respective coating conditions. Normally, small fractures are generated across the scratch length with increasing load at certain periodicity. However, the coatings still stay adhered to the substrate till a second critical load. But when the normal load rises further, coating delaminates forming debris on the sides of the scratch. Scratch tracks are visible even to the naked eye due to the scattering of light by the defects along the width and/or depth of the scratch. While the SD sample has failed in

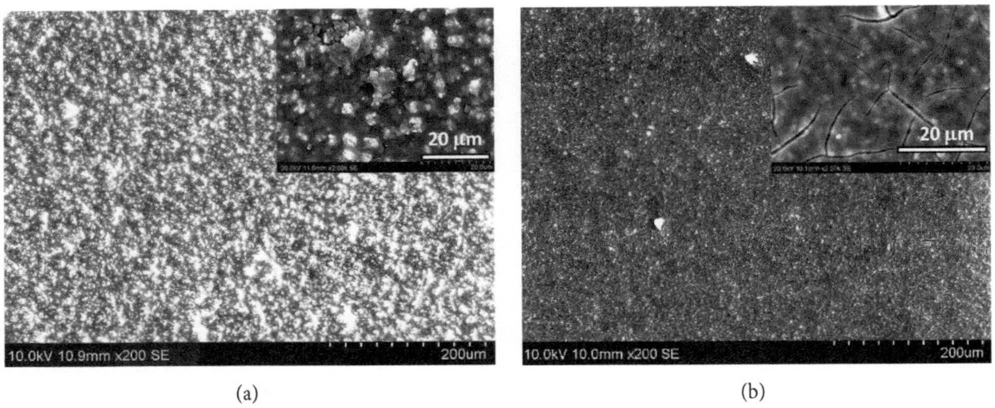

FIGURE 3: SEM images of the surface of coated substrate (a) cured at 130°C and (b) cured at 300°C.

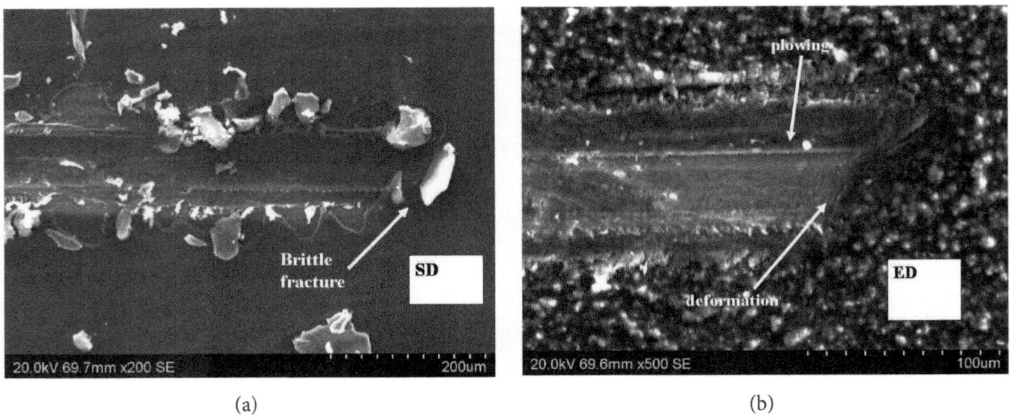

FIGURE 4: SEM images of scratch (a) simple dip coated (SD) and (b) electrolytically deposited (ED) aluminum.

classic brittle fracture as can be discerned from Figure 4(a), ED failed according to classic ductile fracture as seen in Figure 4(b). The results from the scratch tests can be further understood by examining the morphology of the ED samples by SEM analysis. The coatings have a microlevel roughness with extensive porosity due to dendrite-like structure as seen from the inset of Figure 3(a). But these dendrites and the pores are covered with a layer of the silica-zirconia formed from the sol. These dendrites have further submicron and nanoparticle structure resulting from the ED process. Such structure acts like a reinforcement and provides necessary improved tensile strength. Hence, the ED coatings are able to resist the brittle fracture at lower load unlike a normal sol-gel coating, though the coating material is the same. Accordingly, the dendrite structure in case of the former is rougher due to enhanced height of such features. The adhesive properties were tested for ED coatings and were found as 5B (0% removal after adhesion test), which implies that the coating has excellent adhesion to the substrate.

3.4. Corrosion Tests

3.4.1. Neutral Salt Spray Tests (NSST). The photographs of coated and bare substrates after NSST are shown in Figure 5. The substrate panels were scratched prior to exposure to salt spray to accelerate the corrosion process. The bare aluminum panel totally failed in 48 hours, while the coating in the normal dip coated substrate started to peel off after 48 hours and the scratched area was filled with the corrosion product. In case of electrochemically anodized cum deposited substrate (ED), no corrosion was found to occur even after 400 hours. The performance of the ED coating as a barrier coating is far superior to simple dip coated aluminum and is capable of rendering a long-term corrosion protection to the aluminum substrate and, hence, further corrosion resistance property measurements like potentiodynamic polarization and electrochemical impedance studies were carried out only on ED samples and not on the SD samples.

3.4.2. Potentiodynamic Polarization Measurements. The data obtained from the potentiodynamic polarization studies carried out after 1 h and 24 h exposure to 3.5% NaCl, on the electrolytically coated aluminum and bare aluminum substrates are presented in Figure 6 and the results obtained after R_p fitting the data are presented in Table 1. It could be seen that the electrolytically coated Al provided a good barrier protection to the aluminum substrate, since the corrosion current density, i_{corr}, was lower when compared to that for the bare aluminum substrate. The corrosion potentials were lower for the coated aluminum substrates when compared to

(a) (b) (c)

FIGURE 5: Photographs of samples after salt spray test: (a) after 48 hours—bare aluminum substrate, (b) after 48 hours—dip coated substrate with 3 mm/s withdrawal speed, and (c) after 400 hours—electrolytically deposited substrate.

TABLE 1: R_p fit data for the results obtained from potentiodynamic polarization measurements.

Sample	E_{corr} (Volts)	I_{corr} (Amp/cm^2) $\times 10^{-7}$	R_p (Ohms/cm^2)
Bare Al-1 h	−0.806	64.104	4069.4
Bare Al-24 h	−1.059	15.849	16460
ED-1 h	−1.409	5.247	49714
ED-24 h	−1.377	2.513	1.038*E*5

those for the bare Al substrate. The reduction in the open circuit potential (OCP) could be attributed to the effective suppression of the cathodic reaction due to the reason that SiO_2 having a low isoelectric point (IEP = 1.7 − 3.5) leads to a negative surface charge at pH > 2. Since the coatings are exposed to the electrolyte solution 3.5% NaCl that has a pH from 6.5 to 7, which is higher than the IEP, the open circuit potential of the coated substrate is more negative than that of the bare substrate. Though the coatings had lower corrosion potentials than bare Al, they provided a good barrier effect and reduced the corrosion currents.

In case of only the bare Al, after 1 h exposure, the i_{corr} value is higher in magnitude when compared to that of the 24 h exposure, which implies the formation of an oxide layer during longer exposure time. The oxide layer formation is evident by the passive region formation from −1.0 V to −0.70 V (300 mV), which decreases the rate of corrosion during longer periods of exposure to the electrolyte solution as shown in Figure 6.

It has been reported that the presence of Ce^{3+} in an electrolyte solution reduces the rate of oxygen reduction reaction and E_{corr} shifts to more negative values when compared to that without Ce^{3+} ions [23, 24]. In the present case, the Ce^{3+} ions were present in coating itself. In case of the coated substrates exposed to the electrolyte solution for 1 and 24 h, namely, ED-1 h and ED-24 h, the cathodic arm was shifted to more negative potentials and also lower current densities, when compared to the bare aluminum substrate, indicating a reduction in the rate of oxygen reduction reaction as shown in Figure 6. The corrosion resistance is much higher for the ED samples when compared to that for bare Al for both exposure times. A well-established passive region can be observed for ED samples for 1 h exposure itself and it is still improved for 24 h exposure. This shows that the coatings are highly protective. The 900 mV, that is, the difference between E_{corr} (corrosion potential) and E_c (breakdown potential), is associated with the increased corrosion resistance of passive coatings as shown in Figure 6.

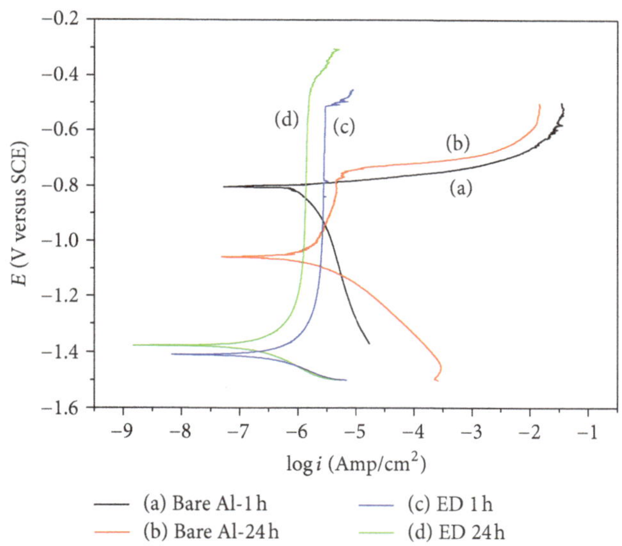

FIGURE 6: Potentiodynamic polarization plots of bare and coated Al (with ED) after 1 h and 24 h exposure to 3.5% NaCl solution.

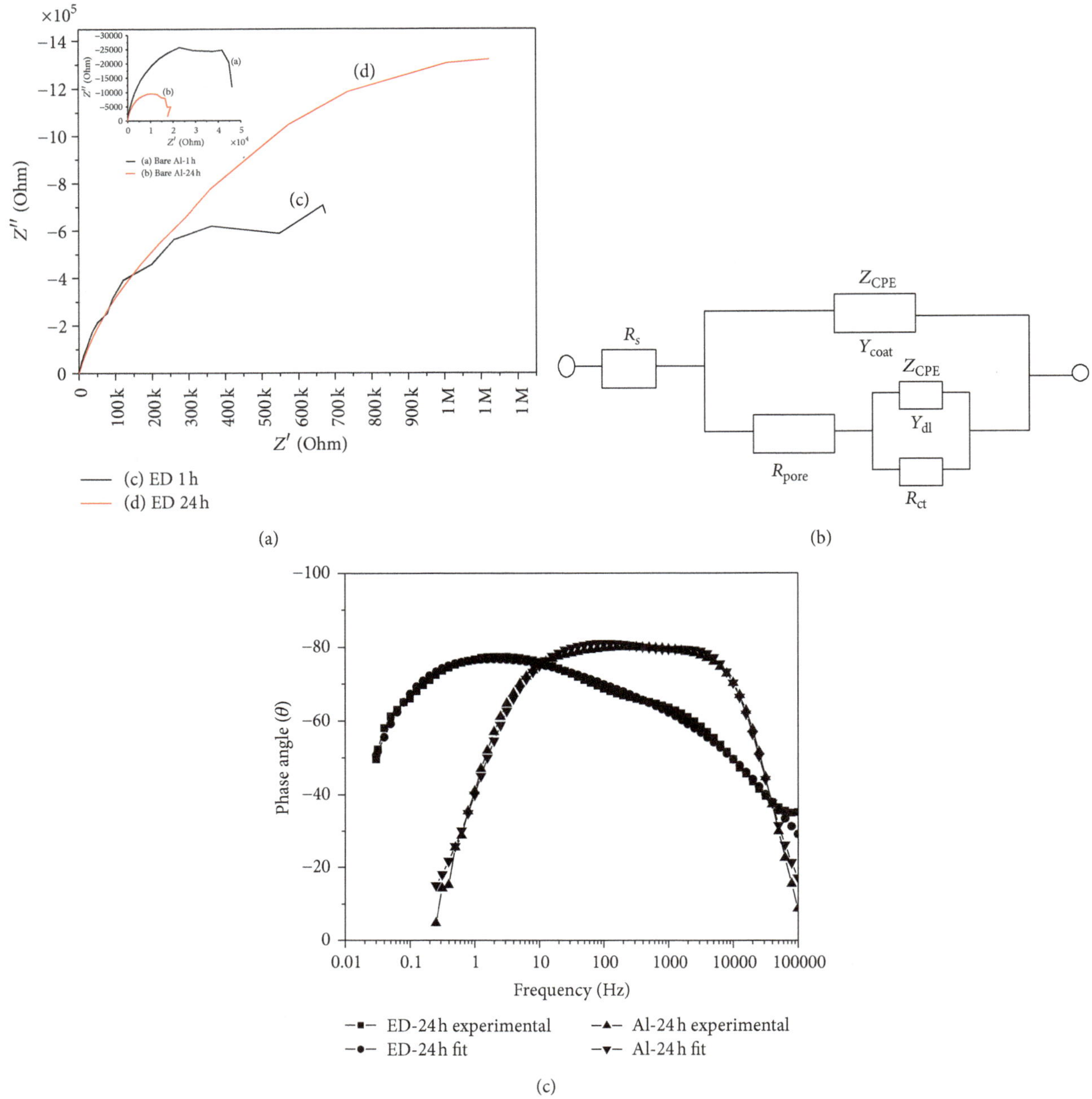

FIGURE 7: (a) Nyquist plots for the coated and bare Al for 1 h and 24 h exposure to 3.5% NaCl, (b) equivalent circuit used for fitting the EIS data of coated substrate, and (c) Bode plot for bare aluminum and ED coating on aluminum after 24 h exposure to 3.5% NaCl.

3.4.3. Electrochemical Impedance Spectroscopy (EIS). EIS is a nondestructive testing technique for evaluation of the barrier properties of the coatings and adhesion to the substrate. The obtained EIS data for the coated substrates are shown as Nyquist plots in Figure 7(a) and were fitted using the equivalent circuit given by two time constants as shown in Figures 7(b) and 7(c).

A constant phase element (CPE) was used instead of an "ideal capacitor" to explain the deviations from ideal behavior. The possible reasons for a nonideal behavior could be due to surface roughness, inhomogeneous reaction rates on the surface, varying thickness, or composition of a coating [25–27]. The time constant at high frequencies is related to the properties of coating and that at medium and lower frequencies is related to the properties of substrate. The impedance of a CPE, that is, Z_{CPE}, could be defined as $Z_{\text{CPE}} = 1/Y(j\omega)^n$, where ω is the angular frequency in rad·s^{-1}, Y is the pseudocapacitance, and n is called the CPE exponent, which is associated with the system inhomogeneity. When $n = 1$, the system behaves like a pure capacitor and $Y = C$. When CPE is used to fit the experimental data, $n < 1$; R_s is the resistance of the electrolyte; R_{pore} is the resistance to charge transfer through pores; Y_{coat} is the pseudo-capacitance of coating; Y_{dl} is pseudocapacitance associated with the double layer formed at the metal-electrolyte interface in parallel with charge transfer resistance R_{ct} describing the corrosion of the

Table 2: Results after fitting the EIS data for the bare and ED aluminum substrates after 1 h and 24 h exposure to 3.5% NaCl.

Sample	R_s ($\Omega\cdot cm^2$)	Y_{coat} ($Ss^n cm^{-2}$)	n	R_{pore} ($\Omega\cdot cm^2$)	Y_{dl} ($Ss^n cm^{-2}$)	n	Rct ($\Omega\cdot cm^2$)	χ^2
Bare Al-1 h	2.37	$7.88E-7$	0.97	70.44	$4.76E-6$	0.83	$6.04E4$	0.003
Bare Al-24 h	1.54	$3.38E-6$	0.96	124.7	$5.89E-6$	0.81	$2.11E4$	0.011
ED-Al-1 h	27.47	$5.56E-7$	0.79	163.5	$3.20E-6$	0.91	$2.02E6$	0.015
ED-Al-24 h	27.34	$1.26E-6$	0.83	637.2	$8.07E-7$	0.86	$4.34E6$	0.011

metal substrate. Table 2 presents the EIS results after circuit fitting of bare and ED samples exposed to 3.5% NaCl for 1 h and 24 h. It can be seen from Figure 7(a) and Table 2 that the R_{ct} values of the ED samples are higher by two orders of magnitude than those of bare samples after 1 h and 24 h exposure which shows that the coating does not allow the electrolyte to penetrate and reach the substrate. Even the R_{pore} values were also found to be higher (refer to Table 2) for coated samples at both exposed timings than those for bare Al substrate, showing less porosity in the coatings. The higher R_{pore} value of ED sample exposed for 24 h is due to closing of the existing pores by Ce^{3+} ions which are converted into $Ce(OH)_3$ and $Ce(OH)_4$ precipitates.

4. Conclusion

The novel process of *in situ* anodizing/sol-gel coating deposition substantially improved the corrosion resistance properties of the aluminum substrate and as expected, this one-step anodizing/deposition process circumvents the deterioration of mechanical properties of substrate due to the process being carried out at room temperature process which does not need high temperature curing. The SEM studies suggest that a low temperature of 130°C is sufficient for curing to obtain crack-free coatings. Since this coating is highly stable in salt spray test, it has considerable potential as future alternatives for nontoxic chromate-free conversion coatings.

Acknowledgments

The authors gratefully acknowledge the constant support and encouragement provided by Dr. G. Sundararajan and Dr. G. Padmanabham, ARCI, Hyderabad, India. The authors also would like to acknowledge Mr. G. V. R. Reddy for the SEM analysis and Mr. A. Ramesh for the technical support.

References

[1] M. P. Ryan, D. E. Williams, R. J. Chater, B. M. Hutton, and D. S. McPhail, "Why stainless steel corrodes," *Nature*, vol. 415, no. 6873, pp. 770–774, 2002.

[2] I. Betova, M. Bojinov, T. Laitinen, K. Mäkelä, P. Pohjanne, and T. Saario, "The transpassive dissolution mechanism of highly alloyed stainless steels: I. Experimental results and modelling procedure," *Corrosion Science*, vol. 44, no. 12, pp. 2675–2697, 2002.

[3] S. A. Kulinich and A. S. Akhtar, "On conversion coating treatments to replace chromating for Al alloys: recent developments and possible future directions," *Russian Journal of Non-Ferrous Metals*, vol. 53, no. 2, pp. 176–203, 2012.

[4] V. Moutarlier, M. P. Gigandet, and J. Pagetti, "Characterisation of pitting corrosion in sealed anodic films formed in sulphuric, sulphuric/molybdate and chromic media," *Applied Surface Science*, vol. 206, no. 1–4, pp. 237–249, 2003.

[5] S. J. Garcia-Vergara, P. Skeldon, G. E. Thompson, and H. Habazaki, "A tracer investigation of chromic acid anodizing of aluminium," *Surface and Interface Analysis*, vol. 39, no. 11, pp. 860–864, 2007.

[6] L. E. Fratila-Apachitei, F. D. Tichelaar, G. E. Thompson et al., "A transmission electron microscopy study of hard anodic oxide layers on AlSi(Cu) alloys," *Electrochimica Acta*, vol. 49, no. 19, pp. 3169–3177, 2004.

[7] X. Liu, J. Gen, A. Yu, G. Wang, L. Song, and X. Zhang, "A study on a submerged membrane bioreactor for municipal wastewater treatment," *Advanced Materials Research*, vol. 531, pp. 415–418, 2012.

[8] A. Jagminas, D. Bigeliend, I. Mikulskas, and R. Tomasiunas, "Growth peculiarities of aluminum anodic oxide at high voltages in diluted phosphoric acid," *Journal of Crystal Growth*, vol. 233, no. 3, pp. 591–598, 2001.

[9] R. Narayanan and S. K. Seshadri, "Phosphoric acid anodization of Ti-6Al-4V—structural and corrosion aspects," *Corrosion Science*, vol. 49, no. 2, pp. 542–558, 2007.

[10] S. J. Garcia-Vergara, P. Skeldon, G. E. Thompson, and H. Habazaki, "Pore development during anodizing of Al-3.5 at.%W alloy in phosphoric acid," *Surface and Coatings Technology*, vol. 201, no. 24, pp. 9506–9511, 2007.

[11] F. L. Coz, L. Arurault, S. Fontorbes, V. Vilar, L. Datas, and P. Winterton, "Chemical composition and structural changes of porous templates obtained by anodising aluminium in phosphoric acid electrolyte," *Surface and Interface Analysis*, vol. 42, no. 4, pp. 227–233, 2010.

[12] M. A. Song-jiang, L. Peng, Z. Hai-hui, F. U. Chao-Peng, and K. Ya-fei, "Preparation of anodic films on 2024 aluminum alloy in boric acid-containing mixed electrolyte," *Transactions of Nonferrous Metals Society of China*, vol. 18, no. 4, pp. 825–830, 2008.

[13] H.-H. Shih and S.-L. Tzou, "Study of anodic oxidation of aluminum in mixed acid using a pulsed current," *Surface and Coatings Technology*, vol. 124, no. 2-3, pp. 278–285, 2000.

[14] M. A. Páez, J. H. Zagal, O. Bustos, M. J. Aguirre, P. Skeldon, and G. E. Thompson, "Effect of benzotriazole on the efficiency of anodizing of Al-Cu alloys," *Electrochimica Acta*, vol. 42, no. 23-24, pp. 3453–3459, 1997.

[15] T. Takenaka, H. Habazaki, and H. Konno, "Formation of black anodic films on aluminum in acid electrolytes containing titanium complex anion," *Surface and Coatings Technology*, vol. 169-170, pp. 155–159, 2003.

[16] A. Da Forno, M. Bestetti, N. Lecis, S. P. Trasatti, and M. Trueba, "Anodic oxidation and silane treatment for corrosion protection of AM60B magnesium alloy," *Materials Science Forum*, vol. 690, pp. 413–416, 2011.

[17] F. Zucchi, V. Grassi, A. Frignani, C. Monticelli, and G. Trabanelli, "Influence of a silane treatment on the corrosion resistance of a WE43 magnesium alloy," *Surface and Coatings Technology*, vol. 200, no. 12-13, pp. 4136–4143, 2006.

[18] K. Watanabe, M. Sakairi, H. Takahashi, K. Takahiro, S. Nagata, and S. Hirai, "Anodizing of aluminum coated with silicon oxide by a sol-gel method," *Journal of the Electrochemical Society*, vol. 148, no. 11, pp. B473–B481, 2001.

[19] X. Du and Y. Xu, "Formation of Al_2O_3-$BaTiO_3$ nanocomposite oxide films on etched aluminum foil by sol-gel coating and anodizing," *Journal of Sol-Gel Science and Technology*, vol. 45, no. 1, pp. 57–61, 2008.

[20] J. G. Kaufman, *Properties of Aluminum Alloy*, ASM International, Materials Park, Ohio, USA, 2008.

[21] D. Wang and G. P. Bierwagen, "Sol-gel coatings on metals for corrosion protection," *Progress in Organic Coatings*, vol. 64, no. 4, pp. 327–338, 2009.

[22] M. L. Zheludkevich, J. Tedim, and M. G. S. Ferreira, "'Smart' coatings for active corrosion protection based on multifunctional micro and nanocontainers," *Electrochimica Acta*, vol. 82, pp. 314–323, 2012.

[23] B. R. W. Hinton, "Corrosion inhibition with rare earth metal salts," *Journal of Alloys and Compounds*, vol. 180, no. 1-2, pp. 15–25, 1992.

[24] E. A. Matter, S. Kozhukharov, M. Machkova, and V. Kozhukharov, "Comparison between the inhibition efficiencies of Ce(III) and Ce(IV) ammonium nitrates against corrosion of AA2024 aluminum alloy in solutions of low chloride concentration," *Corrosion Science*, vol. 62, pp. 22–33, 2012.

[25] C. A. Schiller and W. Strunz, "The evaluation of experimental dielectric data of barrier coatings by means of different models," *Electrochimica Acta*, vol. 46, no. 24-25, pp. 3619–3625, 2001.

[26] W. H. Mulder, J. H. Sluyters, T. Pajkossy, and L. Nyikos, "Tafel current at fractal electrodes: connection with admittance spectra," *Journal of Electroanalytical Chemistry*, vol. 285, no. 1-2, pp. 103–115, 1990.

[27] C.-H. Kim, S.-I. Pyun, and J.-H. Kim, "An investigation of the capacitance dispersion on the fractal carbon electrode with edge and basal orientations," *Electrochimica Acta*, vol. 48, no. 23, pp. 3455–3463, 2003.

16

Corrosion Study of Powder-Coated Galvanised Steel

Manish Kumar Bhadu, Akshya Kumar Guin, Veena Singh, and Shyam K. Choudhary

R&D and Scientific Services, TATA Steel Ltd., Jamshedpur 831001, India

Correspondence should be addressed to Manish Kumar Bhadu; mkbhadu@gmail.com

Academic Editors: C.-H. Hsu and S. Umoren

In general, steel is protected from corrosive environments by conversion coatings, that is, phosphating, chromating, and so forth, and then followed by different layers of paints. Nowadays, strict pollution laws and regulations are creating significant challenges for coating experts to develop an environmentally friendly product. Powder coatings have demonstrated their ability as alternative to traditional solvent-borne coatings. In the present work, polyester-based two coating systems have been investigated and their performances have been evaluated for surface topographical properties by Scanning electron microscope (SEM), and energy dispersive spectrometry (EDX), accelerated corrosion resistance by salt spray test, and impedance property by electrochemical impedance spectroscopy. Coating adhesion with the steel and hardness were evaluated by bond strength, cross cut adhesion, and pencil hardness. This paper explains the results and performance of the coatings by the above two systems.

1. Introduction

Galvanised steel is widely used in construction, automobile, and white goods sector. Zinc coating is the most effective and economic means to protect the steel substrate exposed to atmospheric corroding environments. It protects the steel substrate by acting as a barrier against the corrosive environment and by sacrificially corroding themselves to provide cathodic protection.

Protective ability of galvanised (GI) steel may be enhanced by employing thicker zinc coatings or by painting the metallic substrate [1]. Paints improve the surface life of underlying zinc coating acting as a barrier against zinc reaction with environmental agents. Cracks, crater, and pin holes occurring in the paint are sealed by corroded zinc products. Moreover, corroded zinc products occupy a 20–25% more volume than zinc, while iron oxides (corrosion product of steel) occupy a volume several times larger than the steel; thus, expansive forces are reduced at the zinc-paint interface compared to those at the steel-paint one [2]. The main practical problem concerning painting of zinc-coated surface lies in achieving good bond strength, that is, good adhesion of polymer with GI sheet. Often adhesion looks satisfactory immediately after painting, but it prematurely degrades after water, oxygen, and other corrosive ingredients diffuse through the polymeric coating. Several pretreatment processes were reported to improve coating adhesion. The main function of pretreatment for GI steel surface is to form a very stable passive film which will enhance the adhesion with subsequent polymeric film [3]. There are a number of pretreatment processes like phosphating, chromating, and coating of rust preventing compounds by chemical and physical vapor deposition and diffusion coating, and so forth. Some of these pretreatment processes do not contain a film forming material that is, nonpriming, and hence, they should be compatible with the rest of the painting system [4]. Primers contain a film forming material and are expected to act as the anchorage of the paint system [5]. Some of these coatings are toxic and pose concerns to the environment.

Powder thermoset coatings are solvent-free and unlike the conventional liquid coatings have zero volatile organic content (VOC). Thus, they could offer the coating formulators a robust and a highly promising approach to produce eco-friendly coatings. Moreover, not only powder coatings are easier to apply than solvent-based coatings, but, they also provide a thicker and more uniform coating. If we take into account the economical advantages of powder coatings, like cost savings in water disposal, high yield, and cheap maintenance, the total operating cost of a powder application plant is lower than that of traditional solvent-based liquid paints.

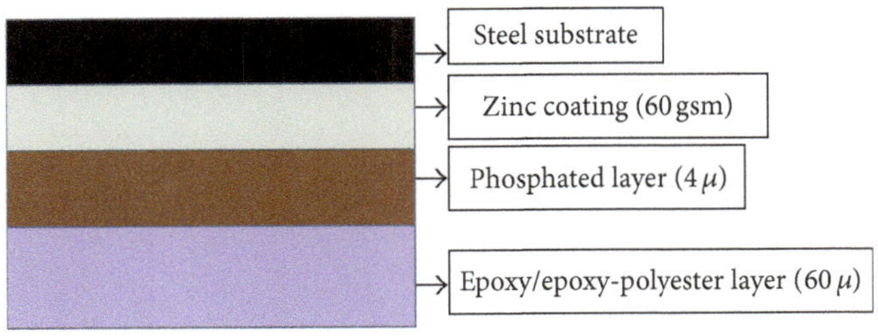

FIGURE 1: Schematic diagram of different coating layers with thickness.

Furthermore, a wide variety of finishes such as structured, wrinkled, metallic, and antique finishes are available with powder coatings [6–9].

Although it is possible to achieve high-gloss and smooth coatings with excellent adhesion, flexibility, and hardness by epoxy powder coating, they however exhibit poor tolerance to heat and light resulting in a pronounced tendency to fade [10]. In order to overcome this shortcoming, polyester/epoxy blends-based coating technology has been developed which shows excellent film smoothness, appropriate mechanical properties, and adherence characteristics [11]. Interestingly, blend coatings show a broader degradation temperature interval, and they volatilize to a substantially less extent compared to pure epoxy and polyester powder coating. Epoxy-polyester powder coating is a hybrid of epoxy and polyester powder coating. These hybrids have properties similar to those of epoxy powders; however, their additional advantage is that they have improved resistance to fading and improved weather resistance [11].

Hybrid powders are now regarded as the main backbone of the powder coating industry. The objective of the present work is to find out a suitable powder coating system for GI surface. A comparative study was made in between polyester and epoxy-polyester powder coating process. Adhesion on GI steel was assessed by suitable standardized adhesion tests. Anticorrosive behaviour was evaluated by salt spray and electrochemical impedance spectroscopy (EIS) method.

2. Experimental Details

Zero-spangle GI (120 g/m^2) sheet was used as the base substrate. These sheets were degreased to remove oil and grease followed by phosphating in tricationic phosphate base solution to obtain a thin coating of dry film thickness (DFT) ~ 4 micron. The polyester and epoxy-polyester powder coating of DFT 60 micron was applied by electrostatic spraying method on phosphate GI sheet referred to as system-1 and system-2, respectively. Coated panels were allowed to stand for 7 days at room temperature for curing before any testing. The microstructure and surface morphology of the coated samples were observed by a scanning electron microscope (SEM) equipped with energy dispersive spectrometry (EDX). The energy used for analysis was 15 KeV.

Corrosion resistance of the samples was evaluated by electrochemical impedance spectroscopy (EIS). The EIS measurements were carried out by using the VersaSTAT MC. A typical three-electrode system was employed in these tests. The samples acted as the working electrode (1 cm^2 of the exposed area), saturated calomel electrode (SCE) as the reference, and graphite as counter electrode. 3.5% NaCl solution was used as an electrolyte in all the measurements.

The EIS measurement was carried out in the frequency range of 100 KHz to 0.01 Hz, and the applied voltage was 5 mV. Salt spray tests were carried out by exposing the scribed samples (6″ × 4″) in a salt spray chamber as per the ASTM B-117 test method. The panels were checked at a regular interval of time, and results were noted down in terms of blisters, creep, and red rust. The adhesion, hardness, and bond strength measurements were performed on the coated steel sample as per the ASTM D3359, ASTM D 3363, and ASTM D 4541, respectively.

3. Results and Discussion

3.1. Morphology. The schematic diagram of the different coating layers on the steel substrate is shown in Figure 1. The cross-sectional SEM photographs of phosphate steel with polyester powder coated and epoxy-polyester powder coated are shown in Figures 2 and 3, respectively. It can be seen that coating formed by epoxy-polyester appears to be uniform, without any surface defects and cracks. From the SEM photograph, it is clearly visible that the epoxy-polyester powder coating is highly dense, and it strongly adheres to surface, whereas some visible cracks appear on polyester-coated sheet and are not so strongly adherent to surface compared to the epoxy-polyester-coated surface.

EDX spectra of polyester powder-coated and epoxy-polyester powder-coated samples are also shown in Figures 2 and 3. The peaks of different pigments materials like zinc, silica, phosphorus, and oxygen are predominant in both cases.

3.2. Bond Strength. Two major pieces of information are obtained from the bond strength test. The first is the pull-off strength, that is, bond strength of coating on substrate, and the second one is about the point where the split occurred in the paint system. The split could be an adhesive break, a cohesive break, a combination of both, or a failure of the glue.

From Table 1, cohesive failure is observed in polyester powder-coated sample, whereas adhesive failure is observed in epoxy-polyester system. Epoxy-polyester system comparatively adheres strongly with phosphated steel substrate and needs more than 5 MPa force for any type of delamination.

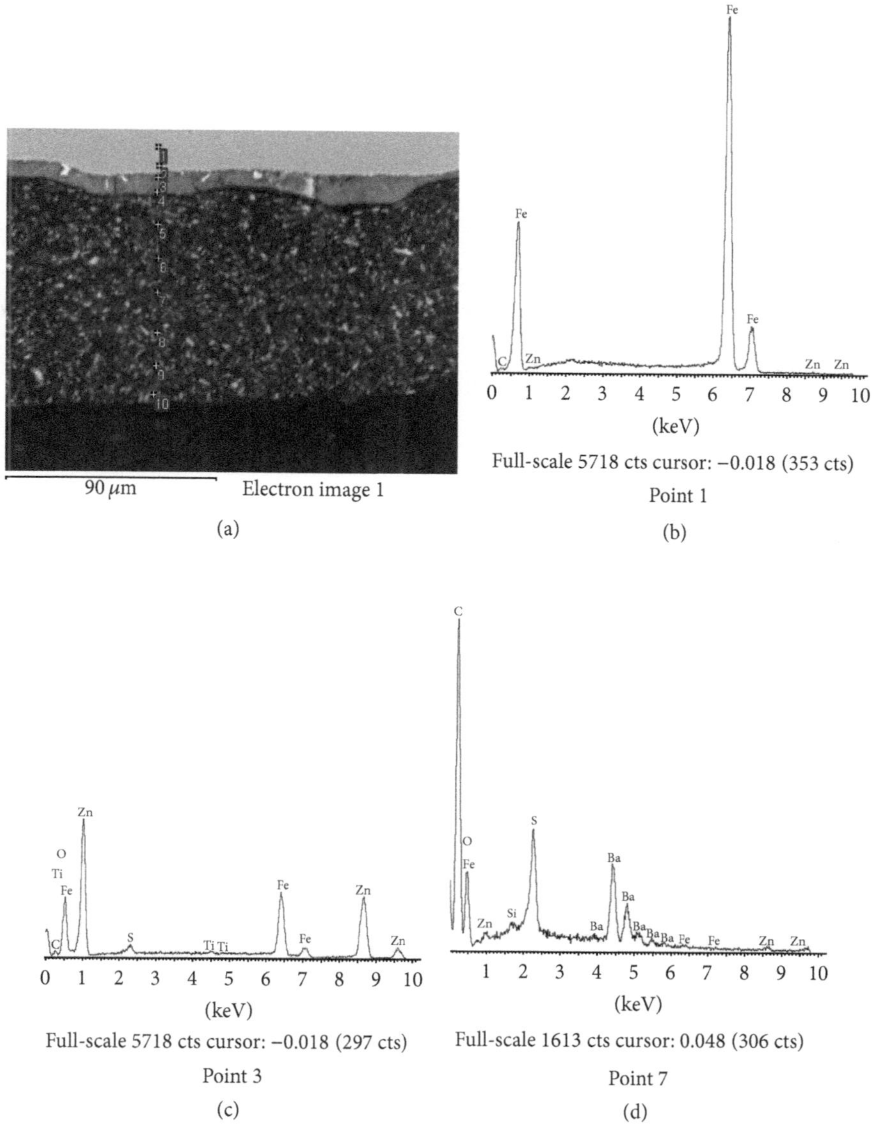

FIGURE 2: (a) Cross-sectional SEM image of polyester powder-coated sheet (system-1). (b), (c), and (d) show EDX analysis of three respective points 1, 3, and 7, shown in (a).

The failure in the case of system-2 is from the glue itself. But polyester coating system shows failure from interface and needs only 4 MPa force for any type of coating delamination. This improvement in system-2 is due to the formation of strong and crack-free bonding of epoxy-polyester hybrid with steel substrate, whereas polyester coating weakly adheres with the presence of internal crack, as noticed from SEM photograph and hence needs minimum force to delaminate from the surface as compared to epoxy-polyester coating [12–14].

3.3. Corrosion Resistance Property by Electrochemical Impedance Spectroscopy. The electrochemical properties of polyester and epoxy-polyester on phosphate steel substrate were examined by EIS measurements. The main role of EIS in the characterization of an organic coating is to provide information about the properties of the protective system, such as defects, adhesion, and barrier properties, as well as to determine the onset and progression of corrosion process on metal substrate underneath the organic coating. For the EIS measurements, elements were selected through a model equivalent circuit (see Figure 4) to represents the systems under study. The systems undergo a charge transfer control according to this circuit, where R_s represents solution resistance, R_p coating resistance, C_c coating capacitance, R_{ct} charge transfer resistance, and C_{dl} double layer capacitance.

The Bode representations of the impedance data have been analyzed with VersaSTAT MC and ZSimpWin software of Princeton applied research. Single slope in the midfrequency range shows the existence of a single time constant, and the impedance data have been analyzed using the equivalent circuit.

The impedance behavior of coating system-1 and coating system-2 on steel substrate after initial study, 144 h, and 264 h

Table 1: Bond strength and nature of failure of galvanised coated substrate.

System	Bond strength (in MPa)	Nature of failure	Photographs
1	4	Failure from top coat and adhesive	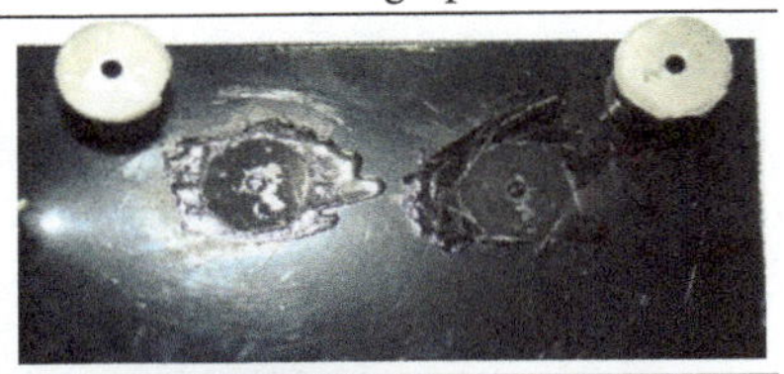
2	5	Failure from top coat and adhesive	

Table 2: Impedance parameters of wash primer and epoxy primer and top coated GI in 3.5% NaCl solution.

Time, h	System-1 resistance (Ω/cm^2)	System-2 resistance (Ω/cm^2)
Initial	8.835×10^6	1.166×10^6
24 h	7.203×10^4	1.446×10^5
96 h	7367	1.185×10^5
144 h	7166	1.633×10^4
264 h	5504	1.576×10^4

Table 3: Cross-hatch adhesion and pencil hardness results of system-1 and system-2.

System	Cross-hatch	Percentage of area removed	Pencil hardness (at 45°)
1	5B	0	7H
2	5B	0	9H

of immersion in 3.5% NaCl solution is shown Figures 5(a)–5(c) with overlay. The individual impedance behaviours with initial time, 24 h, 96 h, 144 h, and 264 h are shown in Figures 6, 7, 8, 9, and 10, respectively. The coating resistance and capacitance values derived from these figures are given in Table 2. It is clear from Figures 6, 7, 8, 9, and 10 and Table 2 that in the beginning of experiment, the coating resistance of polyester- and epoxy-polyester-coated samples is in the same range. Impedance of coating system-1 containing polyester powder coating shows 8.835×10^6 Ω/cm^2 against the 1.166×10^6 Ω/cm^2 resistance of coating system-2.

The coating resistance value of system-1 decreases from 8.835×10^6 Ω/cm^2 to 5.504×10^3 Ω/cm^2 after 264 h of immersion, whereas coating resistance value of system-2 decreases from 1.166×10^6 Ω/cm^2 to 1.633×10^4 Ω/cm^2 in 144 h of immersion and to 1.576×10^4 Ω/cm^2 in 264 h of immersion. Generally, the high impedance value of polyester-coated sample shows a fast reduction in the first 24 h of immersion, due to the development of conductive pathways inside the film. Comparatively, a slow decrease in the impedance value was observed for the epoxy-polyester powder-coated steel sheet, followed by a small recovery after 24 h of immersion [15, 16].

Penetration of water and movement of ionic species through the coating layer may be responsible for the observed decrease in coating resistance value [17]. Other reasons could be a weaker ionic resistance and a lower-cross linking density. The dielectric constants of organic coating and water are about 6 and 80, respectively, at ambient temperature. Therefore, permeation of a small amount of water through the coating can contribute to a relatively large change in the pore resistance. It is known that diffusion of electrolyte, water, and ions through the epoxy-polyester powder-coated sample is much lower than the polyester powder-coated sample. The higher corrosion protection by epoxy-polyester powder coating on steel surface is due to the higher cross-linked density of epoxy-polyester network comparing with polyester system; in addition to high cross-linking density, the additional free hydroxyl group, that is, –OH of epoxy-polyester coating system (as compared to only polyester system), forms a strong bond with phosphate and steel surface. The high cross-linking density and strong adhesion of epoxy group with phosphate

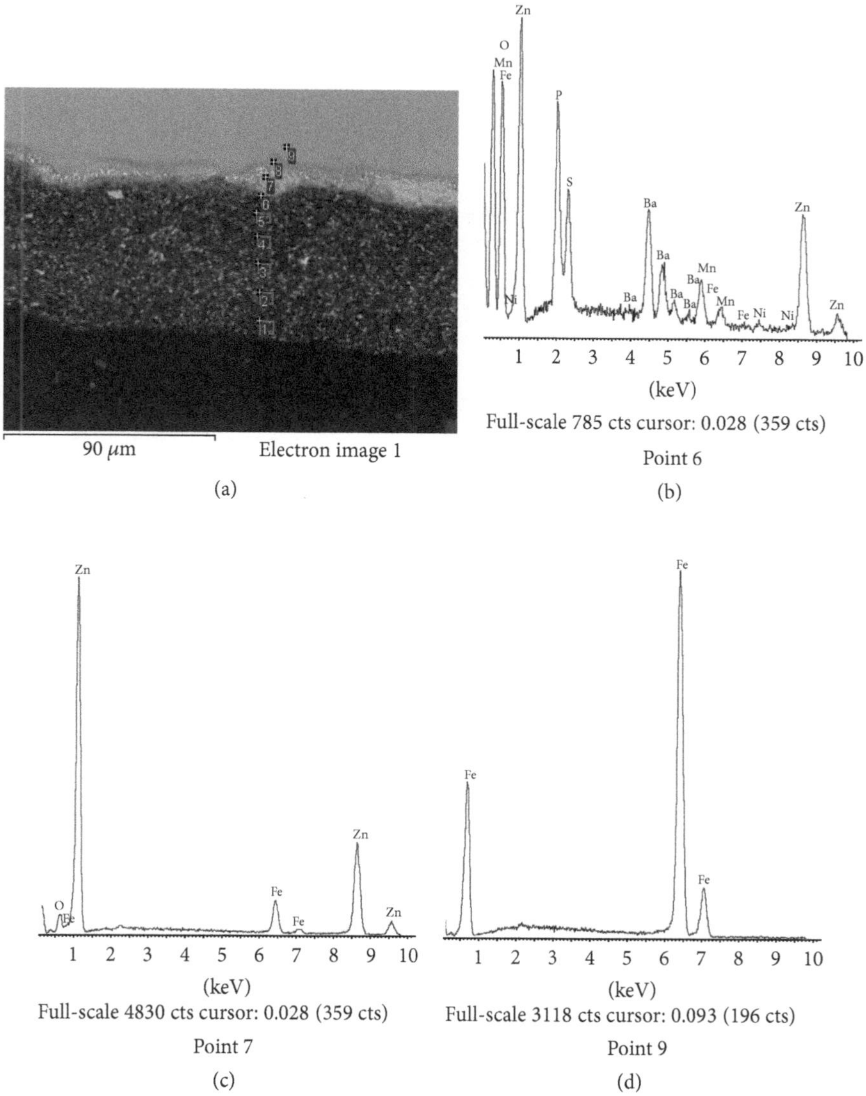

FIGURE 3: (a) Cross sectional SEM image of epoxy-polyester powder-coated sheet (system-2). (b), (c) and, (d) show EDX analysis of three respective points 6, 7, and 9, shown in (a).

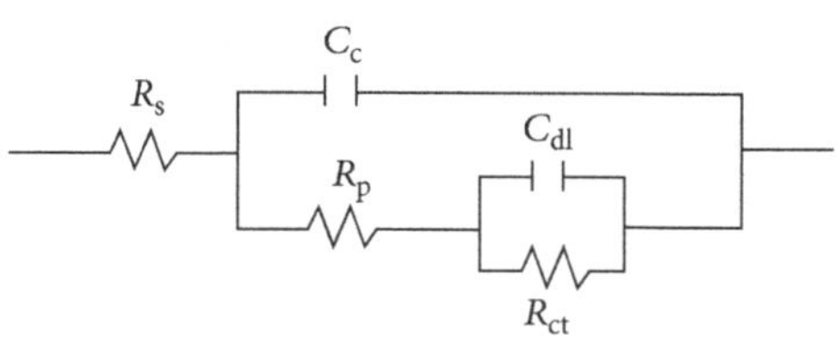

FIGURE 4: Electrical equivalent circuit for the polyester and polyester-epoxy coating on the galvanised sheet.

and steel substrate create an impermeable surface for water and other corrosive ingredients on the coated surface [18–20].

3.4. Corrosion Resistance Property by Salt Sprat Test. Performance of steel coated sheet, after 1608 h exposure in salt spray chamber is shown in Figure 11. It can be seen that epoxy-polyester powder-coated sample (system-2) is found to have less white rust even after 1608 h of exposure in salt spray chamber (ASTM B 117). Whereas coating system-1 containing polyester powder coating fails after 1200 h of exposure in salt spray chamber (ASTM B117). There are a number of blisters in and around the edges of the scribe area in the case of system-1.

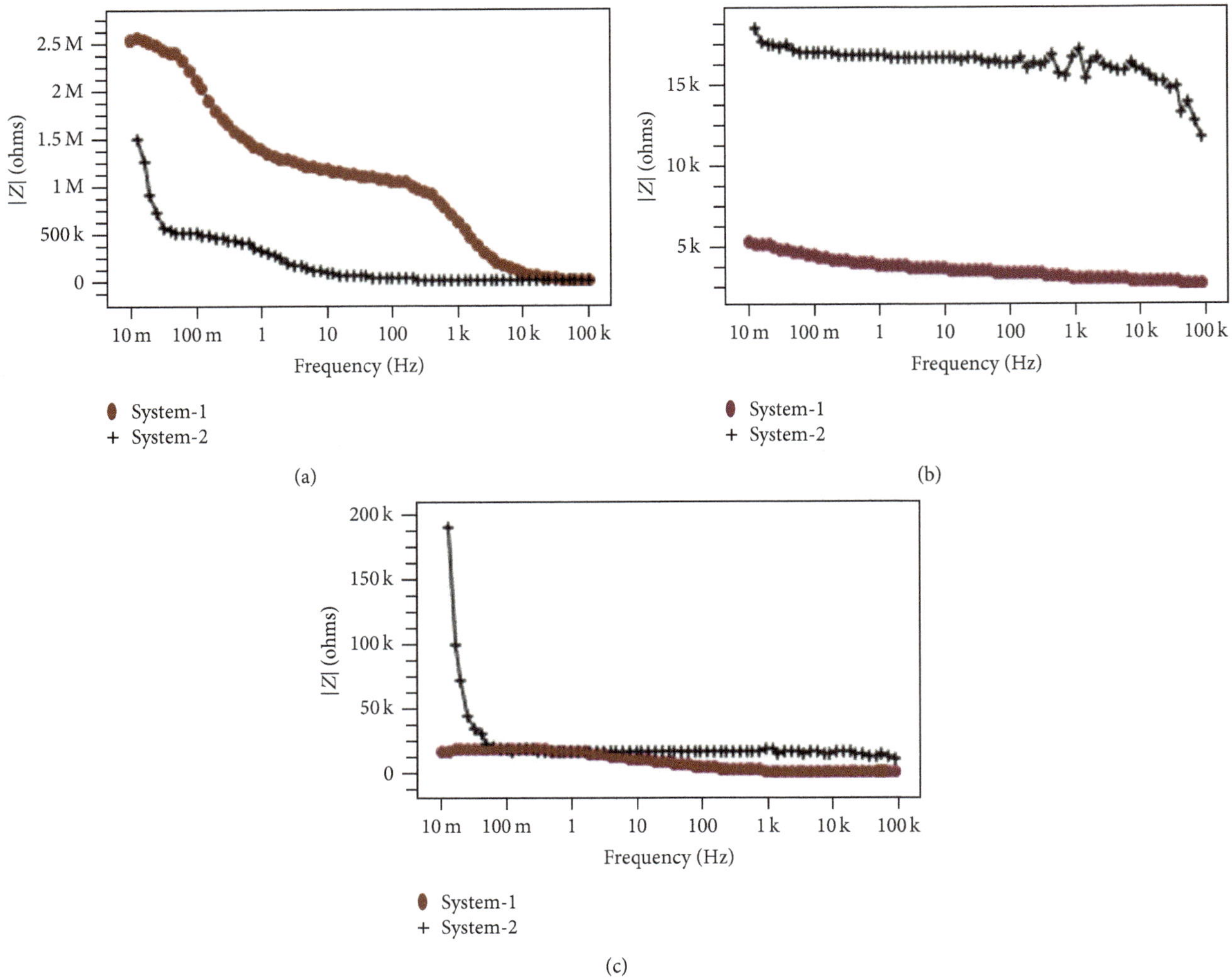

FIGURE 5: (a), (b), and (c) show EIS overlap diagram of samples immersed for time intervals, beginning, 144 h and 264 h respectively.

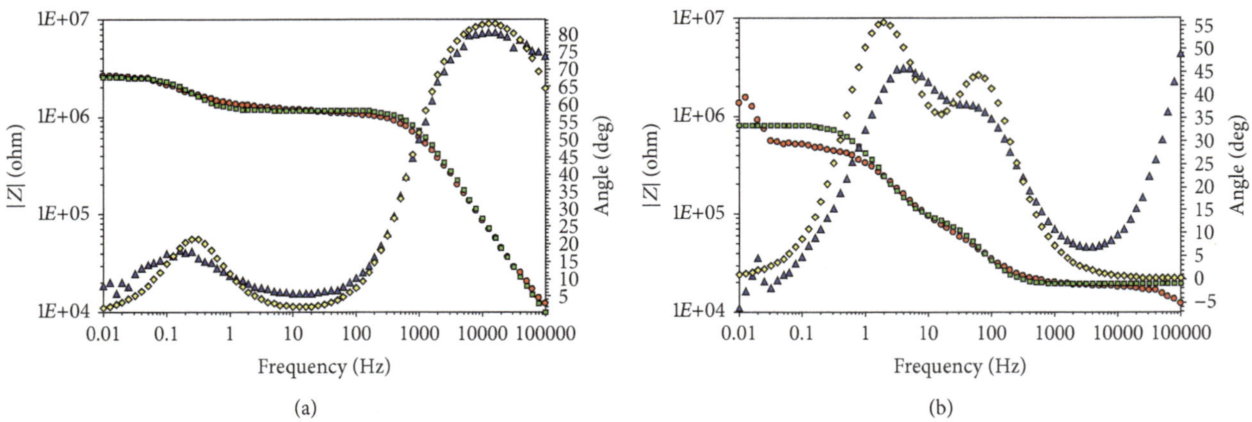

FIGURE 6: Bode plot of (a) system-1 and (b) system-2 at initial time.

3.5. Adhesion and Hardness. Cross hatch and pencil hardness of system-1 and system-2 were performed according to the ASTM D3359 and ASTM D3363-05. The results of these tests are shown in Table 3, which compares the adhesion and hardness properties of the two systems.

The result of the x-cut adhesion test was satisfactory for both the systems providing 5B with no observed flaking in the cross-cut area. Therefore, no adhesion losses appeared in any interfaces. The adhesion of primer to the steel surface and the adhesion of the top coats to the primer, both were satisfactory.

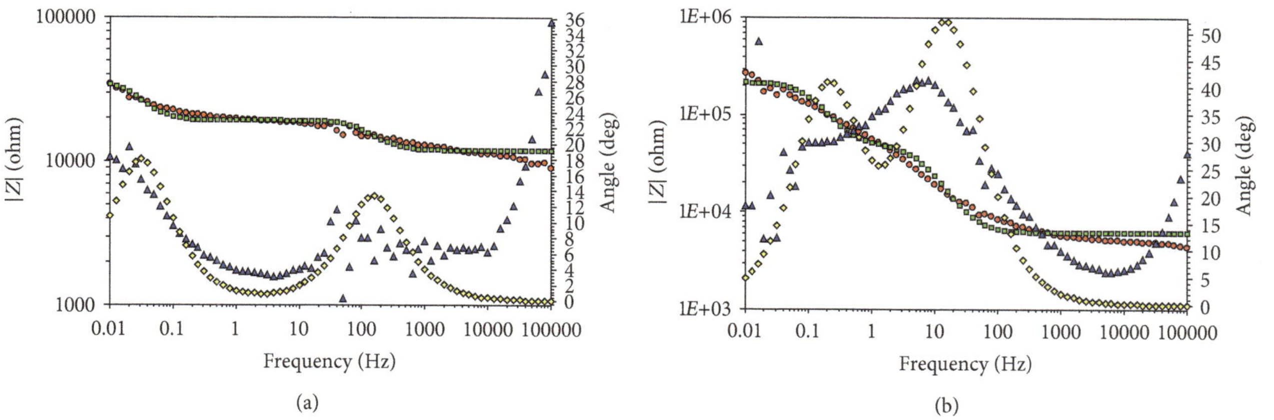

FIGURE 7: Bode plot of (a) system-1 and (b) system-2 at 24 h time.

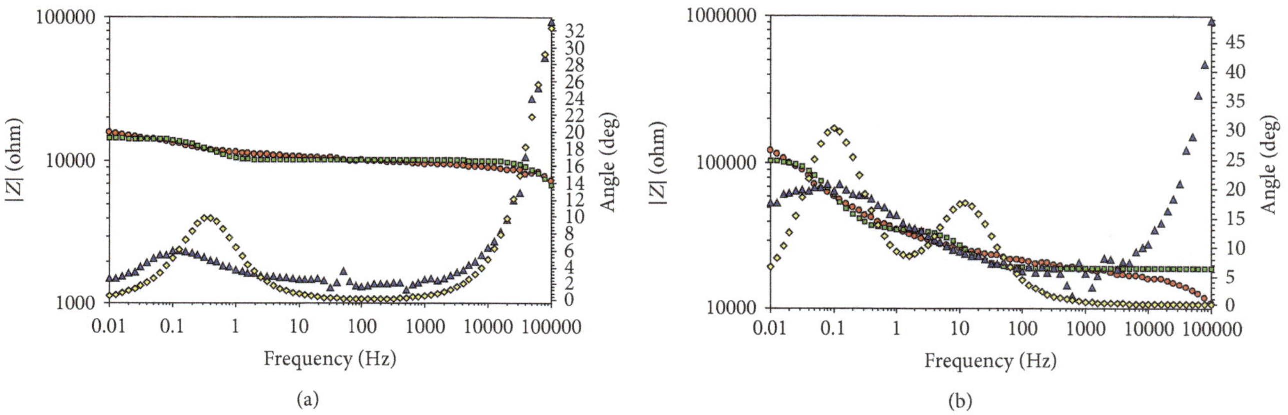

FIGURE 8: Bode plot of (a) system-1 (b) system-2 at 96 h time.

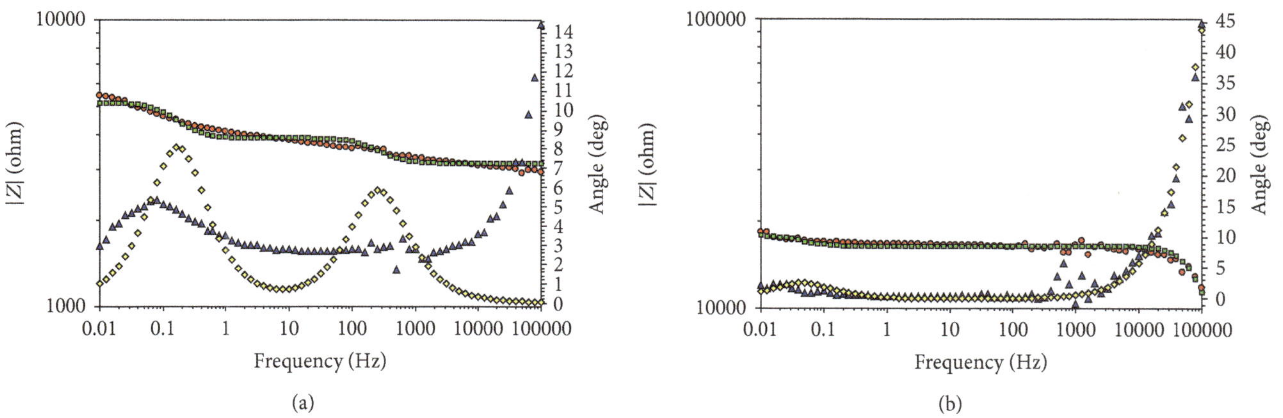

FIGURE 9: Bode plot of (a) system-1 and (b) system-2 at 144 h time.

4. Conclusion

The powder coatings protect galvanised steel substrates by the introduction of a barrier layer with relatively high ohmic resistance between the metallic substrate and the corrosive environment. Also, it is indicated that the cross-link density appears only to affect ionic conductivity of the film, and due to the higher cross-linked density of epoxy-polyester network comparing with polyester system, the permeability of water and ions through the coating film becomes less, and it leads to a more impervious film with a more resistant structure to corrosion. It is indicated that the polyester film is more porous due to the less cross-link density of cured polymer, compared with epoxy-polyester powder-coated samples.

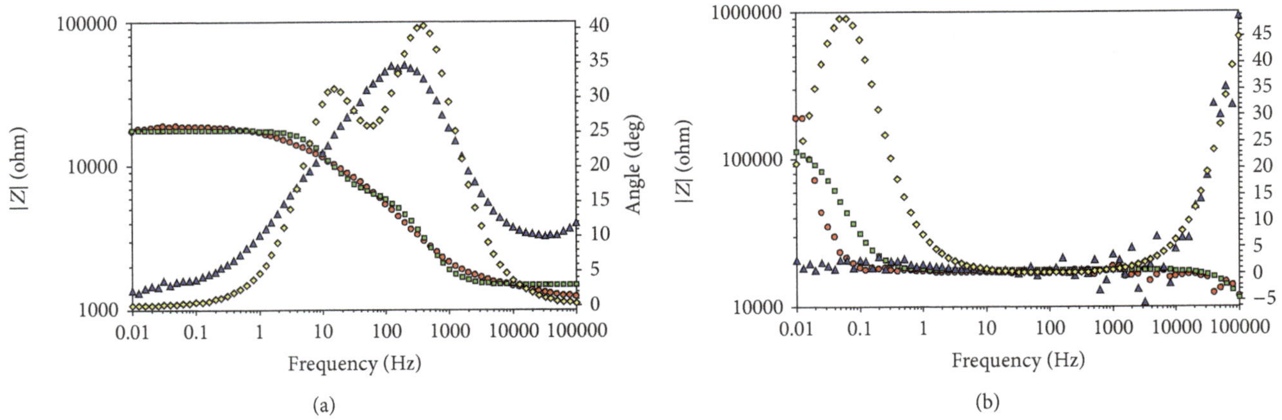

FIGURE 10: Bode plot of (a) system-1 and (b) system-2 at 264 h time.

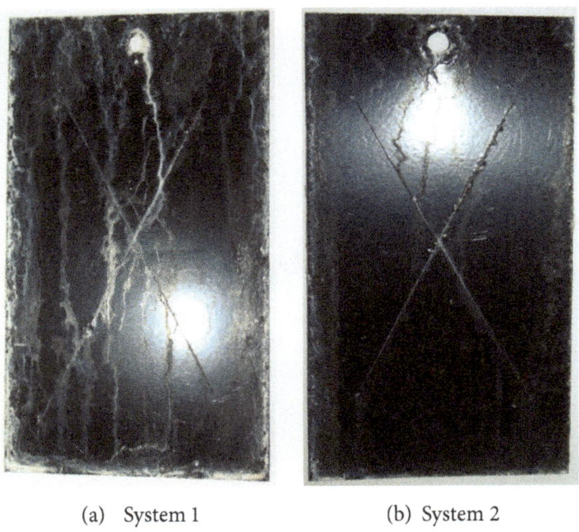

FIGURE 11: Salt spray photographs of coated substrate, (a) polyester-coated sample after 1200 h of salt spray test, (b) epoxy-polyester-coated sample after 1608 h of the salt spray test.

SEM micrograph and bond strength suggest that epoxy-polyester coating is more firmly adherent to the phosphate substrate than polyester coating due to more hydroxyl groups in epoxy-polyester bonding with phosphate substrate. Protective properties of fully cured organic coatings on metallic substrate may be attributed to a barrier and/or an active inhibition mechanism.

Acknowledgment

The authors would like to thank the staff of the corrosion lab for helping them during experiments.

References

[1] A. K. Guin, S. Nayak, T. K. Rout, N. Bandyopadhyay, and D. K. Sengupta, "Corrosion resistance nano-hybrid sol-gel coating on steel sheet," *ISIJ International*, vol. 51, no. 3, pp. 435–440, 2011.

[2] A. S. Khanna, *Introduction to High Temperature Oxidation and Corrosion*, ASM International, 1997.

[3] B. V. Jegdić, J. B. Bajat, J. P. Popić, and V. B. Mišković-Stanković, "Corrosion stability of polyester coatings on steel pretreated with different iron-phosphate coatings," *Progress in Organic Coatings*, vol. 70, no. 2-3, pp. 127–133, 2011.

[4] A. M. P. Simões, R. O. Carbonari, A. R. Di Sarli, B. del Amo, and R. Romagnoli, "An environmentally acceptable primer for galvanized steel: formulation and evaluation by SVET," *Corrosion Science*, vol. 53, no. 1, pp. 464–472, 2011.

[5] R. Mafi, S. M. Mirabedini, M. M. Attar, and S. Moradian, "Cure characterization of epoxy and polyester clear powder coatings using Differential Scanning Calorimetry (DSC) and Dynamic Mechanical Thermal Analysis (DMTA)," *Progress in Organic Coatings*, vol. 54, no. 3, pp. 164–169, 2005.

[6] D. Maetens, "Weathering degradation mechanism in polyester powder coatings," *Progress in Organic Coatings*, vol. 58, no. 2-3, pp. 172–179, 2007.

[7] J. B. Bajat, J. P. Popić, and V. B. Mišković-Stanković, "The influence of aluminium surface pretreatment on the corrosion stability and adhesion of powder polyester coating," *Progress in Organic Coatings*, vol. 69, no. 4, pp. 316–321, 2010.

[8] M. A. J. Batista, R. P. Moraes, J. C. S. Barbosa, P. C. Oliveira, and A. M. Santos, "Effect of the polyester chemical structure on the stability of polyester-melamine coatings when exposed to accelerated weathering," *Progress in Organic Coatings*, vol. 71, no. 3, pp. 265–273, 2011.

[9] F. L. Duivenvoorde, J. Laven, and R. Van der Linde, "Diblock copolymer dispersants in polyester powder coatings," *Progress in Organic Coatings*, vol. 45, no. 2-3, pp. 127–137, 2002.

[10] S. Radhakrishnan, N. Sonawane, and C. R. Siju, "Epoxy powder coatings containing polyaniline for enhanced corrosion protection," *Progress in Organic Coatings*, vol. 64, no. 4, pp. 383–386, 2009.

[11] V. C. Malshe and G. Waghoo, "Weathering study of epoxy paints," *Progress in Organic Coatings*, vol. 51, no. 4, pp. 267–272, 2004.

[12] R. Van der Linde, E. G. Belder, and D. Y. Perera, "Effect of physical aging and thermal stress on the behavior of polyester/TGIC powder coatings," *Progress in Organic Coatings*, vol. 40, no. 1–4, pp. 215–224, 2000.

[13] Technical Data Sheet of Ridoline 1352 BA, of M/s Henkel Chembond India Ltd.

[14] Technical Data Sheet of Granodine, of M/s Henkel Chembond India Ltd.

[15] M. Özcan, I. Dehri, and M. Erbil, "EIS study of the effect of high levels of SO_2 on the corrosion of polyester-coated galvanised steel at different relative humidities," *Progress in Organic Coatings*, vol. 44, no. 4, pp. 279–285, 2002.

[16] R. Naderi, M. M. Attar, and M. H. Moayed, "EIS examination of mill scale on mild steel with polyester-epoxy powder coating," *Progress in Organic Coatings*, vol. 50, no. 3, pp. 162–165, 2004.

[17] R. Mafi, S. M. Mirabedini, R. Naderi, and M. M. Attar, "Effect of curing characterization on the corrosion performance of polyester and polyester/epoxy powder coatings," *Corrosion Science*, vol. 50, no. 12, pp. 3280–3286, 2008.

[18] G. Wuzella, A. Kandelbauer, A. R. Mahendran, and A. Teischinger, "Thermochemical and isoconversional kinetic analysis of a polyester-epoxy hybrid powder coating resin for wood based panel finishing," *Progress in Organic Coatings*, vol. 70, no. 4, pp. 186–191, 2011.

[19] M. Barletta and A. Gisario, "An application of neural network solutions to laser assisted paint stripping process of hybrid epoxy-polyester coatings on aluminum substrates," *Surface and Coatings Technology*, vol. 200, no. 24, pp. 6678–6689, 2006.

[20] R. Naderi, M. M. Attar, and M. H. Moayed, "EIS examination of mill scale on mild steel with polyester-epoxy powder coating," *Progress in Organic Coatings*, vol. 50, no. 3, pp. 162–165, 2004.

17

Corrosion Inhibition and Adsorption Properties of Ethanolic Extract of *Calotropis* for Corrosion of Aluminium in Acidic Media

Sudesh Kumar and Suraj Prakash Mathur

Department of Pure and Applied Chemistry, Maharshi Dayanand Saraswati University, C-698, Panchsheel Nagar, Ajmer, Rajasthan 305009, India

Correspondence should be addressed to Sudesh Kumar; sudeshneyol@gmail.com

Academic Editors: N. Boshkov, C. Gervasi, and Q. Qu

The corrosion inhibition of aluminium in sulfuric acid solution in the presence of different plant parts, namely, leaves, latex, and fruit was studied using weight loss method and thermometric method. The ethanolic extracts of *Calotropis procera* and *Calotropis gigantea* act as an inhibitor in the acid environment. The inhibition efficiency increases with increase in inhibitor concentration. The plant parts inhibit aluminium, and inhibition is attributed, due to the adsorption of the plant part on the surface of aluminium.

1. Introduction

Aluminium is widely used in chemical industries due to its low cost and easy availability for fabrication reaction of vessel, tanks, pipeline, and boiler. Alumninium suffers from severe corrosion in aggressive environments, and it has to be protected. The corrosion of metal in aqueous solution occurs in two steps oxidation and reduction. Oxidation reaction takes place at anode, whereas reduction takes place at cathode. The cathodic reaction may be either by evolution of hydrogen or absorption of oxygen. The use of inhibitors is one of the best methods of protecting metals against corrosion [1, 2]. Most corrosion inhibitors are organic compounds having hetero atoms in their aromatic or long carbon chain. However, there is increasing concern about the toxicity of most corrosion inhibitors. The toxic effect does not only affect living organisms but also poison the environment. Due to toxicity of some corrosion inhibitors, there has been increasing search for green corrosion inhibitors [3]. Natural products of plant origin containing different organic compounds (e.g., alkaloids, tannins, pigments, organic, and amino acids) are known to have inhibitive action [4, 5]. Inhibitors in this class are those that are environment friendly, less polluting, cheap, and easily available and are obtained from natural products such as plant extracts [6]. Several studies have been carried out on the inhibition of corrosion of metals by plant extract [7–13]. Aluminium is a reactive metal to reduce the corrosion problem in these environment inhibitive effects of various naturally occurring substances like *Datura stramonium* [14], *Tannin beet* root, *Tamarind*, *Pomegranate juice*, *Saponin* [15], *Embellica officinalis*, *Terminalia bellerica*, *Mahasudarshana churna*, *Prosopis juliflora* [16], *Caparis decidu* [17], *Adhatoda vasica*, *Vinca rosea*, *Heena* [18], *Eugenia jambolana*, *pomegranate* and *Peels* [19, 20], *Tannins*, *Caffeine*, *Prosopis cineraria* [21], and *Ficus religiosa* [22] that have been evaluated as effective corrosion inhibitors. *Calotropis* is used as a traditional medicinal plant with unique properties. *Calotropis* is used alone or with other medicine to treat common diseases. Generally, organic compounds having heteroatom O, N, and S are found to have basicity and electron density thus assist in corrosion inhibitor. O, N, and S are the active centers for the adsorption on the metal surface [23, 24]. Mass loss is determined gravimetrically, volumetrically, and radiometrically. These all are the direct measures of corrosion, and gravimetrically mass loss methods are most used for inhibitor testing.

2. Extraction of Plants and Experimental Methods

Samples of *Calotropis procera and Calotropis gigantea* obtained from the Maharshi Dayanand Sarswati University Botanical Garden were dried, grinded, and soaked in a solution of ethanol. After 48 hours, the sample was cooled and filtered. The filtrate was subjected to evaporation (in order to leave the sample free of the ethanol) using a rotary evaporator. The stock solution of the extract so obtained was used in preparing 0.5 N, 1.0 N H_2SO_4 for mass loss method and 3 N, 4 N, and 5 N H_2SO_4 for thermometric analysis, respectively.

Aluminium of composition (wt %) Mn (0.6), P (0.36), C (0.15), S (0.07), and Al (98.79) was used for the study. The sheet was mechanically pressed and cut to form different coupons, each of dimensions, 2.5 cm ∗ 1.5 cm ∗ 0.2 cm. Each coupon was degreased by washing with ethanol. The washed sample was dipped in acetone, removed, and allowed to dry in air before use. All reagents used for the study were AR grade, and deionized water was used for their preparation.

Aluminium specimens were cleaned with emery paper. The corrosion products were cleaned with Clark's solution [25]. The weight of the specimens before and after immersion was determined using a Shimadzu balance. Specimens were suspended by glass hook; the degree of surface coverage (θ) is calculated by the formula [26]

$$\theta = \frac{(\Delta M_u - \Delta M_i)}{\Delta M_u}, \quad (1)$$

where θ is surface coverage and ΔM_u, and ΔM_i are the mass loss of the metal in uninhibited and inhibited acids, respectively.

The corrosion rate mmpy (millimeter penetration per year) is obtained by the following equation [27]:

$$\text{Corrosionrate (mmpy)} = \frac{(\text{Mass loss} \times 87.6)}{(\text{Area} \times \text{Time} \times \text{Metal density})}, \quad (2)$$

where mass loss is expressed in mg, area is expressed in cm^2 of metal surface exposed, time is expressed in hours of exposure, metal density is expressed in gm/cm^3, and 87.6 is the conversion factor.

Inhibition efficiency was also determined using a thermometric technique [28]. Each specimen 2.5 cm ∗ 1.5 cm ∗ 0.02 cm was placed in a glass hook and immersed in a beaker containing 50 mL of test solution at 30°C ± 0.5°C and exposed to air. Evaporation losses were made up with deionized water. After the test specimens were cleaned with acetone [29], duplicate experiments were performed in each case, and mean values of the mass loss were calculated.

The acidic solutions (0.5 N, 1 N, 2 N, and 4 N) were prepared by using deionized water. Temperature changes were measured at intervals of one minute using a thermometer with a precision of ±0.5°C. The temperatures increased slowly in the beginning, maximum temperature attained

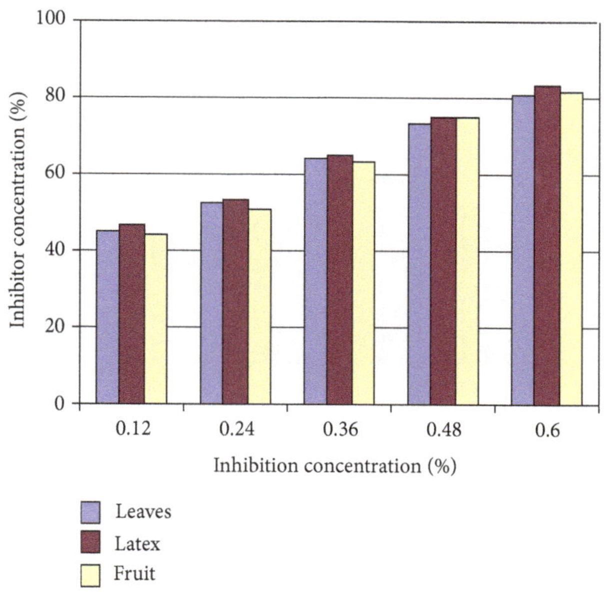

FIGURE 1: Inhibition efficiency versus concentration for Al in 0.5 N sulphuric acid for *Calotropis procera*.

was recorded, and percentage inhibition efficiencies were calculated as [30]

$$\eta = \frac{(RN_f - RN_i)}{RN_f} \times 100, \quad (3)$$

where RN_i and RN_f are the reaction numbers in the presence and absence of inhibitors, respectively, and RN (°C/min) is defined as

$$RN = \frac{(T_m - T_0)}{t}, \quad (4)$$

where T_m and T_0 are the maximum and initial temperatures respectively and t is the time required to reach the maximum temperature.

3. Result and Discussion

The inhibition efficiency (%) calculated from the mass loss measurement for sulfuric acid solution and inhibitors are given in Tables 1, 2, 3, and 4. It is observed that the inhibition efficiency increases with increase in the concentration of inhibition and decreases with increases in acid strength. The corrosion rate decreases with increases in concentration of inhibitor. The maximum efficiency was obtained in low acid concentration. The inhibitors have shown the efficiency in this range. *Calotropis procera* shows minimum 45.00% and maximum 82.98% whereas *Calotropis gigantea* shows minimum 43.26% and maximum 82.82% for 0.5 N sulfuric acid with aluminium, but *Calotropis procera* in 1 N sulfuric acid solution shows minimum 48.58% and maximum 78.86% inhibition efficiency, and *Calotropis gigantea* shows minimum 48.69% and maximum 78.82% inhibition efficiency with latex extract illustrated by Figures 1, 2, 3, and 4.

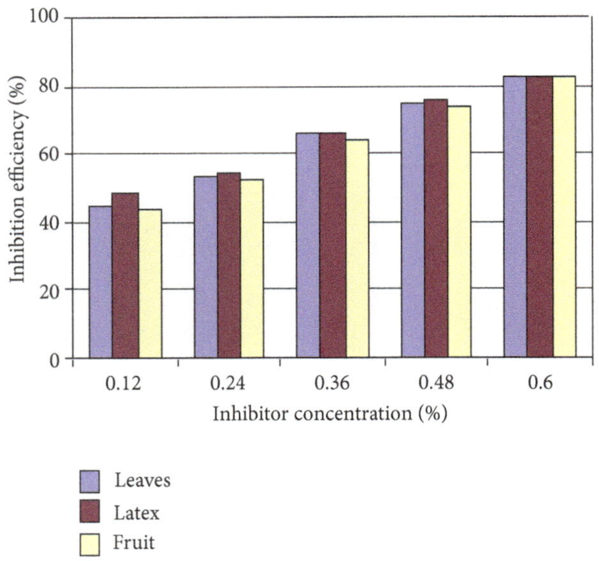

FIGURE 2: Inhibition efficiency versus concentration for AI in 0.5 N sulphuric acid for *Calotropis gigantea.*

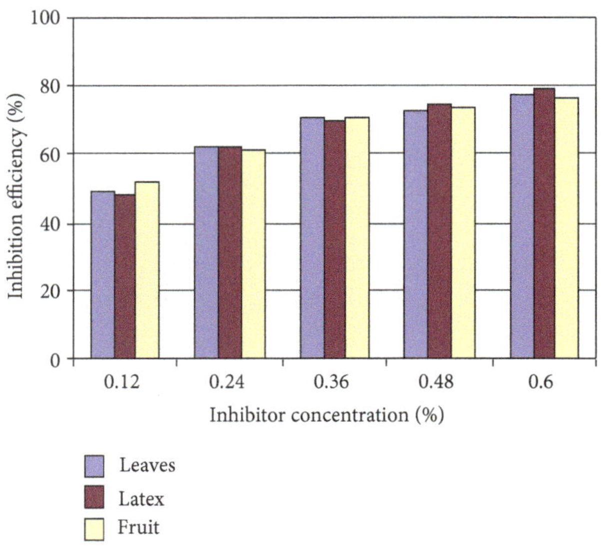

FIGURE 3: Inhibition efficiency versus concentration for AI in 1 N sulphuric acid for *Calotropis procera.*

Inhibition efficiency values were also determined by the thermometric method (Tables 5 and 6). Temperature changes for aluminium in 2 N, 3 N, 4 N, and 5 N sulfuric acid solutions were recorded at various inhibitor concentrations. However, no significance temperature changes were obtained for 1 N sulfuric acid solution. Therefore, the thermometric method was used for 2 N, 3 N, 4 N, and 5 N sulfuric acid solution. *Calotropis procer*a extract shows minimum inhibition efficiency 34.15% and maximum 72.96% for 2 N sulfuric acid, but 3 N sulfuric acid shows minimum 27.83% and maximum 69.03%, while 4 N sulfuric acid shows minimum 36.36% and maximum 67.33%, and 5 N sulfuric acid shows minimum 21.84% and maximum 61.15%, whereas *Calotropis gigantea* extract shows minimum inhibition efficiency 34.75% and

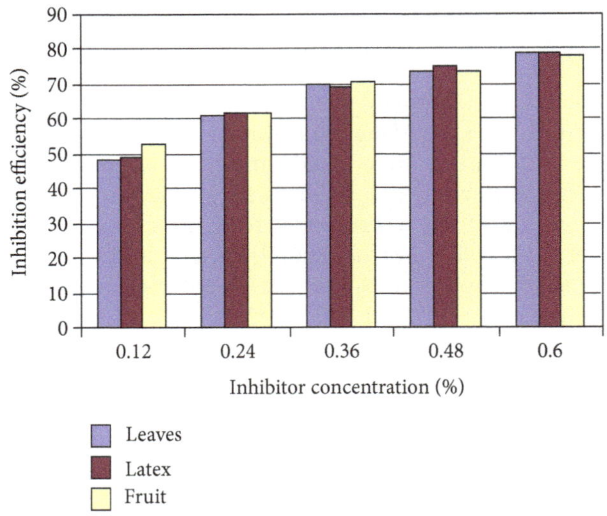

FIGURE 4: Inhibition efficiency versus concentration for AI in 1 N sulphuric acid for *Calotropis gigantea.*

maximum 73.58% for 2 N sulfuric acid, but 3 N sulfuric acid shows minimum 28.83% and maximum 68.56%, while 4 N sulfuric acid shows minimum 37.43% and maximum 67.87%, and 5 N sulfuric acid shows minimum 25.21% and maximum 62.26% (Tables 7 and 8).

Generally, the adsorption of organic molecules involves O, N, and S atoms. This process blocks the active sites and hence decreases the corrosion rate. In the present study, it was assumed that the plant extract is adsorbed on the metal surface and decreases the surface area available for cathodic and anodic reaction to take place. The nitrogen, sulphur and oxygen may be responsible for the adsorption. Organic inhibitors, with active portions, contain generally large C–H chains or rings with positively charged amine nitrogen group at the one end.

Organic corrosion inhibitors may function as follows:

(1) chemisorption of the molecule on the metallic surface;

(2) complexing of the molecule with the metal ion which remains in solid state;

(3) neutralising the corrodent;

(4) adsorbing the corrodent.

They offered large coverage due to the long hydrocarbon chain and by the presence of –OH group, being hydrophilic in nature; the –OH group counteracted the effects of chain length and ensured higher solubility.

Cathodic inhibitors slow down the reaction taking place at cathode (i.e., H_2 evolution).

For example, organic inhibitors like amine substituted urea's heavy metal soap and so forth decreases H_2 evolution process. Aluminium salts deposits on cathode help in decreasing the O_2 absorption process.

4. Absorption of Oxygen

This type of corrosion occurs generally in aqueous solution.

TABLE 1: Mass loss data for aluminium in 0.5 N sulfuric acid with ethanolic extract of leaves, latex, and fruit of *Calotropis procera*.

Inhibitor concentration	Mass loss (gm)	Inhibition efficiency (%)	Corrosion rate (mmpy)	Surface coverage (θ)
Uninhibited	1840		33.1	
Leaves				
0.12	1012	45.00	18.2	0.4500
0.24	872	52.60	15.7	0.5260
0.36	653	64.51	11.7	0.6451
0.48	487	73.53	8.77	0.7353
0.60	348	81.08	6.27	0.8108
Latex				
0.12	987	46.35	17.7	0.4635
0.24	856	53.47	15.4	0.5347
0.36	641	65.16	11.5	0.6516
0.48	465	74.72	9.43	0.7472
0.60	313	82.98	5.64	0.8298
Fruit				
0.12	1026	44.23	18.4	0.4423
0.24	898	51.19	16.1	0.5119
0.36	674	63.36	12.1	0.6336
0.48	465	74.72	8.38	0.7472
0.60	338	81.63	6.09	0.8163

TABLE 2: Mass loss data for aluminium in 0.5 N sulfuric acid with ethanolic extract of leaves, latex, and fruit of *Calotropis gigantea*.

Inhibitor concentration	Mass loss (gm)	Inhibition efficiency (%)	Corrosion rate (mmpy)	Surface coverage (θ)
Uninhibited	1840		33.1	
Leaves				
0.12	1023	44.40	18.4	0.4440
0.24	851	53.73	15.3	0.5373
0.36	634	65.54	11.4	0.6554
0.48	465	74.72	8.38	0.7472
0.60	327	82.22	5.89	0.8222
Latex				
0.12	951	48.31	17.1	0.4831
0.24	833	54.72	15.0	0.5472
0.36	627	65.92	11.3	0.6592
0.48	442	75.97	7.96	0.7597
0.60	316	82.82	5.69	0.8282
Fruit				
0.12	1044	43.26	18.8	0.4326
0.24	875	52.44	15.7	0.5244
0.36	653	64.51	11.7	0.6451
0.48	482	73.80	8.68	0.7380
0.60	319	82.66	5.74	0.8266

TABLE 3: Mass loss data for aluminium in 1.0 N sulfuric acid with ethanolic extract of leaves, latex, and fruit of *Calotropis procera*.

Inhibitor Concentration	Mass loss (gm)	Inhibition efficiency (%)	Corrosion rate (mmpy)	Surface coverage (θ)
Uninhibited	2460		44.3	
Leaves				
0.12	1242	49.51	22.3	0.4951
0.24	935	61.99	16.8	0.6199
0.36	721	70.69	12.9	0.7069
0.48	663	73.04	11.9	0.7304
0.60	548	77.72	9.87	0.7772
Latex				
0.12	1263	48.58	22.7	0.4858
0.24	931	62.15	16.7	0.6215
0.36	748	69.59	13.4	0.6959
0.48	625	74.59	11.2	0.7459
0.60	520	78.86	9.37	0.7886
Fruit				
0.12	1186	51.78	21.3	0.5178
0.24	958	61.05	17.2	0.6105
0.36	712	71.05	12.8	0.7105
0.48	643	73.86	11.5	0.7386
0.60	574	76.66	10.3	0.7666

TABLE 4: Mass loss data for aluminium in 1.0 N sulfuric acid with ethanolic extract of leaves, latex, and fruit of *Calotropis gigantea*.

Inhibitor concentration	Mass loss (gm)	Inhibition efficiency (%)	Corrosion rate (mmpy)	Surface coverage (θ)
Uninhibited	2460		44.3	
Leaves				
0.12	1262	48.69	22.7	0.4869
0.24	957	61.09	17.2	0.6109
0.36	734	70.16	13.2	0.7016
0.48	641	73.94	11.5	0.7394
0.60	523	78.73	9.42	0.7873
Latex				
0.12	1244	49.43	22.4	0.4943
0.24	950	61.38	17.1	0.6138
0.36	763	68.98	13.7	0.6898
0.48	619	74.83	11.1	0.7483
0.60	521	78.82	9.39	0.7882
Fruit				
0.12	1164	52.68	20.9	0.5268
0.24	934	62.03	16.8	0.6203
0.36	727	70.44	13.1	0.7044
0.48	651	73.53	11.7	0.7353
0.60	543	77.92	9.78	0.7792

Table 5: Thermometric reaction for aluminium in 2 N and 3 N sulphuric acid with ethanolic extract of leaves, latex, and fruit of *Calotropis procera*.

Inhibitor concentration	Reaction number (RN) 2 N	Inhibition efficiency (%)	Reaction number (RN) 3 N	Inhibition efficiency (%)
Uninhibited	1.26		1.91	
Leaves				
0.12	0.81	35.65	127	33.63
0.24	0.76	39.21	1.20	37.13
0.36	0.70	44.27	1.15	39.69
0.48	0.51	59.53	0.80	58.07
0.60	0.40	67.83	0.67	64.60
Latex				
0.12	0.83	34.15	1.26	34.05
0.24	0.78	37.94	1.25	34.67
0.36	0.71	43.32	0.95	50.08
0.48	0.54	57.15	0.84	51.33
0.60	0.41	67.43	0.76	62.66
Fruit				
0.12	0.78	38.02	1.38	27.83
0.24	0.71	43.40	1.32	31.07
0.36	0.64	49.01	1.11	41.67
0.48	0.48	61.34	0.75	60.63
0.60	0.34	72.96	0.59	69.03

Table 6: Thermometric reaction for aluminium in 4 N and 5 N sulphuric acid with ethanolic extract of leaves, latex, and fruit of *Ccalotropis procera*.

Inhibitor concentration	Reaction number (RN) 4 N	Inhibition efficiency (%)	Reaction number (RN) 5 N	Inhibition efficiency (%)
Uninhibited	2.48		3.10	
Leaves				
0.12	1.58	36.36	2.27	26.83
0.24	1.55	37.37	2.05	33.86
0.36	1.50	39.38	1.92	38.11
0.48	1.02	58.61	1.60	48.32
0.60	0.89	63.87	1.20	61.15
Latex				
0.12	1.46	41.19	2.42	21.84
0.24	1.40	43.64	2.11	31.89
0.36	1.26	49.12	2.08	32.89
0.48	1.12	54.87	1.84	40.56
0.60	0.94	61.83	1.48	52.19
Fruit				
0.12	1.44	41.95	2.09	32.57
0.24	1.32	44.01	1.94	37.31
0.36	1.22	50.80	1.87	39.47
0.48	1.01	59.17	1.53	50.48
0.60	0.81	67.33	1.18	61.95

Table 7: Thermometric reaction for aluminium in 2 N and 3 N sulphuric acid with ethanolic extract of leaves, latex, and fruit of *Calotropis gigantea*.

Inhibitor concentration	Reaction number (RN) 2 N	Inhibition efficiency (%)	Reaction number (RN) 3 N	Inhibition efficiency (%)
Uninhibited	1.26		1.91	
Leaves				
0.12	0.82	34.75	1.28	33.94
0.24	0.76	39.21	1.19	38.43
0.36	0.69	45.27	1.14	40.31
0.48	0.52	58.53	0.81	59.23
0.60	0.39	68.53	0.67	64.60
Latex				
0.12	0.82	35.19	1.27	35.24
0.24	0.77	38.94	1.15	39.47
0.36	0.72	44.32	0.94	53.08
0.48	0.54	57.25	0.83	57.73
0.60	0.40	68.53	0.72	62.66
Fruit				
0.12	0.79	38.89	1.37	28.83
0.24	0.70	44.60	1.31	32.37
0.36	0.64	49.01	1.10	42.37
0.48	0.47	62.56	0.78	60.63
0.60	0.35	73.58	0.58	68.56

Table 8: Thermometric reaction for aluminium in 4 N and 5 N sulphuric acid with ethanolic extract of leaves, latex, and fruit of *Calotropis gigantea*.

Inhibitor concentration	Reaction number (RN) 4 N	Inhibition efficiency (%)	Reaction number (RN) 5 N	Inhibition efficiency (%)
Uninhibited	2.48		3.10	
Leaves				
0.12	1.57	37.43	2.24	28.83
0.24	1.54	38.37	2.03	35.33
0.36	1.48	41.47	1.95	37.21
0.48	1.01	59.61	1.66	50.48
0.60	0.88	64.87	1.22	60.75
Latex				
0.12	1.48	42.19	2.36	25.21
0.24	1.37	45.34	2.14	31.37
0.36	1.24	50.52	2.08	32.89
0.48	1.09	56.54	1.82	42.56
0.60	0.91	62.83	1.46	53.19
Fruit				
0.12	1.43	41.77	2.10	32.57
0.24	1.33	44.81	1.95	37.22
0.36	1.21	51.83	1.81	43.39
0.48	1.06	58.17	1.55	51.43
0.60	0.82	67.05	1.17	62.26

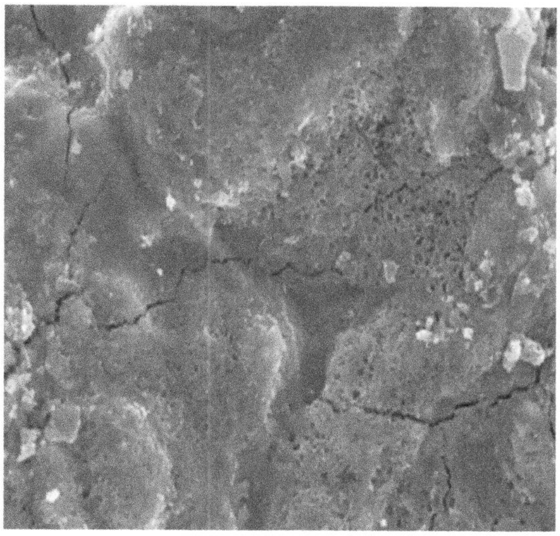

FIGURE 5: Scanning Electronic Microscopy (SEM)—image of aluminum.

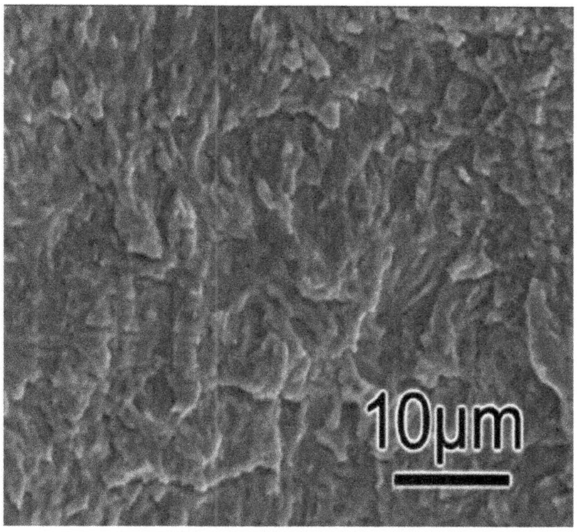

FIGURE 6: Aluminium with 0.5 N sulphuric acid with *Calotropis procera* (SEM).

The corrosion of aluminium and aluminium alloys precede through the following elementary reactions:

Anodic reaction

$$Al \longrightarrow Al^{3+} + 3e^-$$

Cathodic reaction

$$O_2 + 2H_2O + 4e^- \longrightarrow 4OH^- \quad \text{(neutral/alkaline medium)}$$

$$2H^+ + 2e^- \longrightarrow H_2 \quad \text{(acidic medium)} \tag{5}$$

All metals above the hydrogen in electrochemical series have a tendency to get dissolve in acidic solution releasing hydrogen.

It has been observed that the fruit extract of *Calotropis procera* and *Calotropis gigantea* has maximum inhibition efficiency compared to the latex and leaves extracts. This process increased the absorptivity of the fruit extract on the corroding site of the metal. This explained that the higher inhibition efficiency displayed by the latex for mass loss method and fruit extract for thermometric method in 0.60% inhibitors concentration.

5. Surface Analysis

Surface analysis of metal by Scanning Electron Microscopy was carried out on Model-ZEISS EVO 50. The surface morphological characteristics of the blank and inhibited aluminium were analyzed at magnification of 2.0 KX operated at an accelerating voltage of 20 KV. Scanning electron microscopy reveals that plant extract adsorbed on metal surface and increased the smoothness of aluminium that decreases the metal surface for corrosion attack (Figure 6). SEM provides a two-dimensional projection or a two-dimensional image of a sample. SEM of alumnium in sulphuric acid (uninhibited) is showed by Figure 5.

6. Conclusion

From the study, the following conclusions are made:

(i) ethanol extract of *Calotropis procera and Calotropis gigantea* is a good inhibitor for the corrosion of aluminium in H_2SO_4;

(ii) the inhibitor action results in a reduction of surface roughness due to corrosion processes.

Acknowledgments

This work was supported by Material laboratory, Maharshi Dayanand Saraswati University, Ajmer (Rajasthan) India. The authors express their sincere gratitude to the referee for valuable suggestion.

References

[1] N. O. Eddy and E. E. Ebenso, "Adsorption and inhibitive properties of ethanol extracts of Musa sapientum peels as a green corrosion inhibitor for mild steel in H_2SO_4," *African Journal of Pure and Applied Chemistry*, vol. 2, no. 6, pp. 46–54, 2008.

[2] N. O. Eddy and S. A. Odoemelam, *Journal of Materials Science*, vol. 4, no. 1, pp. 10–16, 2008.

[3] A. Al-Sehaibani, "Evaluation of extracts of henna leaves as environmentally friendly corrosion inhibitors for metals," *Materialwissenschaft und Werkstofftechnik*, vol. 31, no. 12, pp. 1060–1063, 2000.

[4] U. J. Ekpe, E. E. Ebenso, and U. J. Ibok, "Inhibitory actions of Azadirachta indica leaf extract on the corrosion of mild steel in H2SO4," *J. W. African Science Association*, vol. 39, pp. 13–30, 1994.

[5] O. K. Abiola and N. C. Oforka, "The corrosion inhibition effect of Cocos Nucifera (coconut) water on mild steel in HCl solution," in *Proceeding's of the Chemical Society of Nigeria, 25th International Conference*, 2002.

[6] G. O. Avwiri and F. O. Igho, "Inhibitive action of Vernonia amygdalina on the corrosion of aluminium alloys in acidic media," *Materials Letters*, vol. 57, no. 22-23, pp. 3705–3711, 2003.

[7] E. S. H. El Ashry, A. El Nemr, S. A. Esawy, and S. Ragab, "Corrosion inhibitors. Part II: quantum chemical studies on the corrosion inhibitions of steel in acidic medium by some triazole, oxadiazole and thiadiazole derivatives," *Electrochimica Acta*, vol. 51, no. 19, pp. 3957–3968, 2006.

[8] A. Y. El-Etre and M. Abdallah, "Natural honey as corrosion inhibitor for metals and alloys. II. C-steel in high saline water," *Corrosion Science*, vol. 42, no. 4, pp. 731–738, 2000.

[9] P. C. Okafor, M. E. Ikpi, I. E. Uwah, E. E. Ebenso, U. J. Ekpe, and S. A. Umoren, "Inhibitory action of Phyllanthus amarus extracts on the corrosion of mild steel in acidic media," *Corrosion Science*, vol. 50, no. 8, pp. 2310–2317, 2008.

[10] M. I. Awad, "Eco friendly corrosion inhibitors: Inhibitive action of quinine for corrosion of low carbon steel in 1 m HCl," *Journal of Applied Electrochemistry*, vol. 36, pp. 1163–1168, 2006.

[11] S. A. Odoemelam and N. O. Eddy, *Journal of Surface Science and Technology*, vol. 24, p. 1, 2008.

[12] S. Rajendran, V. Ganga Sri, J. Arockiaselvi, and A. J. Amalraj, "Corrosion inhibition by plant extracts—an overview," *Bulletin of Electrochemistry*, vol. 21, no. 8, pp. 367–377, 2005.

[13] M. G. Sethuraman and P. B. Raja, "Corrosion inhibition of mild steel by Datura metel in acidic medium," *Pigment and Resin Technology*, vol. 34, no. 6, pp. 327–331, 2005.

[14] G. S. Verma, P. Anthony, and S. P. Mathur, *Journal of the Electrochemical Society of India*, vol. 51, p. 173, 2002.

[15] "The useful plants of India" CSIR New Delhi "The wealth of India-Raw materials" CSIR New Delhi, 1986.

[16] R. Chowdhary, T. Jain, M. K. Rathoria, and S. P. Mathur, "Corrosion inhibition of mild steel by acid extracts of Prosopis Juliflora," *Bulletin of Electrochemistry*, vol. 20, no. 2, pp. 67–75, 2004.

[17] P. Arora, T. Jain, and S. P. Mathur, *Chemistry*, vol. 1, p. 766, 2005.

[18] A. Chetouani and B. Hammouti, "Corrosion inhibition of iron in hydrochloric acid solutions by naturally henna," *Bulletin of Electrochemistry*, vol. 19, no. 1, pp. 23–25, 2003.

[19] A. A. El Hossary, R. H. Saleh, and A. M. Shams El Din, "Corrosion inhibition by naturally occurringsubstances-I. The effect of *Hibiscus subdariffa* (karkade) extract on the dissolution of Al and Zn," *Corrosion Science*, vol. 12, no. 12, pp. 897–904, 1972.

[20] R. M. Saleh and A. A. El Hossary, in *Proceedings of the 13th Seminar on Electrochemistry*, CECRI, 1972.

[21] M. K. Sharma, P. Arora, S. Kumar, S. P. Mathur, and R. Ratnani, "Inhibitive effect of Prosopis cineraria on mild steel in acidic media," *Corrosion Engineering Science and Technology*, vol. 43, no. 3, pp. 213–218, 2008.

[22] T. Jain, R. Chowdhary, and S. P. Mathur, *Journal of the Electrochemical Society of India*, vol. 53, p. 33, 2004.

[23] B. C. Edwards, "The protection of Cu-Ni condenser tubes with high molecular weight water-soluble polymers," *Corrosion Science*, vol. 9, no. 6, pp. 395–402, 1969.

[24] B. Brooks, *Mate Prot*, vol. 7, p. 24, 1968.

[25] G. Waranglen, *Introduction To Corrosion and Protection of Metals*, vol. 236, Chapman and Hall, London, UK, 1985.

[26] J. D. Talati and D. K. Gandhi, "Corrosion of zinc in citric acid containing food colourants," *Indian Journal of Technology*, vol. 29, no. 6, pp. 277–282, 1991.

[27] M. K. Sharma, S. Kumar, R. Ratnani, and S. P. Mathur, "Corrosion inhibition of Aluminium by extracts of Prosopis cineraria in acidic media," *Bulletin of Electrochemistry*, vol. 22, no. 2, pp. 69–73, 2006.

[28] P. B. Raja and M. G. Sethuraman, "Inhibition of corrosion of mild steel in sulphuric acid medium by ITCalotropis procera/IT," *Pigment and Resin Technology*, vol. 38, no. 1, pp. 33–37, 2009.

[29] K. Aziz and A. M. El-Din Shams, *Corrosion Science*, vol. 3, p. 582, 1953.

[30] R. H. Hausler, in *Proceedings of the International Conference on Corrosion Inhibition*, vol. 7, p. 16, Dallas, Tex, USA, 1983.

18

Corrosion of WC-VC-Co Hardmetal in Neutral Chloride Containing Media

C. N. Machio,[1] D. S. Konadu,[2,3] J. H. Potgieter,[4] S. Potgieter-Vermaak,[5,6] and J. Van der Merwe[2]

[1] *Council for Scientific and Industrial Research (CSIR), Materials Science and Manufacturing, P.O. Box 395, Pretoria 0001, South Africa*
[2] *DST/NRF Centre of Excellence in Strong Materials, and School of Chemical and Metallurgical Engineering, University of Witwatersrand, Johannesburg, Wits 2050, South Africa*
[3] *Materials Science and Engineering Department, University of Ghana, Legon, Accra, Ghana*
[4] *School of Research, Enterprise and Innovation, Manchester Metropolitan University, Oxford Road, Manchester M1 5GD, UK*
[5] *School of Science and the Environment, Manchester Metropolitan University, Oxford Road, Manchester M1 5GD, UK*
[6] *School of Chemistry, University of the Witwatersrand, Johannesburg, Private Bag X3, Wits 2050, South Africa*

Correspondence should be addressed to C. N. Machio; cmachio@csir.co.za

Academic Editors: G. Marginean and R. Wang

The effect of varying VC content on the corrosion behavior of WC-10 wt% Co hardmetals in sodium chloride (NaCl) and synthetic mine water (SMW) solutions has been investigated using anodic polarization scans and surface analytical methods. It is shown that the polarization behavior is active-pseudopassive in NaCl and active in SMW regardless of the VC content, while the corrosion resistance is poorer and independent of VC content in NaCl but better at high VC contents in SMW. The corrosion behavior of samples is explained using the effect of VC on the chemical composition of the binder.

1. Introduction

Corrosion is an important surface degradation process in some, if not all, applications of tungsten-carbide-(WC-) cobalt- (Co-) based hardmetals. Applications like tools for machining of metals [1] and for wear resistance in the mining industry [1] expose the WC-Co to fluids that can be corrosive, leading to a reduced useful life. The Co binder is the least corrosion resistant constituent [2], and efforts to improve WC-Co corrosion resistance have involved altering the chemical composition of the binder by introducing more corrosion resistant elements. Nickel [3], and chromium [4] especially, and recently, ruthenium [5] have been shown to markedly improve the corrosion resistance of WC-Co-based hardmetals.

The corrosion resistance of WC-Co can also be improved by some transition carbides, which have been usually added in small amounts. Chromium carbide (Cr_3C_2), for example, added to WC-Co in amounts of about 0.5 wt.% markedly improves corrosion resistance [6]. The effect of vanadium carbide (VC) has not been as straight forward as that of chromium carbide. Earlier studies [6] found small additions of VC to WC-Co had at best a neutral effect on the corrosion resistance. Subsequent studies by other researchers, using higher amounts of VC have indicated otherwise. For example, studies in hydrochloric and sulphuric acids have indicated that 10 wt% VC improves the corrosion resistance of high Co content WC-Co [7]. This result has been confirmed by the current authors [8] who have shown that the corrosion resistance, determined from the corrosion current density at the free corrosion potential, increases with increasing VC content. The current article is extending the corrosion study of the materials to neutral chloride electrolytes to determine whether the improved corrosion resistance is solution independent. It has happened that observations made on the corrosion of WC-Co in acidic electrolytes do not appear to be reproduced in tests that use neutral chloride containing solutions [9].

TABLE 1: Nominal chemical composition and magnetic properties of test specimens.

Specimen	Approx. vol% Co	Magnetic saturation ($4\pi\sigma$) (G cm^3 g^{-1})
WC-10Co	16	172
WC-0.4VC-10Co	16	185
WC-10VC-12Co	17	230
WC-27VC-11Co	13	194

TABLE 2: Synthetic mine water composition [15].

Compound	Concentration (mg/L)
Na_2SO_4	1237
$CaCl_2$	1038
$MgSO_4$	199
NaCl	1380

2. Experimental

2.1. Materials. The nominal composition (wt.%) of the test materials and their magnetic saturation levels are given in Table 1. The microstructures of the specimens are fully described elsewhere [8]. The VC content took values of 0, 0.4, 10, and 27 wt%. However, the VC, at the 10 and 27 wt.% content, occurred as the double carbide (W,V)C because WC dissolves in VC during sintering [10]. For easier referencing, the samples will be referred to using their VC content. The target binder content was 10 wt% Co for all the specimens, but samples 10 VC and 27 VC had 12 and 11 wt% Co, respectively. Comparison between the samples is still possible because the effect of cobalt content on the corrosion behavior of WC-Co hardmetals is known. For example, it is known that the current densities at (free) open circuit potential are a function of Co content, and in acidic solutions, the current densities increase with increasing Co content [11].

The magnetic saturation of the specimens increased with the introduction of VC and with Co content (Table 1). Cobalt is the magnetic phase and the higher values with higher Co contents are expected. The magnetic saturation provides a measure of the solute content of the binder and is a quick technique for determining the quality of sintered WC-Co [12]. Generally, and within limits, the magnetic saturation decreases linearly with an increase of W content of the binder [13, 14].

2.2. Electrochemical Measurements. Electrochemical measurements were carried out using the conventional setup of a three-electrode cell in which the test specimen is the working electrode, an inert material, in the case here, graphite rod, is the counter electrode (CE) and a reference electrode against which all potentials are reported. The reference electrode used was a silver/silver chloride electrode (SSE) (in saturated KCl) and was connected to the test cell via a Luggin probe capillary. All tests were done at room temperature and the electrolytes were neither deaerated nor aerated during testing.

The electrolytes used were 1 M sodium chloride (NaCl) and synthetic mine water (SMW) whose composition is given in Table 2. They were chosen to represent the mining environment where WC-Co is used to combat wear [23]. The composition of the SMW given in Table 2 is the most aggressive mine water found in the gold mines of South Africa [15, 24].

Test samples were polished to a 1 μm surface finish after being made electrically conductive by attaching an aluminum sticker tape to one face in a cold-mounted mold. After immersion in the electrolytes, the open circuit potential (OCP) was allowed to stabilize for two hours. Potentiodynamic anodic polarisation measurements were done using an Autolab potentiostat/galvanostat connected to a personal computer with a General Purpose Electrochemical System (GPES) software. As in an earlier study [8], potential was varied from −600 mV to +1200 mV at a scan rate of 2 mV/s for all the samples. This scan rate was similar to those used in literature [25, 26]. The corrosion potential (E_{corr}), corrosion current density (i_{corr}), and corrosion rates were retrieved from the corrosion measurement data using the intersection of the anodic and cathodic Tafel lines.

Chronoamperometry measurements were performed immediately on specimens at the end of their potential sweeps to investigate the corrosion products on their surfaces. The measurements were done for eight hours at potentials selected from the anodic polarisation curves.

Samples were characterised for phases before and after corrosion testing using a Phillips PW 1710 X-ray diffractometer. Surfaces of specimens after chronoamperometric tests were analyzed using a Raman spectrometer (Senterra, Bruker Optics) coupled with a Peltier cooled CCD detector (576 × 288 pixels) equipped with a 532 nm wavelength laser. The vanadium content of the test electrolytes after chronoamperometric testing was determined using a Spectro Genesis inductively coupled plasma—optical emitting spectrometer (ICP-OES).

3. Results

3.1. Potentiodynamic Polarisation Behaviour. Figures 1 and 2 show the potentiodynamic polarisation curves of the samples in 1 M NaCl and SMW, respectively. As was the case in a previous study, the anodic polarization behavior for each specimen was repeatable (Figure 3). The anodic polarization behavior in NaCl, where all samples showed pseudopassive behavior, is similar to that observed in HCl acid solutions [8]. The pseudopassive behavior was more pronounced for the base alloy WC-10Co, the specimen with the least magnetic saturation (Table 1).

The pseudopassive behavior observed for WC-10Co hardmetal in the NaCl test solution is in contrast with other literatures [27] where dilute (0.1 M) NaCl solutions were used. This is because the higher Cl^- anion concentration used in this study compared to literature favours passivation [28].

Compared to NaCl, anodic dissolution occurred continuously, without any pseudopassive behavior, for all specimens

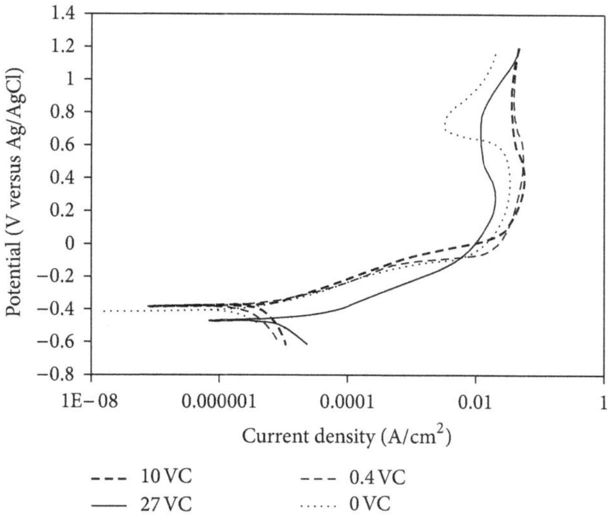

FIGURE 1: Potentiodynamic curves of the samples in NaCl.

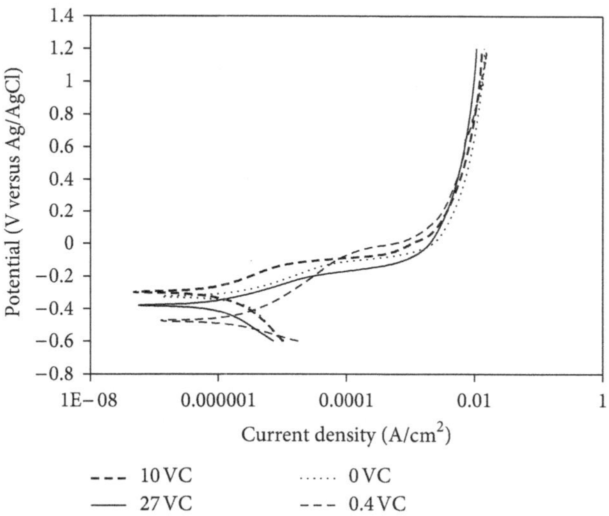

FIGURE 2: Potentiodynamic curves of the samples in SMW.

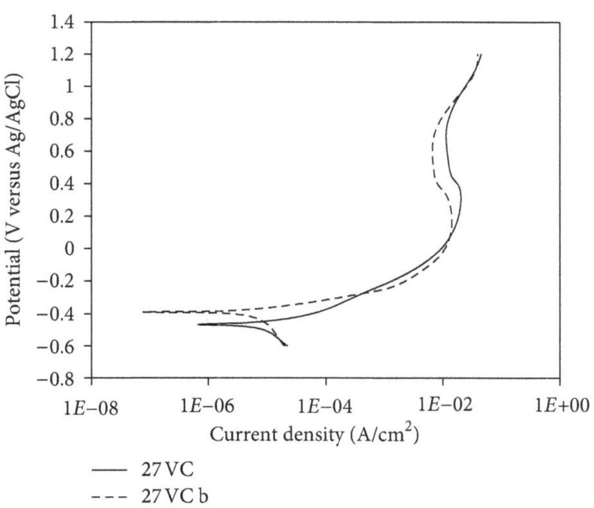

FIGURE 3: Repeatability of 27 wt% VC in 1 M NaCl.

during potentiodynamic polarization in SMW (Figure 2). The behavior of the base alloy was in line with literature [9].

3.2. Corrosion Parameters. Table 3 presents the corrosion parameters of the specimens in NaCl and SMW. Vanadium carbide additions of 0.4 wt% VC, levels used for grain refinement, simultaneously made the E_{corr} more negative, and the corrosion current density higher than that of the base alloy in both electrolytes. A similar trend was obtained in acidic solutions [8]. Higher VC contents made E_{corr} nobler in NaCl compared to the base alloy. However, the corrosion current density increased and appeared to be independent of the VC content, an observation that was different from that found for HCl and H_2SO_4 acidic solutions which had a similar potentiodynamic polarization behavior and where current density decreased with increasing VC content [8].

There was no observable trend of the effect of increasing VC content on E_{corr} in SMW: 10 wt% VC increased E_{corr} to a more nobler value, while 27 wt% VC decreased it to a lesser noble value, compared to the base alloy. However, the corrosion current densities decreased to values lower than those of the base alloy, the decrease tempered by the cobalt content.

3.3. Chronoamperometry Studies. Chronoamperometric tests were done at 0.8 V (v SSE) for the base alloy and 0.7 V (v SSE) for the WC-VC-Co hardmetals in 1 M NaCl based on pseudopassivation at these potentials. Due to the continuous dissolution of all the samples during anodic polarization in SMW, a potential of 1.1 V (v SSE) was used for all the samples. Figure 4 shows the variation of current density for tests in the NaCl solution. The base alloy passivated instantaneously and had the least steady state current density, unlike the VC specimens. VC additions at the level used for grain refinement (0.4 wt% VC) led to fluctuations in the current density curve, implying some pitting corrosion. The current for the high VC content alloys decreased quickly and then increased before decreasing again. Decreasing current densities are associated with a build-up of corrosion products on the test surface [29] while increases suggest oxidation events [17], that are caused by the fracture and/or fall off of the corrosion product that had formed on the surface to expose the underlying surface.

There was no instantaneous passivation for any specimen in SMW (Figure 5). In addition to the base alloy, 27VC also attained a steady state current density. As was the case in NaCl (Figure 4), the base alloy had the least steady state current. Current fluctuations were observed for specimen 0.4VC, an indication of pitting, but the fluctuations were less pronounced than in NaCl (Figure 4).

3.4. Surface Phases after Corrosion. The results of the XRD patterns of the samples before corrosion have been published elsewhere [8]. Figures 6 and 7 are XRD patterns showing the change of surface phases after chronoamperometric testing in NaCl and SMW, respectively. Cobalt peaks, observed in all the specimens before corrosion, were absent after corrosion. This occurrence was in line with literature where

Table 3: Electrochemical parameters of the WC-VC-Co samples in NaCl and SMW.

	NaCl		SMW	
	E_{corr} (V) (v SSE)	i_{corr} ($\mu A/cm^2$)	E_{corr} (V) (v SSE)	i_{corr} ($\mu A/cm^2$)
WC-10Co	−0.399	0.03	−0.313	0.18
WC-0.4VC-10Co	−0.469	0.20	−0.474	0.21
WC-10VC-12C0	−0.370	0.36	−0.296	0.15
WC-27VC-11Co	−0.361	0.28	−0.379	0.08

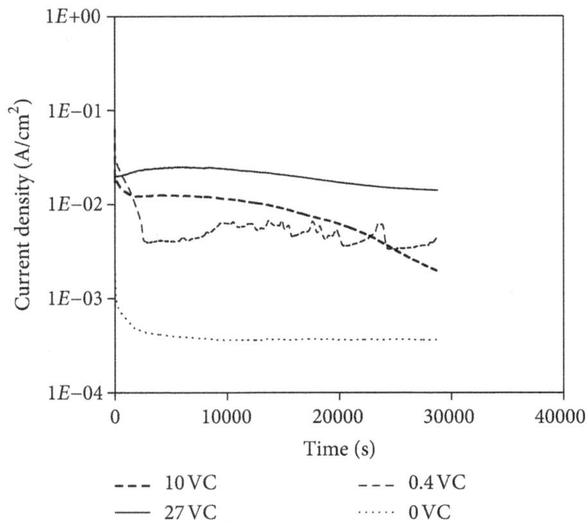

Figure 4: Chronoamperometry curves of the samples in NaCl.

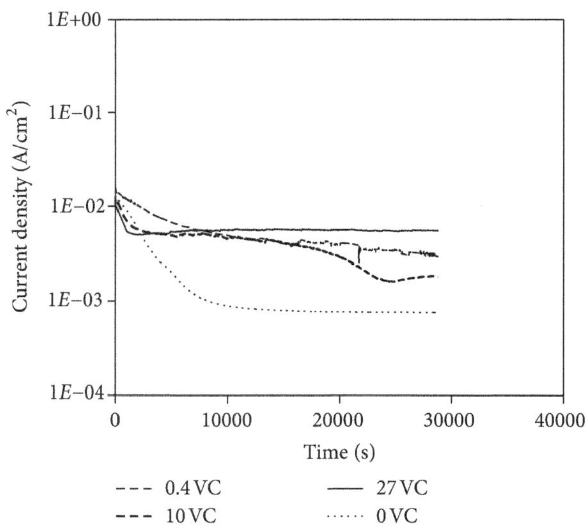

Figure 5: Chronoamperometry curves of samples in SMW.

the Co binder phase in WC-Co-based hardmetals corrodes preferentially [2]. Tungsten oxides appeared to have formed for some specimens, but the XRD analysis was inconclusive, necessitating further analysis using Raman spectroscopy (Figures 8 and 9). In contrast, there were no XRD peaks for vanadium-based corrosion phases on the surfaces of the VC-containing specimens after corrosion in any of the test electrolyte (Figures 6 and 7).

Some of the VC-containing specimens did not have XRD peaks for VC (which occurs as (V,W)C) after corrosion (10VC and 27VC) specimens in NaCl and 10VC in SMW as presented earlier in [8]). These peaks were present after corrosion in acid solutions, where corrosion rates were at least an order of magnitude higher than [8]. It is possible that in the current case, the (V,W)C grains fell out during specimen handling due to the dissolution of the binder, as has been observed in literature [30].

Figures 8 and 9 show the Raman spectra obtained on the samples after testing in NaCl and SMW, respectively. The corrosion products for all test specimens were tungsten-oxide based, even though they had not been detected by XRD analysis in some cases, especially in SMW (Figure 7). Tables 4 and 5 allocate the Raman bands to their possible sources based on observations in literature. The literature of WO_3-based films indicates that Raman bands below 190 cm^{-1} are caused by lattice vibration of crystalline WO_3; bands between 190 and 400 cm^{-1} are due to bending of O–W–O bonds; those in the range 500–900 cm^{-1} are characteristic of stretching O–W–O bonds; and bands in the vicinity of 950, characteristic of hydrated WO_3, are caused by the stretching of terminal W=O bonds [16, 17, 19–21, 31, 32].

No Raman band matches were found for possible vanadium-based corrosion products, for example, vanadium oxide, vanadium oxy-chloride ($VOCl_3$), whose Raman bands are available in literature [33, 34]. However, during the chronoamperometric tests, light green solids were observed to form in the solutions. Inductive coupled plasma optical emitting spectrometry (ICP-OES) analysis of the electrolytes indicated the presence of vanadium cations probably V(II) (Table 6) indicating that the green solids that formed were probably VCl_2, which is known to be green [35].

The formation of WO_3 as a film accounts for WC-VC-Co pseudopassive behavior in acidic test media [8], and in the current study, the observation of pseudopassivity in the neutral 1 M NaCl test solution (Figure 1). Its formation on the specimens tested in SMW was however not accompanied by pseudopassive behaviour. The difference in pseudopassive behavior between the NaCl and SMW could have been caused by the difference in Cl^- anion concentrations, where the higher concentrations in NaCl encouraged passivation compared to the low concentration obtained in the SMW [28]. It is also possible that the WO_3 film formed in SMW had poorer adhesion properties.

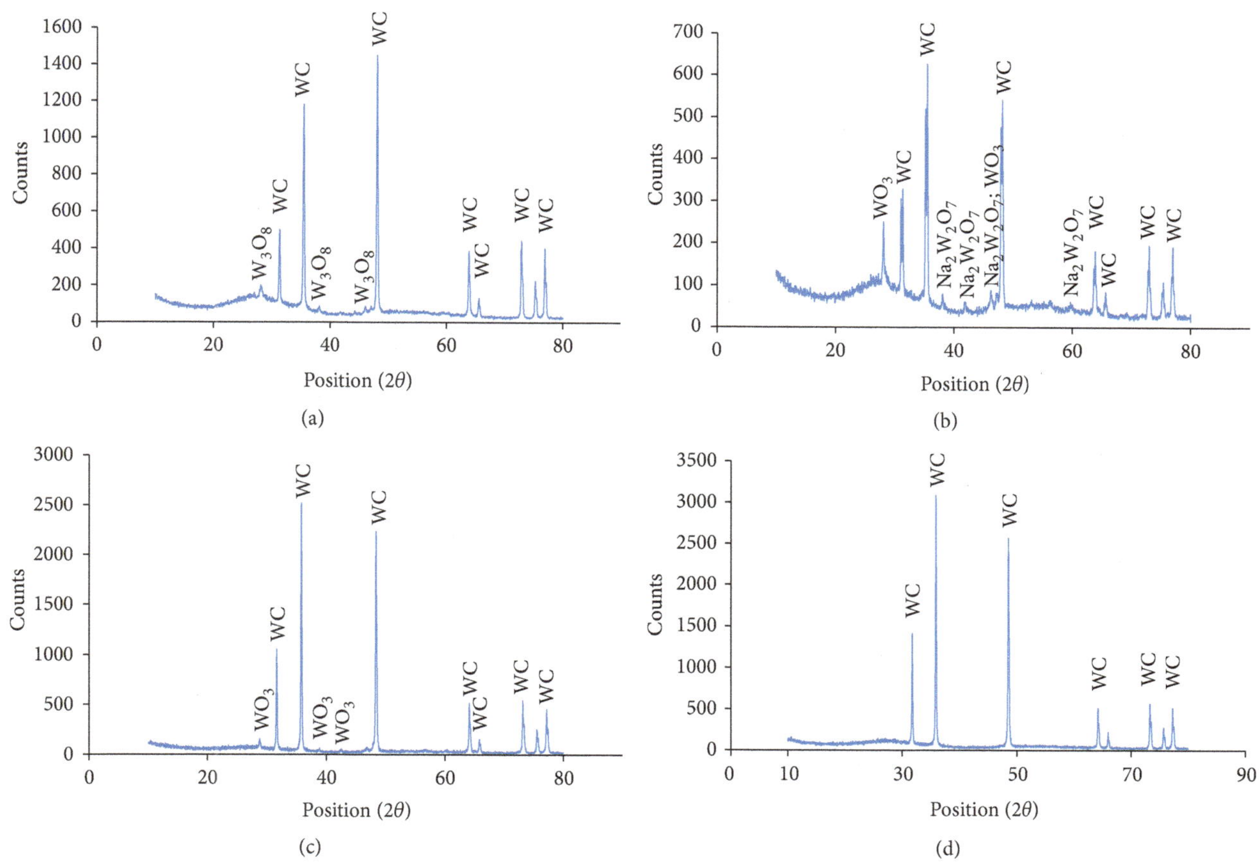

FIGURE 6: XRD pattern of the corrosion products of the samples on (a) WC-10Co, (b) WC-0.4VC-10Co, (c) WC-10VC-12Co, and (d) WC-27VC-11Co in NaCl.

TABLE 4: Raman shifts after corrosion in NaCl.

Specimen	Raman bands	Possible phase	References
WC-10C	133 & 249	WO_3	[16]
	814.5	WO_3	[17]
	679 & 958.5	$WO_3 \cdot 2H_2O$	[16]
WC-0.4VC-10Co	131	WO_3	[16]
	697	WO_3	[18]
	813.5	WO_3	[17]
	881.5	$WO_3 \cdot nH_2O$	[19]
	968.5	$WO_3 \cdot 2H_2O$	[16]
WC-10VC-12Co	131 & 803	WO_3	[16]
	261 & 962	$WO_3 \cdot 2H_2O$	[20]
WC-27VC-11Co	132.5, 262.5, 700, & 805.5	WO_3	[16]

4. General Discussion

The effect of increasing amounts VC on the corrosion resistance of WC-Co appears to have been mixed. The VC reduces the extent of the pseudopassive behavior in NaCl and does not change it in SMW; it does not alter the corrosion current density in NaCl, where the current density increases with Co content, while it reduces it in SMW; chronoamperometric current densities stabilize at higher values for specimens with VC. The trends in the results reported here suggest that VC impairs the corrosion resistance of WC-Co except in SMW where it reduces the current density. These results can be explained from two perspectives.

In general, it can be expected that VC additions affect the corrosion resistance of WC-Co in two ways. In the first instance, vanadium atoms from the VC can go into solution in the Co binder and either on their own and/or in conjunction with W atoms that also go in solution in the Co, alter the corrosion resistance of the Co binder. Solute atoms of W improve the corrosion resistance of Co [22]. This effect can

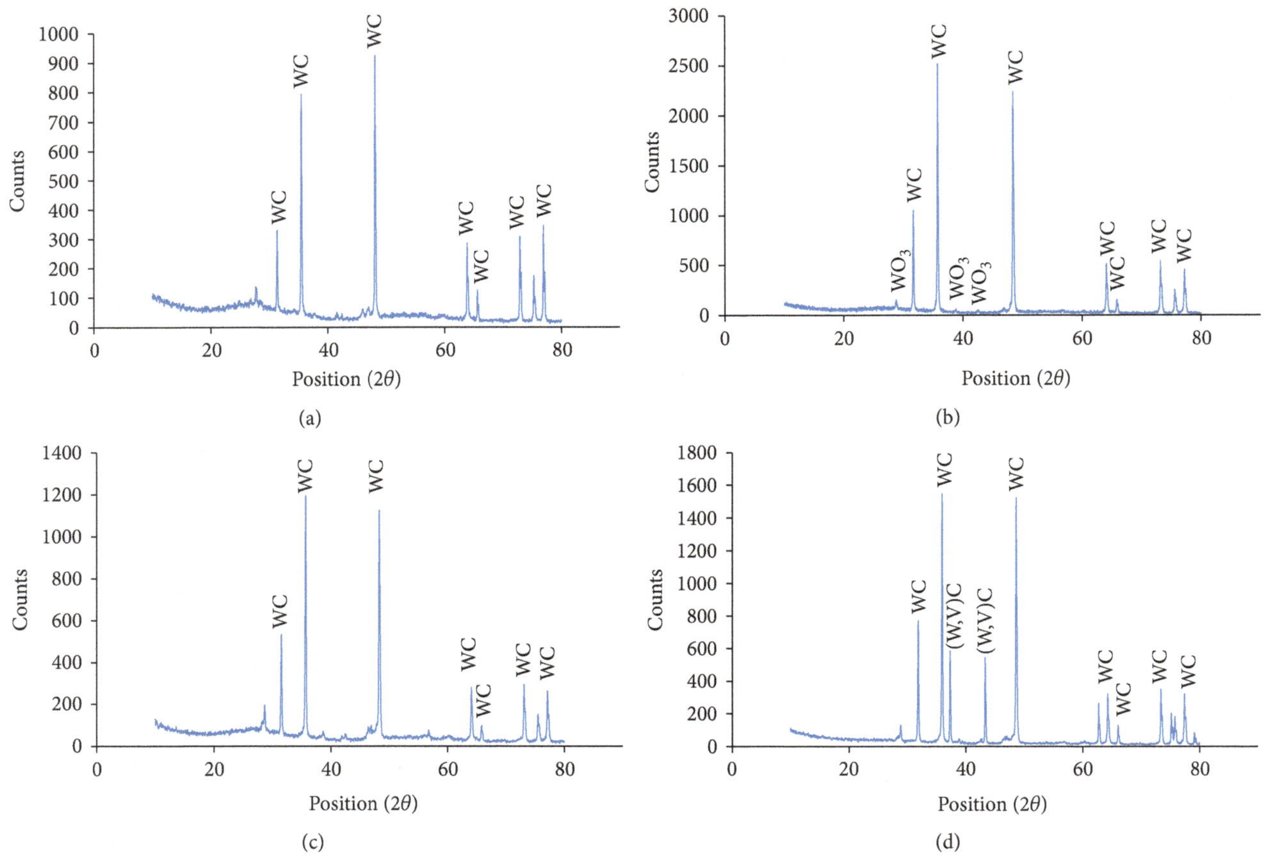

FIGURE 7: XRD pattern of the corrosion products of the samples on (a) WC-10Co, (b) WC-0.4VC-10Co, (c) WC-10VC-12Co, and (d) WC-27VC-11Co in SMW.

TABLE 5: Raman bands after corrosion in SMW.

Specimen	Raman bands	Possible phase	Reference
WC-10Co	133, 262.5, 704.5 & 808	WO_3	[16, 21]
WC-0.4VC-10Co	139	WO_3	[21]
	743	Cubic WO_3	[20]
	891	WO_3	[19]
	955	$WO_3 \cdot 2H_2O$	[16]
WC-10VC-12Co	129, 263, & 802	WO_3	[21]
	980	$WO_3 \cdot nH_2O$	[22]
WC-27VC-11Co	130, 263, 708, & 803	WO_3	[21]
	327	$CaWO_4$	[19]

TABLE 6: ICP-OES results of VC containing corrosion product solutions after dissolution in SMW and NaCl.

Sample	Vanadium content (ppm)	
	SMW	NaCl
WC-0.4VC	46.7 ± 1.3	83.1 ± 3.5
WC-10VC	15.1 ± 3.7	24.9 ± 2.7
WC-27VC	58.4 ± 3.1	2.0 ± 0.3

either be augmented or curtailed by the presence of VC through its effect on the quantity of W atoms going into the Co binder. The W solute content in Co can be determined from the magnetic saturation. It is known, broadly, that the magnetic saturation of conventional WC-Co increases with decreasing amounts of W atoms in Co [13, 14]. Judging from the magnetic saturations of the test specimens, which increase with VC content (Table 1), the introduction of VC reduced the amount of W atoms in the Co binders of these specimens.

The reduction of the W solute content of the binders of the VC specimens deduced above explains a number of the observations in this study. Some authors have used the W solute content, captured in the magnetic saturation,

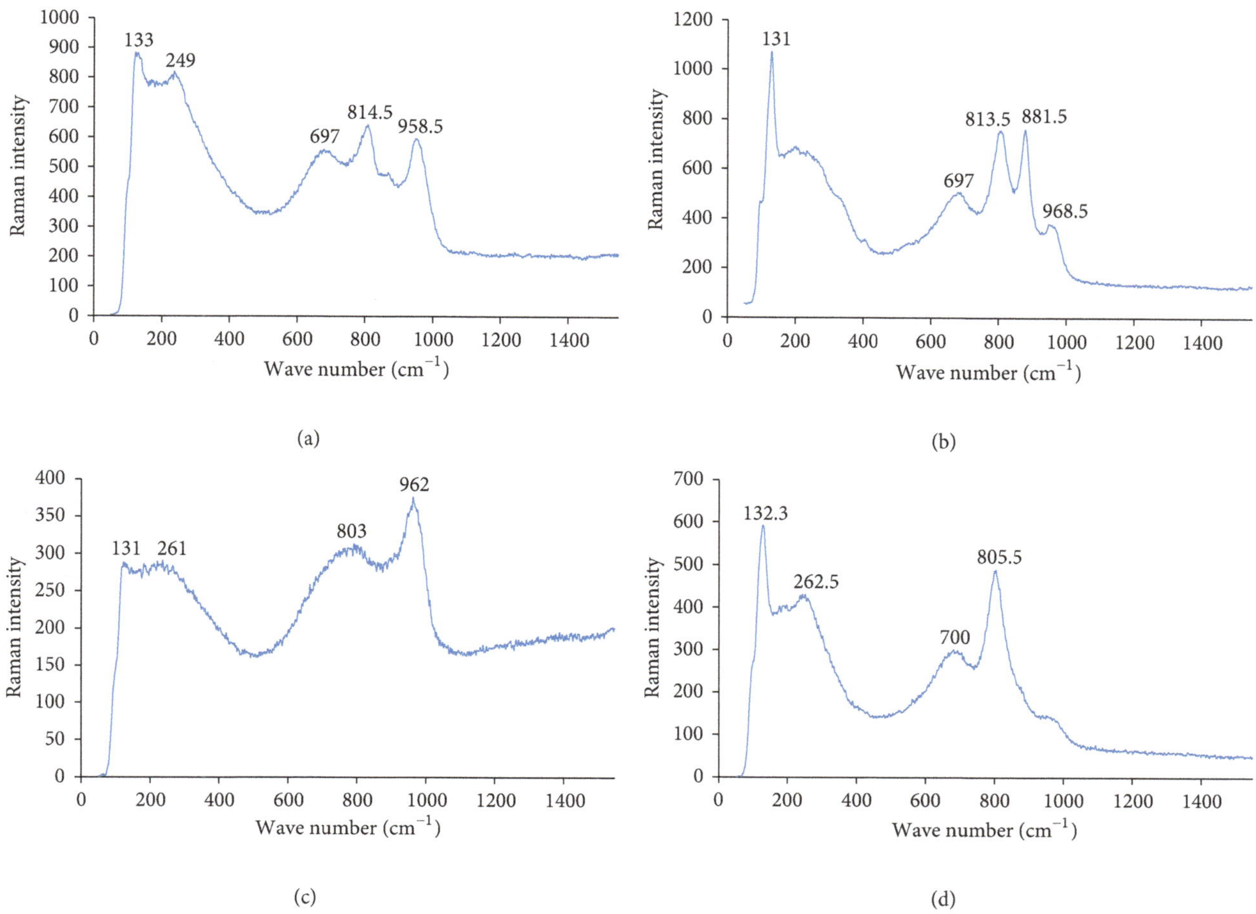

FIGURE 8: Raman spectra of corroded product on (a) WC-10Co, (b) WC-0.4VC-10Co, (c) WC-10VC-12Co, and (d) WC-27VC-11Co in NaCl.

to explain the corrosion behavior of WC-Co. According to literature, a low magnetic saturation, implying a higher W solute content of the Co binder, is associated with a more pronounced pseudopassive behavior in sulphuric acid [36]. The pronounced pseudopassive behavior observed for the base alloy in NaCl in the current study (Figure 1) can therefore be explained on the basis of its lower magnetic saturation (Table 1) compared to the other test specimens. By extension, the variation of magnetic saturation also explains the variation of corrosion current densities, i_{corr}, in NaCl (Table 3), which increase with increasing magnetic saturation, that is, decreasing W solute content of the binder. This variation of current density associated with changing W solute content in Co is, as has already been mentioned, in line with literature, where lower Co binder W solute content is associated with poor corrosion resistance [22].

The above thinking where magnetic saturation/Co binder W solute content explains corrosion behavior needs to be qualified, because it does not appear like this is a universal phenomenon: the relationship appears to be solution-specimen dependent. For example, corrosion of WC-Co in nitric acid is not explained by the magnetic saturation of the specimens [37]. The solution can play a role in accentuating the effect of the magnetic saturation. For example, the pronounced pseudopassive behavior observed for corrosion of the base alloy in NaCl in this study could have been aided by the effect of higher Cl^- ion concentration, which also encourages pseudopassive behavior [28].

The specificity of the solution-specimen dependence of the effect of magnetic saturation on corrosion, referred to above, appears to partly apply to the corrosion behavior in SMW, where all specimens continuously dissolved during potentiodynamic polarization (Figure 2) and also where, generally, corrosion current densities, i_{corr}, decreased even as the magnetic saturation was increasing (Table 3). The Cl^- ion concentration in the SMW was much less, preventing the solution from predisposing specimens to pseudopassive behavior. The variation of i_{corr}, in SMW can be loosely correlated with the variation of the volume fraction of the binder, meaning that the effect of VC on corrosion in SMW was probably through its effect on the volume fraction of the binder.

The second possible impact of the VC on the corrosion of WC-Co hardmetal would be expected to be through its effect on the corrosion product. It has been suggested that vanadium is naturally passivating [38]. However, this tendency did not manifest itself in this study: for example, the potentiodynamic polarization behavior of the specimens was

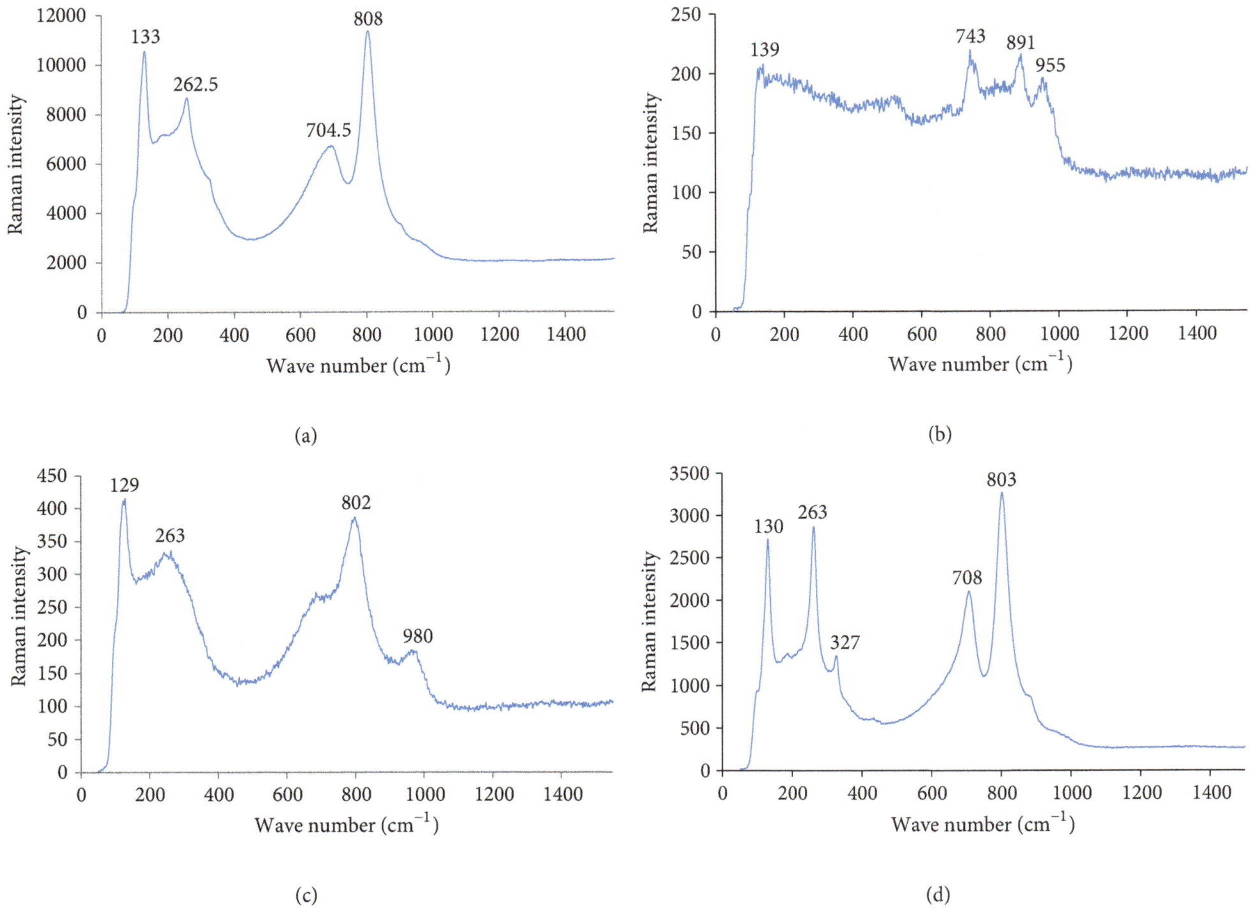

FIGURE 9: Raman spectra of after corrosion in SMW (a) WC-10Co, (b) WC-0.4VC-10Co, (c) WC-10VC-12Co, and (d) WC-27VC-11Co.

not changed in a positive way, for example, the emergence of true passivity, by the presence of VC (Figures 1 and 2). Also, no vanadium-based compound was detected on the corroded surfaces of the specimens in this study either by XRD studies (Figures 6 and 7) or by Raman spectroscopy (Figures 8 and 9). These two analysis techniques showed that no solid film of a vanadium compound formed on the surfaces of the test specimens after corrosion. It appears that a vanadium-based corrosion compound formed during corrosion, however, it dissolved in the test solutions (as shown by the presence of Vanadium ions in the test electrolytes (Table 6)), and, in some cases, precipitated out of the electrolytes and did not participate in forming a barrier to curtail corrosion. This failure to form a corrosion barrier explains the higher current densities observed for the VC specimens during chronoamperometric testing (Figures 4 and 5).

The results of the 0.4 wt% VC specimen deserve some emphasis, given the importance of VC-grain refinement of WC-Co materials. Fine-grained WC-Co materials are important for wear applications. While the wear resistance is high, the current results have shown that the corrosion resistance is poorer than for any other specimen in the study. This specimen had a higher corrosion current density, in both electrolytes (Table 3), and was the only specimen that pitted during chronoamperometric testing (Figures 4 and 5). Even though this has been reported in the literature of the corrosion of WC-Co in neutral chloride solutions [6, 9], the behavior had not been effectively explained. This detrimental effect stems, primarily, from the effect of VC on the Co binder W solute content: the VC restricts the amount of W that goes into solution in the Co binder during sintering by forming a (V,W)C film around the WC grains. This is the mechanism by which VC grain refiners are known to effect WC grain refinement [38]. The effect of low W solute content on corrosion has already been explained.

5. Conclusion

Investigations into the corrosion behaviour of WC-VC-Co hardmetals using potentiodynamic scans, XRD, and Raman spectroscopy measurements in 1 M NaCl and synthetic mine water (SMW) were undertaken. The observations can be summarized as follows.

(i) Specimens exhibit active-pseudopassive behavior in NaCl but the intensity of the pseudopassivation is impaired by the presence of VC. In SMW, all specimens actively corrode during anodic polarization with no impact whatsoever from VC. The negative impact in NaCl and the lack of impact in SMW are caused, primarily, by the fact that the VC does not participate in the formation of a protective film on the surfaces of the corroding samples, and, partly by the fact that the VC prevents W atoms from going into solution in the Co binder, that is, the VC fails to lower the magnetic saturation of samples.

(ii) Additions of VC to WC-Co at levels used for WC grain size refinement increase corrosion current densities and predispose specimens to pitting corrosion. This occurs because the VC reduces the amount of W atoms dissolving in the Co binder (increases the magnetic saturation compared to the base alloy).

(iii) The effect of high VC content on the corrosion resistance of WC-Co is solution dependent: high VC contents are not beneficial for corrosion resistance in NaCl, but greatly improve corrosion resistance in SMW.

Acknowledgments

This work was funded by South Africa's DST's Centre of Excellence in Strong Materials, based at the University of the Witwatersrand, Johannesburg, and the National Research Foundation.

References

[1] H. Engqvist, U. Beste, and N. Axén, "Influence of pH on sliding wear of WC-based materials," *International Journal of Refractory Metals and Hard Materials*, vol. 18, no. 2, pp. 103–109, 2000.

[2] M. H. Ghandehari, "Anodic behavior of cemented WC-6% co alloy in phosphoric acid solutions," *Journal of the Electrochemical Society*, vol. 127, no. 10, pp. 2144–2147, 1980.

[3] W. J. Tomlinson and C. R. Linzell, "Anodic polarization and corrosion of cemented carbides with cobalt and nickel binders," *Journal of Materials Science*, vol. 23, no. 3, pp. 914–918, 1988.

[4] A. Human, I. Northrop, S. Luyckx, and M. James, "A comparison between cemented carbides containing cobalt- and nickel-based binders," *Journal of Hard Materials*, vol. 2, no. 3, pp. 245–256, 1991.

[5] J. H. Potgieter, N. Thanjekwayo, P. Olubambi, N. Maledi, and S. S. Potgieter-Vermaak, "Influence of Ru additions on the corrosion behaviour of WC-Co cemented carbide alloys in sulphuric acid," *International Journal of Refractory Metals and Hard Materials*, vol. 29, no. 4, pp. 478–487, 2011.

[6] W. J. Tomlinson and N. J. Ayerst, "Anodic polarization and corrosion of WC-Co hardmetals containing small amounts of Cr_3C_2 and/or VC," *Journal of Materials Science*, vol. 24, no. 7, pp. 2348–2352, 1989.

[7] S. Broccardo, *An investigation into the corrosion resistance of WC-VC-Co hardmetals [M.Sc. thesis]*, University of the Witwatersrand, 2003.

[8] D. S. Konadu, J. V. D. Merwe, J. H. Potgieter, S. Potgieter-Vermaak, and C. N. Machio, "The corrosion behaviour of WC-VC-Co hardmetals in acidic media," *Corrosion Science*, vol. 52, no. 9, pp. 3118–3125, 2010.

[9] A. M. Human and H. E. Exner, "The relationship between electrochemical behaviour and in-service corrosion of WC based cemented carbides," *International Journal of Refractory Metals and Hard Materials*, vol. 15, no. 1-3, pp. 65–71, 1997.

[10] H. E. Exner, "Physical and chemical nature of cemented carbides," *International metals reviews*, vol. 24, no. 4, pp. 149–173, 1979.

[11] A. M. Human and H. E. Exner, "Electrochemical behaviour of tungsten-carbide hardmetals," *Materials Science and Engineering A*, vol. 209, no. 1-2, pp. 180–191, 1996.

[12] M. T. Laugier, "Coercivity, hardness and microstructure in WC-Co composites," *Journal of Materials Science Letters*, vol. 4, no. 2, pp. 211–216, 1985.

[13] D. L. Tillwick and I. Joffe, "Precipitation and magnetic hardening in sintered WC-Co composite materials," *Journal of Physics D*, vol. 6, no. 13, pp. 1585–1596, 1973.

[14] B. Roebuck, E. A. Almond, and A. M. Cottenden, "The influence of composition, phase transformation and varying the relative F.C.C. and H.C.P. phase contents on the properties of dilute CoWC alloys," *Materials Science and Engineering*, vol. 66, no. 2, pp. 179–194, 1984.

[15] C. N. Machio, *Preparation, characterisation and testing of WC-VC-Co HP/HVOF thermal spray coatings [Ph.D. thesis]*, University the Witwatersrand, 2006.

[16] P. Delichere, P. Falaras, M. Froment, A. Hugot-Le Goff, and B. Agius, "Electrochromism in anodic WO_3 films I: preparation and physicochemical properties of films in the virgin and coloured states," *Thin Solid Films*, vol. 161, pp. 35–46, 1988.

[17] B. Bozzini, G. Pietro De Gaudenzi, A. Fanigliulo, and C. Mele, "Electrochemical oxidation of WC in acidic sulphate solution," *Corrosion Science*, vol. 46, no. 2, pp. 453–469, 2004.

[18] T. Kubo and Y. Nishikitani, "Deposition temperature dependence of optical gap and coloration efficiency spectrum in electrochromic tungsten oxide films," *Journal of the Electrochemical Society*, vol. 145, no. 5, pp. 1729–1734, 1998.

[19] G. N. Kustova, Y. A. Chesalov, L. M. Plyasova, I. Y. Lin, and A. I. Nizovskii, "Vibrational spectra of $WO_3 \cdot nH_2O$ and WO_3 polymorphs," *Vibrational Spectroscopy*, vol. 55, no. 2, pp. 235–240, 2011.

[20] J. A. Horsley, I. E. Wachs, J. M. Brown, G. H. Via, and F. D. Hardcastle, "Structure of surface tungsten oxide species in the WO_3/Al_2O_3 supported oxide system from X-ray absorption near-edge spectroscopy and Raman spectroscopy," *Journal of Physical Chemistry*, vol. 91, no. 15, pp. 4014–4020, 1987.

[21] K. Nonaka, A. Takase, and K. Miyakawa, "Raman spectra of sol-gel-derived tungsten oxides," *Journal of Materials Science Letters*, vol. 12, no. 5, pp. 274–277, 1993.

[22] A. M. Human, B. Roebuck, and H. E. Exner, "Electrochemical polarisation and corrosion behaviour of cobalt and Co(W,C) alloys in 1 N sulphuric acid," *Materials Science and Engineering A*, vol. 241, no. 1-2, pp. 202–210, 1998.

[23] C. Colin, L. Durant, N. Favrot, J. Besson, G. Barbier, and F. Delannay, "Processing of functional-gradient WC-Co cermets by powder metallurgy," *International Journal of Refractory Metals and Hard Materials*, vol. 12, no. 3, pp. 145–152, 1993.

[24] C. Allen, "Corrosion of galvanised steel in synthetic mine waters," *British Corrosion Journal*, vol. 26, no. 2, pp. 93–101, 1991.

[25] S. Yeo, D. J. Kim, and J. W. Park, "Enhanced corrosion resistance of WC-Co with an ion beam mixed silicon carbide coating," *International Journal of Refractory Metals and Hard Materials*, vol. 29, no. 5, pp. 582–585, 2011.

[26] S. Hochstrasser-Kurz, Y. Mueller, C. Latkoczy, S. Virtanen, and P. Schmutz, "Analytical characterization of the corrosion mechanisms of WC-Co by electrochemical methods and inductively coupled plasma mass spectroscopy," *Corrosion Science*, vol. 49, no. 4, pp. 2002–2020, 2007.

[27] B. Schnyder, C. Stössel-Sittig, R. Kötz et al., "Investigation of the electrochemical behaviour of WC-Co hardmetal with electrochemical and surface analytical methods," *Surface Science*, vol. 566–568, no. 1–3, pp. 1240–1245, 2004.

[28] V. Mitrovic-Scepanovic, B. MacDougall, and M. J. Graham, "The effect of Cl^- ions on the passivation of Fe26Cr alloy," *Corrosion Science*, vol. 27, no. 3, pp. 239–247, 1987.

[29] B. Bozzini, G. P. De Gaudenzi, A. Fanigliulo, and C. Mele, "Anodic behaviour of WC-Co type hardmetal," *Materials and Corrosion*, vol. 54, no. 5, pp. 295–303, 2003.

[30] D. Zur Megede and E. Heitz, "Das korrosionsverhalten von Hartmetall-Verbundwerkstoffen in chloridhaltigen wäßrigen lösungen," *Materials and Corrosion*, vol. 37, no. 5, pp. 207–214, 1986.

[31] M. F. Daniel, B. Desbat, J. C. Lassegues, B. Gerand, and M. Figlarz, "Infrared and Raman study of WO3 tungsten trioxides and $WO_3{\cdot}xH_2O$ tungsten trioxide tydrates," *Journal of Solid State Chemistry*, vol. 67, no. 2, pp. 235–247, 1987.

[32] D. S. Kim, M. Ostromecki, and I. E. Wachs, "Surface structures of supported tungsten oxide catalysts under dehydrated conditions," *Journal of Molecular Catalysis A*, vol. 106, no. 1-2, pp. 93–102, 1996.

[33] F. D. Hardcastle and I. E. Wachs, "Determination of vanadium-oxygen bond distances and bond orders by Raman spectroscopy," *Journal of Physical Chemistry*, vol. 95, no. 13, pp. 5031–5041, 1991.

[34] M. A. Vuurman, D. J. Stufkens, A. Oskam, G. Deo, and I. E. Wachs, "Combined Raman and IR study of $MOx\text{-}V_2O_5/Al_2O_3$ ($MOx = MoO_3$, WO_3, NiO, CoO) catalysts under dehydrated conditions," *Journal of the Chemical Society—Faraday Transactions*, vol. 92, no. 17, pp. 3259–3265, 1996.

[35] C. K. Gupta and N. Krishnamurthy, *Extractive Metallurgy of Vanadium*, Elsevier Science Publishers B.V., Amsterdam, The Netherlands, 1992.

[36] S. Sutthiruangwong and G. Mori, "Corrosion properties of Co-based cemented carbides in acidic solutions," *International Journal of Refractory Metals and Hard Materials*, vol. 21, no. 3-4, pp. 135–145, 2003.

[37] E. Breval, J. P. Cheng, D. K. Agrawal et al., "Comparison between microwave and conventional sintering of WC/Co composites," *Materials Science and Engineering A*, vol. 391, no. 1-2, pp. 285–295, 2005.

[38] I. Sugiyama, Y. Mizumukai, T. Taniuchi et al., "Formation of (W, V)Cx layers at the WC/Co interfaces in the VC-doped WC-Co cemented carbide," *International Journal of Refractory Metals and Hard Materials*, vol. 30, no. 1, pp. 185–187, 2012.

19

Corrosion Inhibition of Cu-Ni (90/10) Alloy in Seawater and Sulphide-Polluted Seawater Environments by 1,2,3-Benzotriazole

Venkata Appa Rao Boyapati and Chaitanya Kumar Kanukula

Department of Chemistry, National Institute of Technology Warangal (NIT Warangal), Warangal, Andhra Pradesh 506 004, India

Correspondence should be addressed to Venkata Appa Rao Boyapati; boyapativapparao@rediffmail.com

Academic Editors: W. Fuerbeth and S. J. Lee

The inhibiting effect of 1,2,3-benzotriazole (BTAH) against the corrosion of Cu-Ni (90/10) alloy in seawater and seawater polluted with inorganic sulphide was studied by electrochemical impedance studies (EISs), potentiodynamic polarization studies, and cyclic voltammetric (CV) and weight-loss studies. Surface examination studies were carried out by X-ray photo electron spectroscopy (XPS) and scanning electron microscopy (SEM)/energy dispersive X-ray analysis (EDX). EIS studies have been carried out in seawater and 10 ppm of inorganic sulphide containing seawater in the absence and presence of BTAH at different concentrations, different immersion periods, and at different temperatures. Appropriate equivalent circuit model was used to calculate the impedance parameters. The potentiodynamic polarization studies inferred that BTAH functions as a mixed inhibitor. The impedance, polarization, and weight-loss studies showed that the inhibition efficiency of BTAH is in the range between 99.97 and 99.30% under different conditions. Cyclic voltammeric studies show the stability of the protective BTAH film even at anodic potentials of +550 mV versus Ag/AgCl. All these studies infer that BTAH functions as an excellent inhibitor for Cu-Ni (90/10) alloy in seawater and sulphide-polluted seawater. XPS and SEM-EDX studies confirm the presence of protective BTAH film on the alloy surface.

1. Introduction

Copper-nickel alloys are extensively used in marine applications because of their good electrical and thermal conductivities, corrosion resistance, and ease of fabrication of the equipment [1]. The 90/10 copper-nickel alloy is a material of selection for condensers and heat exchangers, where seawater is used as a coolant and in desalination plants [2, 3]. This alloy is resistant to stress corrosion cracking by ammonia and sulphide ions [4] and has good resistance to biofouling due to the release of copper ions [5, 6]. This alloy is also resistant to pitting and crevice corrosion in quiet seawater [7]. The corrosion resistance of this alloy is related to the performance of the passive film, which is mainly composed of Cu_2O [8–11], though other copper (II) based compounds such as atacamite and cupric oxide are also present in the film [6] after long exposure. The cupric species generally overlies the cuprous species. However, in the sulphide containing seawater, the corrosion rate of Cu-Ni (90/10) alloy is increased as the sulphide ions interfere with the film formation and produce a nonprotective black layer containing cuprous oxide and sulphide ions. The pollution of seawater with sulphide ions at the coastal areas can occur due to industrial waste water discharges into sea and also due to biological and bacteriological processes taking place in seawater. Thus, corrosion of Cu-Ni (90/10) alloy in seawater polluted with sulphide ions is a serious problem, which has drawn the attention of the researchers in this field. Sayed et al. [12] studied the corrosion behaviour of Cu-Ni (90/10) alloy in 3.5% sodium chloride solution in the absence and presence of different concentrations of sulphide ions in the range from 100 to 1000 ppm by potentiodynamic polarization studies and surface analysis by scanning electron microscopy and X-ray diffraction technique. They found that the corrosion rate of the alloy increased in the presence of sulphide ions and at a very high concentration of S^{2-}, the alloy suffered from pitting corrosion. Hack [13] reported that the required concentration of S^{2-} to increase the corrosion susceptibility of Cu-Ni (90/10) alloy may be as low as 0.01 ppm. De Sanchez and Schiffrin [14] studied the disruptive effect of sulphide ions on the protective film formed on Cu-Ni (90/10) alloy in seawater polluted with sulphides.

Organic inhibitors have been reported in the literature to protect Cu-Ni alloys from corrosion in different environments [15, 16]. The heterocyclic organic compounds, namely, azoles, were reported to show marked inhibition efficiency on corrosion of Cu-Ni alloys [17, 18]. Omar et al. [19] studied the corrosion inhibition of Cu-Ni (90/10) alloy in 0.5% hydrochloric acid and 0.5% sulfuric acid solutions by using Schiff bases as corrosion inhibitors. Badawy et al. [20] studied the corrosion inhibition of Cu-5Ni and Cu-65Ni alloys in 0.6 mol dm^{-3} chloride solution using amino acids as inhibitors. Benmessaoud et al. [15] studied the inhibiting effect of 2-mercaptobenzimidazole against the corrosion of 70Cu-30Ni alloy in aerated 3% sodium chloride solution polluted with ammonia (pH = 9.25) by potentiodynamic polarization studies and electrochemical impedance studies. 1,2,3-benzotriazole is a well-known corrosion inhibitor for copper. Babic et al. [21] studied the inhibiting effect of benzotriazole against corrosion of Cu-Ni (90/10) alloy in 1 M sodium acetate solution by cyclic voltammetry, photopotential measurements, and impedance spectroscopy. Allam et al. [22] studied the effect of benzotriazole on the corrosion of Cu-Ni (90/10) alloy in 3.4% sodium chloride solution containing 2 ppm of sulphide ions by weight-loss measurements and X-ray diffraction technique. Maciel and Agostinho [23] employed the potentiodynamic polarization studies on the Cu-Ni (90/10) alloy rotating disc electrode in deaerated 0.5 mol L^{-1} sulfuric acid solution containing Fe(III) ions as oxidant and benzotriazole as inhibitor and showed that benzotriazole at a concentration $>1.0 \times 10^{-3}$ mol L^{-1} afforded an excellent inhibition. Nevertheless, no studies have been reported in the literature on the corrosion behaviour of Cu-Ni (90/10) alloy in seawater and sulphide-polluted seawater in the presence of 1,2,3-benzotriazole at different immersion periods and at different temperatures. Studies using cyclic voltammetry have not been reported on benzotriazole as inhibitor for Cu-Ni (90/10) alloy in seawater and sulphide-polluted seawater. A comprehensive study, employing electrochemical impedance, potentiodynamic polarization, cyclic voltammetry, weight-loss studies, XPS, and SEM-EDX on corrosion inhibition of Cu-Ni (90/10) alloy in seawater and sulphide containing seawater environments by 1,2,3-benzotriazole (BTAH), has been attempted in the present work.

2. Experimental

2.1. Materials. The composition of Cu-Ni (90/10) alloy used in the present study is given in Table 1. The composition of synthetic seawater is given in Table 2 [24]. All the chemicals used in the preparation of synthetic seawater were of AnalaR grade. A stock solution of 1000 ppm of inorganic sulphide containing seawater was prepared by dissolving 2.44 g of AnalaR sodium sulphide in synthetic seawater and making the solution up to the mark in 1000 mL standard measuring flask. From this stock solution, 10 ppm of sulphide containing seawater was prepared. 1,2,3-benzotriazole obtained from Sigma-Aldrich was used as such. The cupronickel alloy specimens were taken from the same Cu-Ni (90/10) alloy sheet. For electrochemical studies, specimens with the dimensions of 4.0 × 1.0 × 0.2 cm were used by exposing 1 cm^2 surface area, while the remaining area was insulated with epoxy resin. For weight-loss studies also, 4.0 × 1.0 × 0.2 cm specimens were used. For XPS and SEM-EDX studies, 1.0 × 1.0 × 0.2 cm and 0.5 × 0.5 × 0.2 cm specimens were used, respectively. All these specimens were polished with 1/0, 2/0, 3/0, and 4/0 grade emery papers consecutively. Later these specimens were polished to mirror finish with alumina and water slurry on the micro polishing cloth, which is fixed on the rotating disc polishing machine. Then, these specimens were washed with triple distilled water, degreased with acetone and dried by blowing N_2 gas for 20 minutes. These specimens were tested in different corrosive environments.

2.2. Electrochemical Studies. Electrochemical impedance studies were carried out in a three-electrode cell assembly (in accordance with ASTM specifications) using an electrochemical work station, model IM6e Zahner-Elektrik, Germany. The Cu-Ni (90/10) alloy of the surface area of 1 cm^2 was used as the working electrode in the absence and presence of BTAH in seawater and sulphide-polluted seawater environments. A platinum electrode was used as the counter electrode and the reference electrode was Ag/AgCl/1.0 M KCl electrode. The impedance studies were carried out at the open circuit potential (OCP) in the frequency range from 60 kHz to 10 mHz under the excitation of sinusoidal wave of ±5 mV amplitude.

The potentiodynamic polarization studies and cyclic voltammetric studies were carried out by using the same three-electrode cell assembly and the electrochemical work station as described under impedance studies. The potentiodynamic polarization studies were conducted in the potential range of −0.7 V to +0.7 V versus Ag/AgCl with a scan rate of 1 mVs^{-1}. The cyclic voltammetric studies were carried out in the potential range of −1.0 V to +1.0 V versus Ag/AgCl with a scan rate of 10 mVs^{-1}. All the electrochemical studies were conducted at different immersion periods in the range from 1 to 48 h. The impedance studies were carried out at three different temperatures, namely, 30, 40, and 60°C.

2.3. Gravimetric Measurements. In all the gravimetric experiments, the polished Cu-Ni (90/10) alloy specimens were weighed and immersed in duplicate, in 500 mL seawater and 10 ppm of inorganic sulphide containing seawater in the absence and presence of BTAH at two different concentrations, for a period of 30 days. Then these specimens were reweighed after dipping for 2 to 3 min in hydrochloric acid (1 + 1) or sulfuric acid (1 + 10) at room temperature. Later, these specimens were scrubbed with a bristle brush under

TABLE 1: Composition of the Cu-Ni (90/10) alloy.

Element	Cu	Ni	Fe	Mn	Pb	Al	Others in trace amounts
Composition (%)	88.512	9.882	1.086	0.412	0.046	0.038	0.024

TABLE 2: Composition of the synthetic seawater.

Compound name	NaCl	KCl	$MgCl_2{\cdot}6H_2O$	$CaCl_2$	$MgSO_4{\cdot}7H_2O$	$NaHCO_3$	KBr	H_3BO_3	$SrCl_2{\cdot}6H_2O$	NaF
$g\,L^{-1}$	27.24	1.40	10.11	2.28	13.92	0.39	0.20	0.026	0.04	0.006

running water, degreased with acetone, and dried. During the studies, only those results were taken into consideration, in which the difference in the weight loss of the two specimens immersed in the same solution did not exceed 0.1 mg. Accuracy in weighing up to 0.01 mg is recommended by ASTM G31 guidelines [25]. The immersion period of 30 days was fixed in view of the considerable magnitude of the corrosion rate obtained in the absence of any inhibitor after this immersion period. Under these conditions of accuracy, the relative standard error in corrosion rate determinations is of the order of 2% or less [26]. Corrosion rates (CRs) of alloy specimens in the absence and presence of BTAH are expressed in mmpy. Inhibition efficiencies of the inhibitor were calculated by using the formula:

$$\text{IE}\ (\%) = \frac{[(\text{CR})_0 - (\text{CR})_I]}{(\text{CR})_0} \times 100, \tag{1}$$

where $(\text{CR})_0$ and $(\text{CR})_I$ are the corrosion rates in the absence and presence of BTAH, respectively.

2.4. Surface Analytical Techniques. The surface analysis of Cu-Ni (90/10) alloy in the absence and presence of BTAH was carried out using the X-ray photoelectron spectrometer, ESCA Kratos model AXIS-165, with Mg Kα radiation (1253.6 eV) and energy resolution of 0.1 eV. Computer deconvolution was applied to detect the elemental peaks of copper, oxygen, carbon, nitrogen, and sulfur present on the surface of the alloy. SEM-EDX studies were carried out using JEOL JSM-6390LV/LGS Scanning Electron Microscope, USA, and JEOL Model JED-2300, Energy Dispersive X-ray Analyzer, USA.

3. Results and Discussion

3.1. Electrochemical Impedance Studies. Figures 1(a) and 1(b) show the Nyquist and Bode plots of Cu-Ni (90/10) alloy in synthetic seawater in the absence of BTAH at different immersion periods in the range from 1 to 48 h. The corresponding impedance parameters are given in Table 3. The Nyquist plots exhibit diffusion impedance at all the immersion periods. The diameter of semicircular region increases with immersion time. This means that there is an increase in the charge transfer resistance (R_{ct}) with the increase in immersion time. The total impedance Bode plot also indicates that the impedance increases with the immersion time. The phase angle Bode plots reveal that the phase maximum is the same at 1- and 3-hour immersion periods and it shifts to a lower frequency at higher immersion periods. There is only one maximum in all these plots and the angle corresponding to phase maximum is found to increase from 51° at 1-hour immersion time to 56° at 48-hour immersion time. All the above results infer that with the increase in immersion time, the film is becoming more and more protective. In these figures the $|Z|$ versus f curves exhibit three distinctive segments. In the high frequency extreme region, the value of $|Z|$ tends to become very small and the corresponding phase angle also falls off very rapidly with increasing frequency. This region is typical of a resistive behaviour and corresponds to solution resistance. In the medium frequency region, a linear relationship exists between $|Z|$ and f and there is a broad maximum in the phase angle versus frequency plot. This region corresponds to the charge transfer at the metal/electrical double layer. In the low frequency segment, the resistive behaviour of the electrode increases and the value of $|Z|$ should remain constant with a further decrease in frequency. But such behaviour is not found. $|Z|$ still increases, with a decrease in frequency. This corresponds to the diffusion impedance because of the diffusion of oxygen and chloride from the bulk of the solution to the interface at the instant of the immersion of the electrode in synthetic seawater and also of $CuCl_2^-$ species from the interface into the bulk of the solution at higher immersion periods, besides diffusion of oxygen and chloride from solution to the interface [27, 28]. In order to account for the above results, the metal/solution interface is modeled as per the equivalent circuit shown in Figure 2(a) [24] and the best fit is obtained for the experimental results with this model. In this model constant phase element (CPE) is substituted for the double layer capacitance to give a more accurate fit [29]. The CPE is a special element whose admittance value is a function of the angular frequency (ω) and whose phase is independent of frequency. Its admittance and impedance components are given in the following equations, respectively:

$$Y_{\text{CPE}} = Y_0 (j\omega)^n, \tag{2}$$

$$Y_{\text{CPE}} = \frac{1}{\left[Y_0 (j\omega)^n\right]}. \tag{3}$$

Y_0 is the magnitude of the CPE, j is the imaginary number ($j = (-1)^{1/2}$), and ω (= $2\pi f$) is the angular frequency [30]. The factor "n" is the adjustable parameter that usually lies between 0 and 1 [31]. The CPE describes an ideal capacitor when n is equal to 1 [29, 31]. Both the charge transfer resistance of the double layer (R_{ct}) and the film resistance (R_{film}) are found to increase with immersion period. The "n"

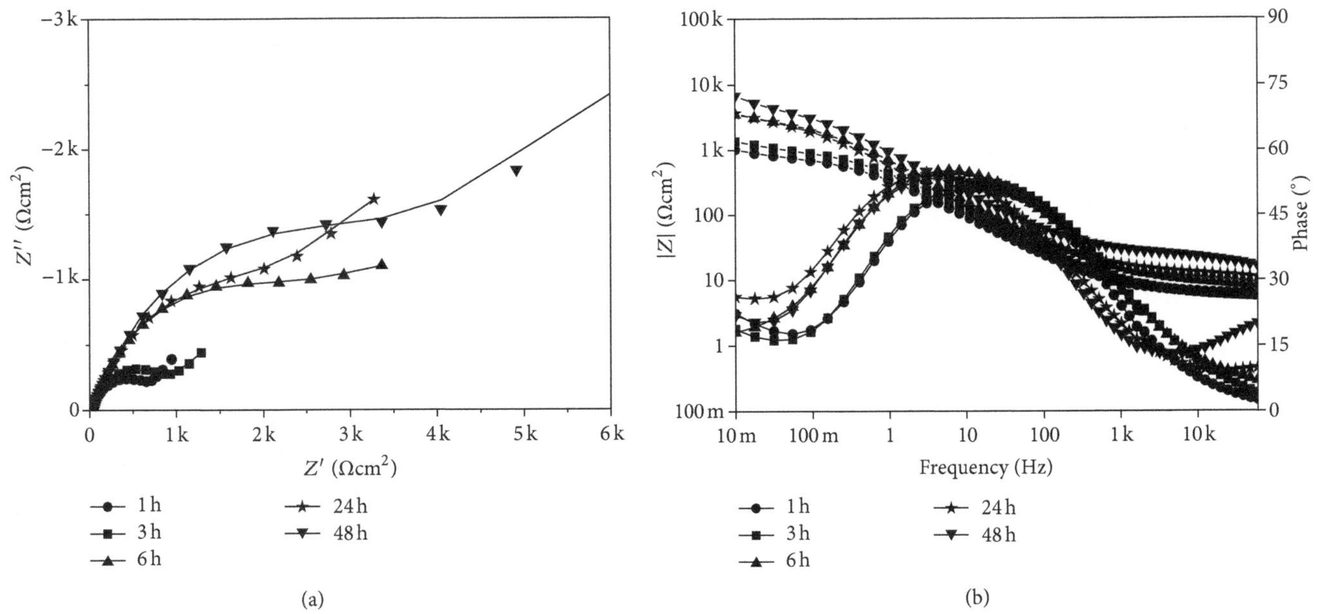

FIGURE 1: (a) Nyquist and (b) Bode plots of Cu-Ni (90/10) alloy in synthetic seawater at different immersion periods. -●-: 1 h, -■-: 3 h, -▲-: 6 h, -★-: 24 h, and -▼-: 48 h. (Temperature: 30°C).

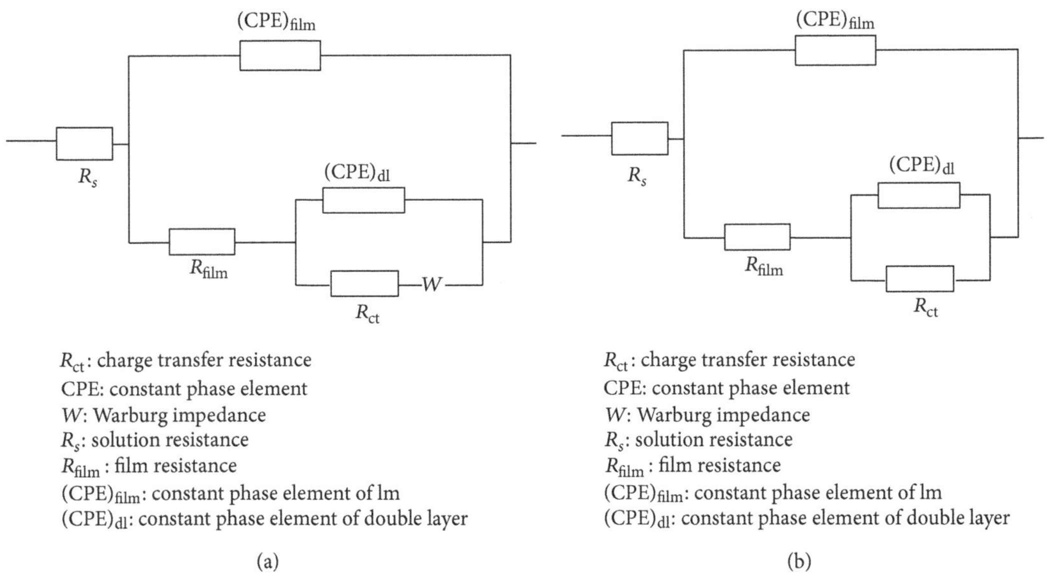

FIGURE 2: Equivalent circuits used to calculate the impedance parameters.

values corresponding to both the CPE of double layer and of the film are found to increase slightly with the increase in immersion period.

Figures 3(a) and 3(b) show the Nyquist and Bode plots, respectively, of the Cu-Ni (90/10) alloy in seawater containing 1,2,3-benzotriazole inhibitor at various concentrations, namely, 0.82, 1.67, 2.50, and 3.33 mM at a constant immersion period of 48 h. The Nyquist plots are not perfect semicircles. With increasing concentration, the diameter of the semicircle is found to increase. Though there is absence of Warburg-like behaviour, the plots show a combination of a semicircle and a linear (horizontal) pattern. These results in the low frequency region indicate the resistive behaviour of the alloy. The Bode plots are quite interesting. The phase angle Bode plots show that there is a shift of frequency corresponding to phase maximum from about 5 Hz in the absence of BTAH to a much higher frequency of the order of kHz in the presence of 0.82 mM BTAH. The angle corresponding to phase maximum is also increased from 56° to about 75°. A close examination of these plots reveals that there is a possibility of two time constants. It is interesting to note that with a further increase in the concentration of the inhibitor to 3.33 mM, there is a shift in the phase maximum to a lower frequency side and phase angle is increased to about

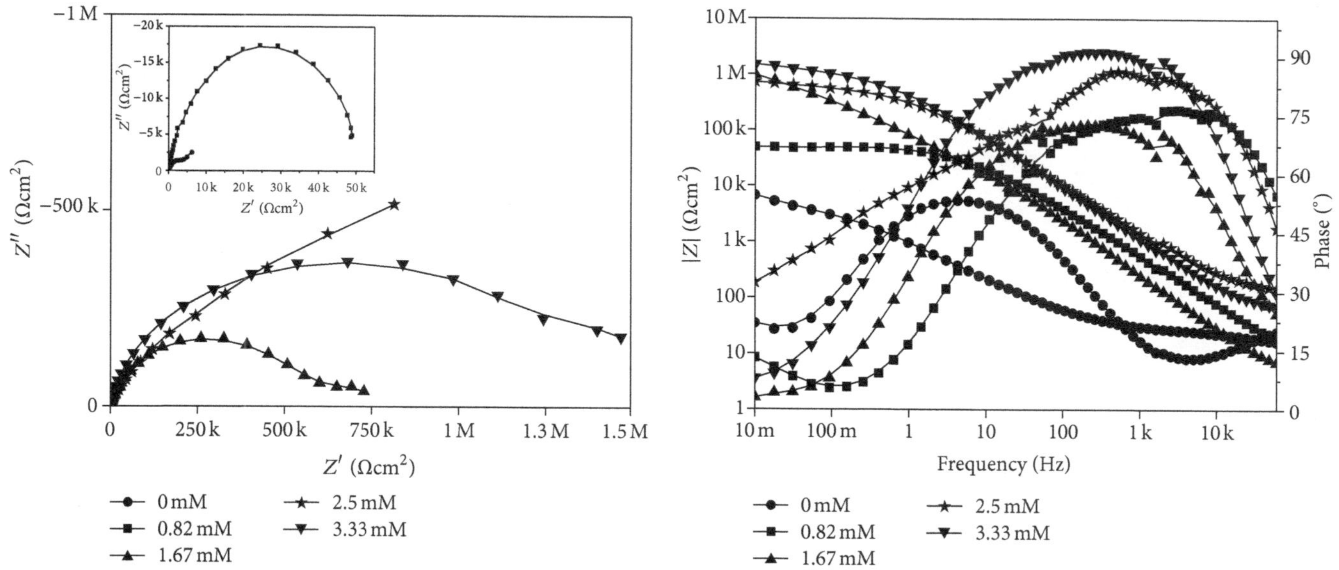

(a) (b)

FIGURE 3: (a) Nyquist and (b) Bode plots of Cu-Ni (90/10) alloy in synthetic seawater in the presence of different concentrations of BTAH at 48-hour immersion period. -•-: 0 mM, -■-: 0.82 mM, -▲-: 1.67 mM, -★-: 2.50 mM, and -▼-: 3.33 mM. (Temperature: 30°C).

TABLE 3: Impedance parameters of Cu-Ni (90/10) alloy in seawater at different immersion periods. (Temperature: 30°C).

Immersion period (h)	R_{ct} (kΩ cm^2)	$(CPE)_{dl}$ (μF cm^{-2})	n_1	R_f (kΩ cm^2)	$(CPE)_{film}$ (μF cm^{-2})	n_2
1	0.722	28.41	0.64	0.002	36.44	0.60
3	1.004	14.50	0.65	0.005	34.81	0.64
6	3.168	13.30	0.66	0.007	27.34	0.64
24	4.174	12.32	0.67	0.008	15.47	0.65
48	4.720	12.20	0.67	0.010	15.12	0.66

80°. The total impedance is increased to a very great extent in the presence of the inhibitor and with the increase in the concentration of the inhibitor. There is more broadening of the phase maximum with the increase in the concentration of the inhibitor. This result reveals that the protective nature of the inhibitor film is increased with the increase in the concentration of the inhibitor. The electrochemical system may be modeled as consisting of metal/protective surface film/electrical double layer/electrolyte solution. Therefore, for all the systems, in the presence of BTAH, the equivalent circuit shown in Figure 2(b) is used. The experimental data fitted with this equivalent circuit better than with other equivalent circuits reported in the literature [16, 20, 24]. The corresponding impedance parameters are given in Table 4. It is interesting to note that both R_{ct} and R_{film} are increased with the increase in the concentration of the inhibitor. An increase in R_{ct} from 4.72 to 1046 kΩ cm^2 and an increase in R_{film} from 0.01 to 557.8 kΩ cm^2 are obtained by the addition of 3.33 mM of BTAH to seawater. Inhibition efficiencies are calculated from the R_{ct} values in the absence and presence of BTAH using

$$\text{IE }(\%) = \left(R^1_{ct} - \frac{R_{ct}}{R^1_{ct}}\right) \times 100. \quad (4)$$

In (4), R^1_{ct} and R_{ct} are the charge transfer resistances in the presence and absence of BTAH, respectively. An IE% of 99.55 is obtained at 3.33 mM concentration of BTAH. Both the $(CPE)_{dl}$ and $(CPE)_{film}$ values are reduced to extremely small values of 0.034 and 0.060 μF cm^{-2}, respectively, in the presence of BTAH and the corresponding "n" values are increased to 0.94 and 0.78, respectively. These results reveal that the concentration of BTAH molecules in the protective film is increased and the Cu_2O formation is much less. BTAH molecules replaced almost all the water molecules and the ions in the electrical double layer.

As a protective film with a high charge transfer resistance is formed when the concentration of BTAH is 3.33 mM, it was of interest to investigate the effect of immersion time in the range from 1 to 48 h, on the impedance spectra. The corresponding Nyquist and Bode plots are shown in Figures 4(a) and 4(b), respectively. The impedance parameters are given in Table 5. The Nyquist plots show the absence of Warburg impedance at all immersion periods and the increase in the diameter of the semicircular part of the plots (in the intermediate frequencies) and also the increase in the linear (horizontal) part of the plots in the lower frequency region with increase in the immersion period. The Bode plots reveal

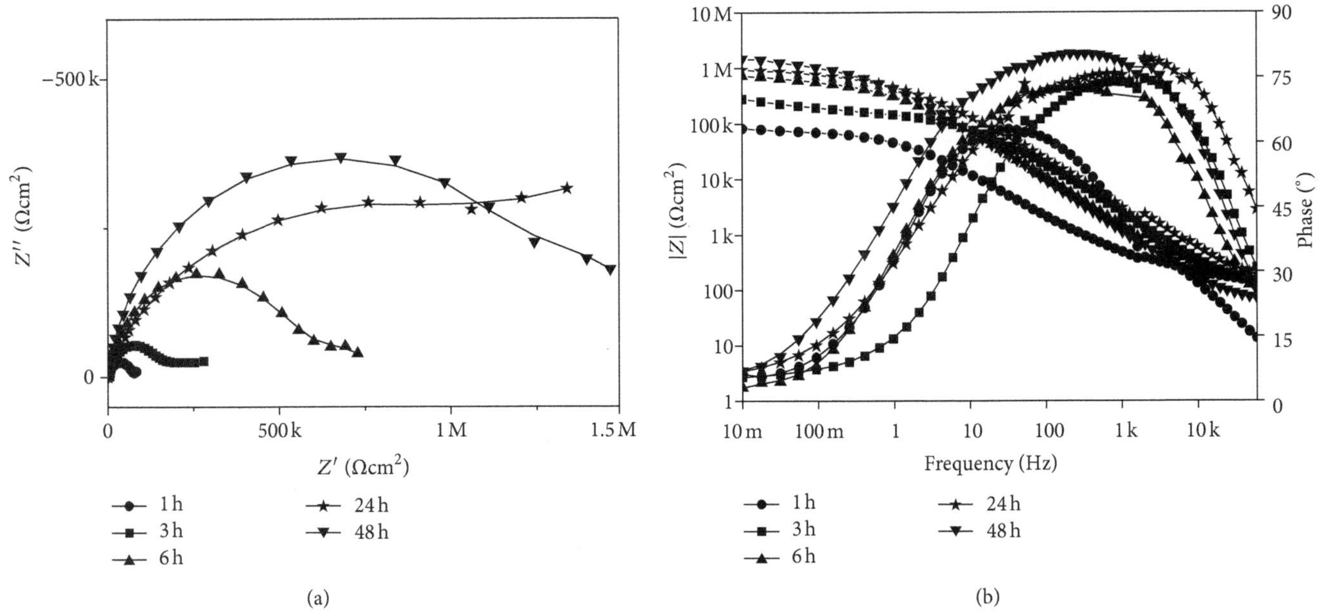

FIGURE 4: (a) Nyquist and (b) Bode plots of Cu-Ni (90/10) alloy in synthetic seawater in the presence of 3.33 mM BTAH at different immersion periods. -●-: 1 h, -■-: 3 h, -▲-: 6 h, -★-: 24 h, and -▼-: 48 h. (Temperature: 30°C).

TABLE 4: Impedance parameters of Cu-Ni (90/10) alloy in seawater in the presence of different concentrations of BTAH at 48-hour immersion period. (Temperature: 30°C).

Concentration of BTAH (mM)	R_{ct} (kΩ cm^2)	(CPE)$_{dl}$ (μF cm^{-2})	n_1	R_{film} (kΩ cm^2)	(CPE)$_{film}$ (μF cm^{-2})	n_2	IE (%)
0	4.72	12.20	0.67	0.01	15.12	0.66	—
0.82	44.8	0.096	0.73	5.34	0.290	0.68	89.48
1.67	362.5	0.058	0.84	80.3	0.137	0.72	98.70
2.50	620.3	0.045	0.91	250.6	0.072	0.73	99.24
3.33	1046.0	0.034	0.94	557.8	0.060	0.78	99.55

some interesting features. After 1-hour immersion time, the phase angle Bode plot reveals two phase maxima, one in the high frequency region and one in the mid frequency region. With increase in immersion period, the broadening of the phase maximum and increase in phase angle maximum are quite significant. After 48-hour immersion period, the total impedance Bode plot is linear, almost throughout the frequency region. With increase in immersion time, the slopes of the $|Z|$ versus f plots are found to increase in the lower frequency region while there is almost a merger of these plots somewhere in the middle frequency region after the immersion period ≥3 h.

After the formation of the protective film on the Cu-Ni (90/10) alloy by immersing it in seawater containing 3.33 mM BTAH for 48 h, the effect of the increase of temperature on the impedance behaviour of this prefilmed alloy was studied in seawater containing 3.33 mM BTAH for an immersion period of 48 h at two different temperatures, namely, 40 and 60°C. The nature of Nyquist and Bode plots are similar to those plots at 30°C (plots not shown). However, the total impedance and phase angle maximum values in the Bode plots are reduced with the increase in temperature. Nevertheless, in the absence of BTAH also, the total impedance and phase angle maximum values are reduced with the increase in temperature. The impedance parameters at different temperatures for bare Cu-Ni (90/10) alloy and prefilmed Cu-Ni (90/10) alloy after an immersion period of 48 h in seawater containing 3.33 mM BTAH are given in Table 6. Both R_{ct} and R_{film} of the prefilmed alloy are found to decrease with the increase in temperature. Nevertheless, these values are still far higher in comparison to the corresponding values of the bare alloy. Hence, inhibition efficiencies >99% are obtained even at 40 and 60°C. The increase in (CPE)$_{dl}$ values is not as much as the increase in the (CPE)$_{film}$ values. Both the n_1 and n_2 values are decreased to some extent with the increase in temperature. All these results are indicative of the deviation of double layer and also of the film from more capacitative to less capacitative behaviour.

Figures 5(a) and 5(b) show the Nyquist and Bode plots of Cu-Ni (90/10) alloy in seawater environment containing 10 ppm of inorganic sulphide ions in the absence of BTAH at different immersion periods. The corresponding impedance parameters are given in Table 7. The Nyquist plots show that the real and imaginary parts of the complex impedance decrease to a great extent in comparison to the behaviour of the alloy in the absence of sulphide ions at all

TABLE 5: Impedance parameters of Cu-Ni (90/10) alloy in seawater in the presence of 3.33 mM BTAH at different immersion periods. (Temperature: 30°C).

Immersion period (h)	R_{ct} (kΩ cm^2)	$(CPE)_{dl}$ (μF cm^{-2})	n_1	R_{film} (kΩ cm^2)	$(CPE)_{film}$ (μF cm^{-2})	n_2	IE (%)
1	66.26	1.732	0.68	14.55	2.546	0.68	98.91
3	142.8	0.064	0.72	117.0	1.321	0.70	99.30
6	379.9	0.055	0.74	227.3	0.141	0.77	99.17
24	882.4	0.022	0.86	762.0	0.077	0.78	99.53
48	1046.0	0.034	0.94	557.8	0.060	0.78	99.55

TABLE 6: Impedance parameters of Cu-Ni (90/10) alloy in seawater and prefilmed alloy in the presence of 3.33 mM BTAH in seawater at different temperatures. (Immersion period: 48 h).

Specimen	Environment	Temperature	R_{ct} (kΩ cm^2)	$(CPE)_{dl}$ (μF cm^{-2})	n_1	R_{film} (kΩ cm^2)	$(CPE)_{film}$ (μF cm^{-2})	n_2	IE (%)
Bare alloy	Seawater	40	2.972	8.38	0.71	0.004	72.52	0.57	—
Prefilmed alloy	Seawater + 3.33 mM BTAH	40	1231.0	0.310	0.86	237.5	0.493	0.80	99.76
Bare alloy	Seawater	60	0.315	50.42	0.58	0.002	210.0	0.55	—
Prefilmed alloy	Seawater + 3.33 mM BTAH	60	542.0	0.520	0.82	473.0	0.748	0.78	99.94

TABLE 7: Impedance parameters of Cu-Ni (90/10) alloy in seawater containing 10 ppm of inorganic sulphide at different immersion periods. (Temperature: 30°C).

Immersion period (h)	R_{ct} (kΩ cm^2)	$(CPE)_{dl}$ (μF cm^{-2})	n_1	R_{film} (kΩ cm^2)	$(CPE)_{film}$ (μF cm^{-2})	n_2
1	0.500	93.02	0.62	0.001	106.5	0.52
3	0.512	104.2	0.64	0.002	111.2	0.58
6	0.661	161.6	0.64	0.002	174.5	0.60
24	0.871	200.2	0.64	0.002	246.7	0.63
48	0.992	210.0	0.64	0.003	322.3	0.63

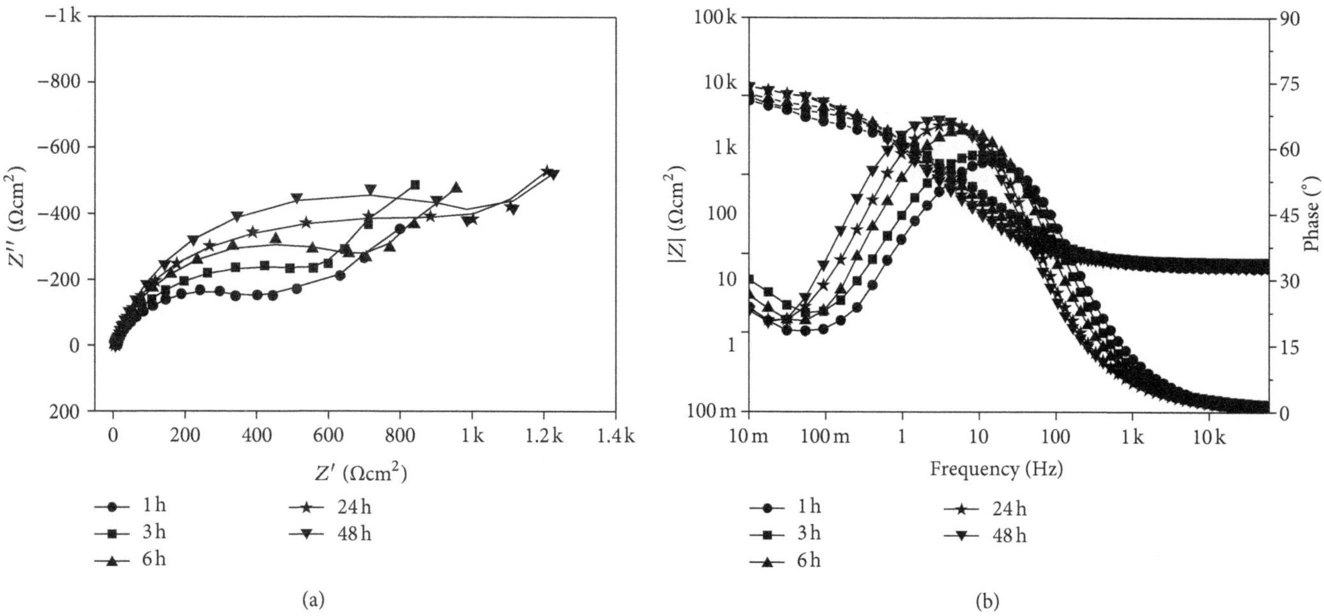

FIGURE 5: (a) Nyquist and (b) Bode plots of Cu-Ni (90/10) alloy in synthetic seawater polluted with 10 ppm sulphide at different immersion periods. -●-: 1 h, -■-: 3 h, -▲-: 6 h, -★-: 24 h, and -▼-: 48 h. (Temperature: 30°C).

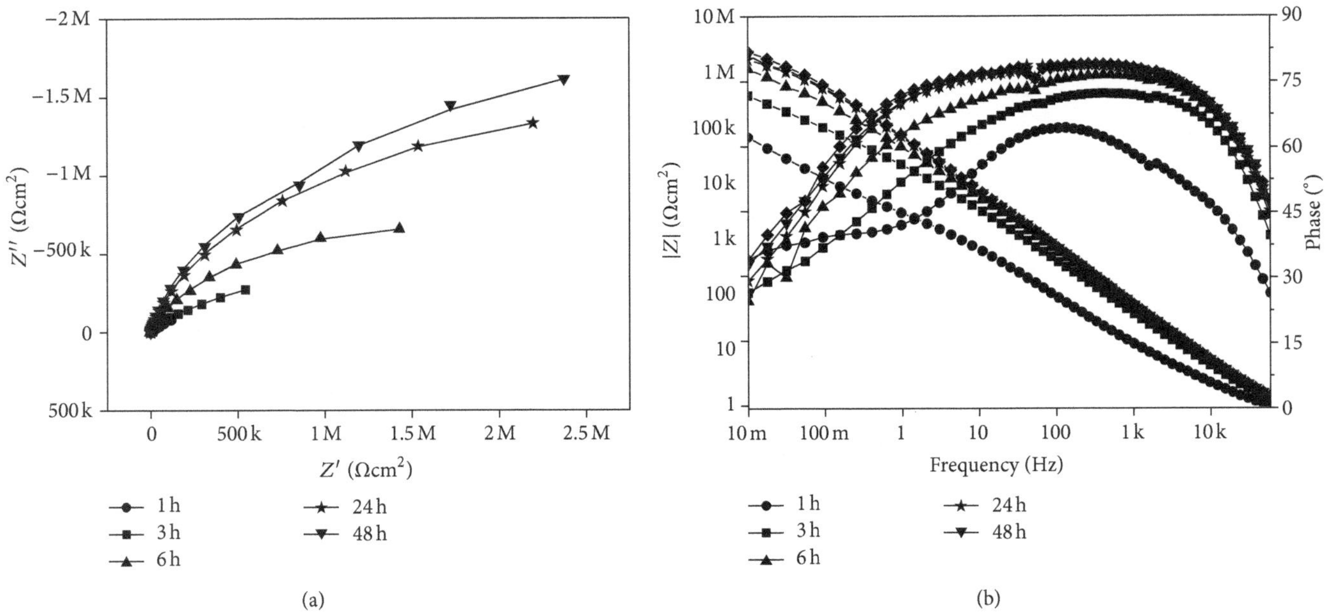

FIGURE 6: (a) Nyquist and (b) Bode plots of Cu-Ni (90/10) alloy in synthetic seawater polluted with 10 ppm sulphide in the presence of 3.33 mM BTAH at different immersion periods. -•-: 1 h, -■-: 3 h, -▲-: 6 h, -★-: 24 h, and -▼-: 48 h. (Temperature: 30°C).

immersion periods. The Bode plots show some interesting features. There is almost a merger of the plots for different immersion periods at higher frequencies. The phase angle Bode plots showed a clear shift of phase angle maximum towards lower frequencies and broadening of the maximum with the increase in immersion period. However, the phase angle maximum region is narrower in comparison to the corresponding plots in the absence of sulphide. The total impedance Bode plots show the linear behaviour in relatively less range of frequency in the midregion in comparison to the plots in the absence of sulphide ions. The total impedance is also found to be much less in comparison to the system containing no sulphides. Both R_{ct} and R_{film} are increased with the increase in immersion period. But these values are less than those in the absence of sulphides ions in seawater. The $(CPE)_{dl}$ and $(CPE)_{film}$ values are increased in the presence of sulphide ions. That means, by the addition of sulphide ions, the protective nature of the film is reduced. However, both the double layer and the film still show a combination of capacitative and resistive behaviours. Such a large increase in both $(CPE)_{dl}$ and $(CPE)_{film}$ in the presence of only 10 ppm S^{2-} ions shows the significant effect of S^{2-} in modifying especially the inner layer due to the strong specific adsorption of sulphide [32, 33].

Figures 6(a) and 6(b) show the Nyquist and Bode plots of Cu-Ni (90/10) alloy in seawater environment containing 10 ppm of sulphide ions in the presence of 3.33 mM BTAH at different immersion periods. The corresponding impedance parameters are given in Table 8. The Nyquist plots show an increase in both the real and imaginary parts of the complex impedance in the presence of inhibitor. The Bode plots have become much more broadened and the phase angle maxima are increased at all the immersion periods. The phase angle maximum is shifted to much higher frequencies. After 24- and 48-hour immersion periods, the total impedance plots show linearity from the highest frequency of 60 kHz to a low frequency of 10 mHz. All these plots clearly exhibit the capacitative behaviour of both the double layer and of the protective film. Both R_{ct} and R_{film} values are found to increase enormously in the presence of the inhibitor and also with an increase in immersion period. No wonder that the inhibition efficiencies are in the range of 98.81 to 99.97% within the range of immersion period studied. After 48 h immersion period, R_{ct} and the film resistance, respectively, are 2516 and 937.5 kΩ cm^2. The capacitance of the double layer $(CPE)_{dl}$ is reduced from 210.0 μF cm^{-2} to an extremely small value of 0.026 μF cm^{-2} in the presence of BTAH even after 1-hour immersion period. This result indicates the absence of all the ions present in seawater and sulphide ions and presence of only BTAH molecules in the double layer.

Once again the prefilmed alloy is immersed in seawater containing 10 ppm sulphide and 3.33 mM BTAH to investigate the effect of the increase of temperature on the impedance behaviour. The impedance parameters are given in Table 9. Both R_{ct} and R_{film} are reduced with increase in temperature. While $(CPE)_{dl}$ is increased only slightly with the increase in temperature; the increase in $(CPE)_{film}$ is quite significant. Both n_1 and n_2 are reduced to some extent. Nevertheless, the inhibition efficiencies are greater than 99% in the entire temperature range. Thus, impedance studies reveal that BTAH functions quite well as inhibitor in the entire temperature range studied.

3.2. Potentiodynamic Polarization Studies. The potentiodynamic polarization curves for Cu-Ni (90/10) alloy in synthetic seawater in the potential range from −0.7 to +0.7 V versus Ag/AgCl are given in Figure 7 and the corrosion parameters are tabulated in Table 10. The anodic polarization

TABLE 8: Impedance parameters of Cu-Ni (90/10) alloy in seawater containing 10 ppm of inorganic sulphide in the presence of 3.33 mM BTAH at different immersion periods. (Temperature: 30°C).

Immersion period (h)	R_{ct} (kΩ cm^2)	$(CPE)_{dl}$ (μF cm^{-2})	n_1	R_{film} (kΩ cm^2)	$(CPE)_{film}$ (μF cm^{-2})	n_2	IE (%)
1	42.13	0.577	0.61	8.399	1.621	0.64	98.81
3	390.1	0.086	0.81	47.98	0.602	0.74	99.87
6	712.0	0.043	0.82	114.6	0.476	0.77	99.91
24	1361.0	0.028	0.88	686.2	0.438	0.77	99.94
48	2516.0	0.026	0.90	937.5	0.239	0.82	99.97

TABLE 9: Impedance parameters of Cu-Ni (90/10) alloy in seawater containing 10 ppm of inorganic sulphide and prefilmed alloy in the presence of 3.33 mM BTAH in seawater containing 10 ppm of inorganic sulphide at different temperatures. (Immersion period: 48 h).

Specimen	Environment	Temperature	R_{ct} (kΩ cm^2)	$(CPE)_{dl}$ (μF cm^{-2})	n_1	R_{film} (kΩ cm^2)	$(CPE)_{film}$ (μF cm^{-2})	n_2	IE (%)
Bare alloy	Seawater + 10 ppm sulphide	40	0.390	168.8	0.58	0.002	434.6	0.57	—
Prefilmed alloy	Seawater + 10 ppm sulphide + 3.33 mM BTAH	40	612.7	0.788	0.79	250.9	0.920	0.73	99.94
Bare alloy	Seawater + 10 ppm sulphide	60	0.184	256.1	0.54	0.002	525.6	0.54	—
Prefilmed alloy	Seawater + 10 ppm sulphide 3.33 mM BTAH	60	398.8	1.033	0.73	223.5	2.621	0.76	99.95

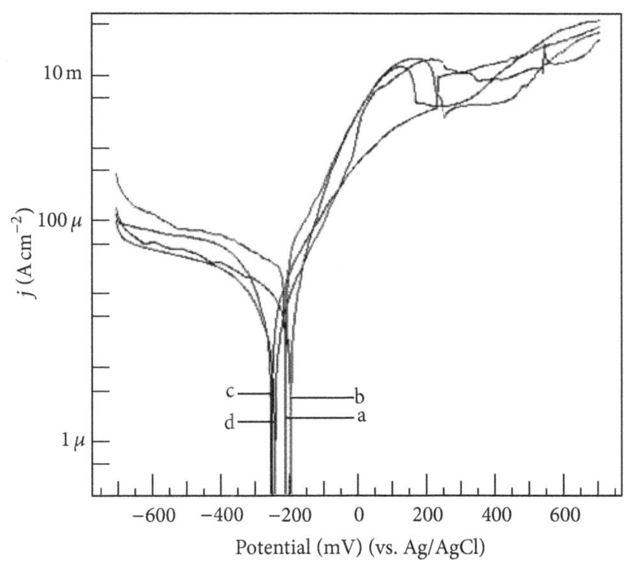

FIGURE 7: Potentiodynamic polarization curves of Cu-Ni (90/10) alloy in synthetic seawater at different immersion periods. a: 1 h, b: 12 h, c: 24 h, and d: 48 h. (Temperature: 30°C).

TABLE 10: Corrosion parameters of Cu-Ni (90/10) alloy obtained by potentiodynamic polarization studies in seawater in the absence of BTAH at different immersion periods. (Temperature: 30°C).

Immersion period (h)	E_{corr} (mV)	i_{corr} (μA cm^{-2})	β_c (mV dec^{-1})	β_a (mV dec^{-1})
1	−219.0	23.3	−165.0	86.8
12	−239.1	4.98	−105.0	90.8
24	−241.0	4.25	−78.1	84.8
48	−251.3	5.98	−115.0	68.2

curves show three distinct regions, namely, Tafel region at lower overpotentials extending to the peak current density due to the dissolution of copper to Cu^+ ions, a region of decreasing current density until a minimum (i_{min}) is reached due to the formation of CuCl, and a region of the sudden increase in current density leading to a limiting value (i_{lim}), due to the formation of $CuCl_2^-$. These three regions have been reported in the literature in the polarization curves of copper in 3.5% sodium chloride solution [34] and also in the case of Cu-Ni (90/10) alloy [35]. With the increase in immersion period from 1 to 48 h, the corrosion potential (E_{corr}) is shifted towards more negative value and the shift in cathodic Tafel slope (β_c) is greater than the shift in anodic slope (β_a). However, the corrosion current density (i_{corr}) is found to decrease significantly up to 12-hour immersion period and then increase slightly up to 48 h. These results infer that with the increase in immersion period from 1 to 48 h, there is a decrease in the corrosion rate of the alloy due to the formation of copper oxide film, and this has been reported by many investigators [7, 36–38]. Another interesting feature is that while there is only one anodic peak after 1- and 12-hour immersion periods, there are two anodic peaks after 24- and 48-hour immersion periods at 150 and 250 mV versus Ag/AgCl electrode, respectively. The larger peak at 150 mV is

TABLE 11: Corrosion parameters of Cu-Ni (90/10) alloy obtained by potentiodynamic polarization studies in seawater in the presence of different concentrations of BTAH at 48-hour immersion period. (Temperature: 30°C).

Concentration of BTAH (mM)	E_{corr} (mV)	i_{corr} (μA cm^{-2})	β_c (mV dec^{-1})	β_a (mV dec^{-1})	IE (%)
0	−251.3	5.98	−115.0	68.2	—
0.82	−217.9	0.141	−95.0	124.0	97.64
1.67	−173.4	0.056	−60.5	81.8	99.06
2.50	−141.4	0.043	−85.2	103.0	99.28
3.33	−117.4	0.042	−70.4	83.7	99.30

TABLE 12: Corrosion parameters of Cu-Ni (90/10) alloy obtained by potentiodynamic polarization studies in seawater in the presence of 3.33 mM BTAH at different immersion periods. (Temperature: 30°C).

Immersion period (h)	E_{corr} (mV)	i_{corr} (μA cm^{-2})	β_c (mV dec^{-1})	β_a (mV dec^{-1})	IE (%)
1	−207.7	0.128	−96.1	97.6	99.45
12	−210.0	0.070	−74.7	108.0	98.59
24	−195.2	0.067	−81.1	147.0	98.42
48	−117.4	0.042	−70.4	83.7	99.30

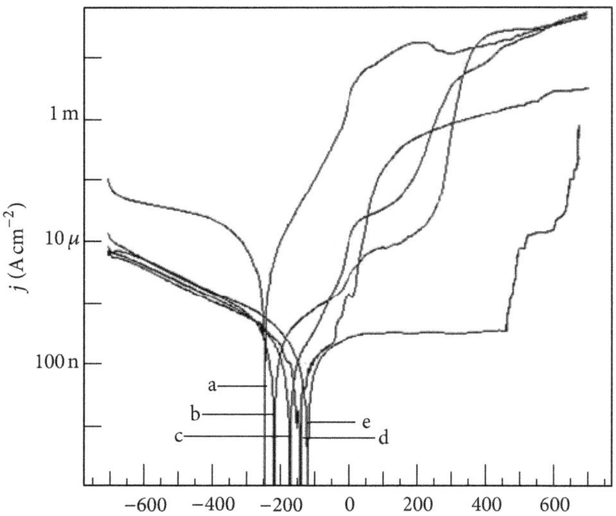

FIGURE 8: Potentiodynamic polarization curves of Cu-Ni (90/10) alloy in synthetic seawater in the presence of different concentrations of BTAH at 48-hour immersion period. a: 0 mM, b: 0.82 mM, c: 1.67 mM, d: 2.50 mM, and e: 3.33 mM. (Temperature: 30°C).

due to the oxidation of Cu to Cu^+ ions and the smaller peak at 250 mV is due to the oxidation of Cu^+ to Cu^{2+} ions.

The effect of the addition of different concentrations of BTAH to seawater on the corrosion rate of the alloy at a fixed immersion period of 48 h is shown in Figure 8. It is interesting to note that with the addition of the inhibitor, E_{corr} is shifted to a less negative value and as the concentration of inhibitor is increased, E_{corr} is shifted to less and less negative values. However, there is a shift in both the anodic and cathodic Tafel slopes, which indicates that BTAH functions as a mixed inhibitor. The corrosion current density is significantly reduced even with the addition of 0.82 mM BTAH and it decreases further with an increase in the concentration of the inhibitor to 2.50 mM and then remains almost the same at 3.33 mM. Another significant effect is on anodic peaks in the presence of the inhibitor. The two characteristic anodic peaks appeared in the absence of inhibitor are practically absent in the presence of inhibitor. The current density values in the entire anodic region are greatly reduced in the presence of the inhibitor in comparison to the decrease in current densities in the cathodic region. Another interesting feature is to be noted in the anodic polarization curve at 3.33 mM concentration of the inhibitor. The anodic region up to +500 mV versus Ag/AgCl has become passive with extremely low current densities of <1 μA cm^{-2}. It may be noted that the passivity breaks down at a potential >500 mV versus Ag/AgCl. The corrosion parameters are given in Table 11. The inhibition efficiencies calculated using (5) are found to be in the range of 99.30 to 97.64% within the concentration range studied:

$$\text{IE (\%)} = \left(\frac{i_{corr} - i^1_{corr}}{i_{corr}} \right) \times 100. \quad (5)$$

In (5), i_{corr} and i^1_{corr} are the corrosion current densities of Cu-Ni (90/10) alloy in synthetic seawater in the absence and presence of BTAH, respectively.

The potentiodynamic polarization curves of the alloy in seawater in the presence of 3.33 mM of BTAH at different immersion periods are depicted in Figure 9. The corrosion parameters are shown in Table 12. It is interesting to note that the protective nature of the film is evident after 1-hour immersion period itself by a large reduction in corrosion current density from 5.98 to 0.128 μA cm^{-2} and a passive anodic region up to +400 mV versus Ag/AgCl with a small current density of the order of 1 μA cm^{-2}. Nevertheless, with the increase in immersion period up to 48 h, i_{corr} is further reduced to 0.042 μA cm^{-2} and the passive anodic region is extended up to +500 mV versus Ag/AgCl. E_{corr} is shifted to less negative potentials with the increase in immersion period. There is a shift in both the anodic and cathodic Tafel slopes. BTAH shifts E_{corr} from −207.7 mV versus Ag/AgCl after 1-hour immersion period to a less negative potential of

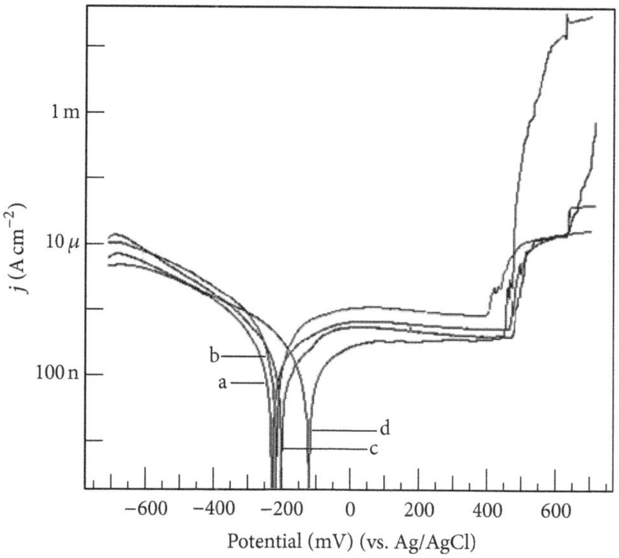

FIGURE 9: Potentiodynamic polarization curves of Cu-Ni (90/10) alloy in synthetic seawater in the presence of 3.33 mM BTAH at different immersion periods. a: 1 h, b: 12 h, c: 24 h, and d: 48 h. (Temperature: 30°C).

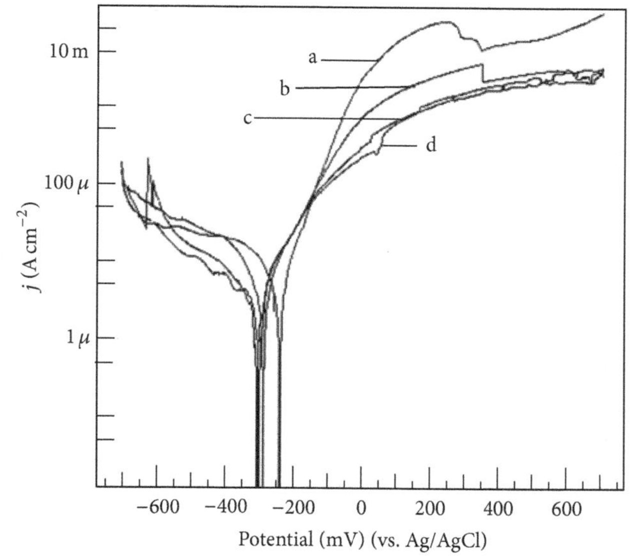

FIGURE 10: Potentiodynamic polarization curves of Cu-Ni (90/10) alloy in synthetic seawater polluted with 10 ppm sulphide at different immersion periods. a: 1 h, b: 12 h, c: 24 h, and d: 48 h. (Temperature: 30°C).

−117.4 mV versus Ag/AgCl after 48-hour immersion period. This can be possibly interpreted by the inhibitive effect on the anodic reaction of the protective film in terms of the mixed potential theory [39]. Similar behaviour was presented by Allam and Ashour in the case of Cu-Ni (90/10) alloy in 3.4% sodium chloride +2 ppm sulphide in the presence of different concentrations of benzotriazole [40].

The effect of the addition of 10 ppm sulphide to seawater (in the absence of inhibitor) on the polarization behaviour of the alloy after various immersion periods is shown in Figure 10. The corresponding corrosion parameters are shown in Table 13. The corrosion potential is shifted to more negative potentials by the addition of 10 ppm sulphide and the corrosion current density is somewhat reduced [12, 32, 33]. In the presence of sulphide, there is only one anodic peak after 12-hour immersion period and no anodic peaks are observed after 24- and 48-hour immersion periods. With the increase in immersion time, the shift in β_c is much higher than the shift in β_a. This is due to the depletion of sulphide with the increase in immersion time due to the formation of a thick cuprous sulphide film at the surface [41]. According to De Sanchez and Schiffrin [14] and Kato et al. [42]; the main effect of copper sulphide is to accelerate the charge transfer of oxygen reduction, thereby controlling the reduction current density by a diffusive transport through the liquid layer adjacent to the electrode, and this is manifested as the increase in β_c.

When 3.33 mM BTAH is added to the environment containing seawater in the presence of 10 ppm S^{2-} (Figure 11), the corresponding polarization curves of the alloy at different immersion periods show some interesting features. The corresponding corrosion parameters are shown in Table 14. After 1-hour immersion period itself, the corrosion current density is drastically reduced and E_{corr} is shifted to less negative potentials. Beyond the Tafel region, there is a passive region up to +10 mV versus Ag/AgCl, and then there is an active region up to +120 mV versus Ag/AgCl, again passive region up to +390 mV versus Ag/AgCl, and finally dissolution leading to limiting current. With the increase in immersion period up to 12 h, the passive region extends from +150 mV to +390 mV and then the dissolution leading to a limiting current. The same trend is observed after 24- and 48-hour immersion times except that there is a further reduction in the corrosion current density. E_{corr} is shifted to more

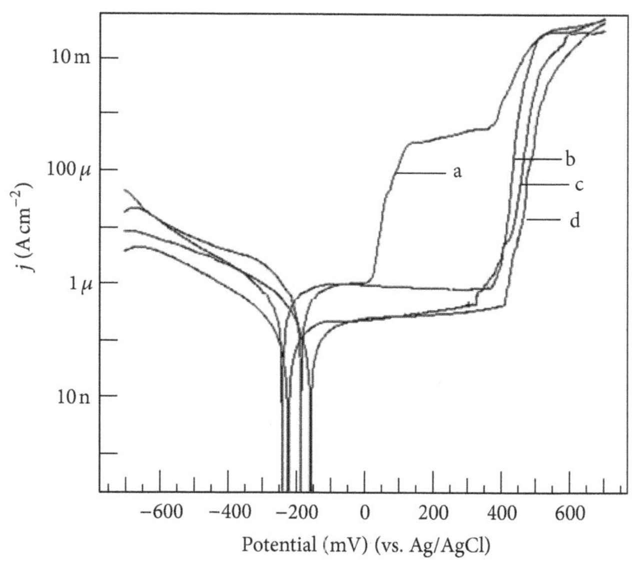

FIGURE 11: Potentiodynamic polarization curves of Cu-Ni (90/10) alloy in synthetic seawater polluted with 10 ppm sulphide in the presence of 3.33 mM BTAH at different immersion periods. a: 1 h, b: 12 h, c: 24 h, and d: 48 h. (Temperature: 30°C).

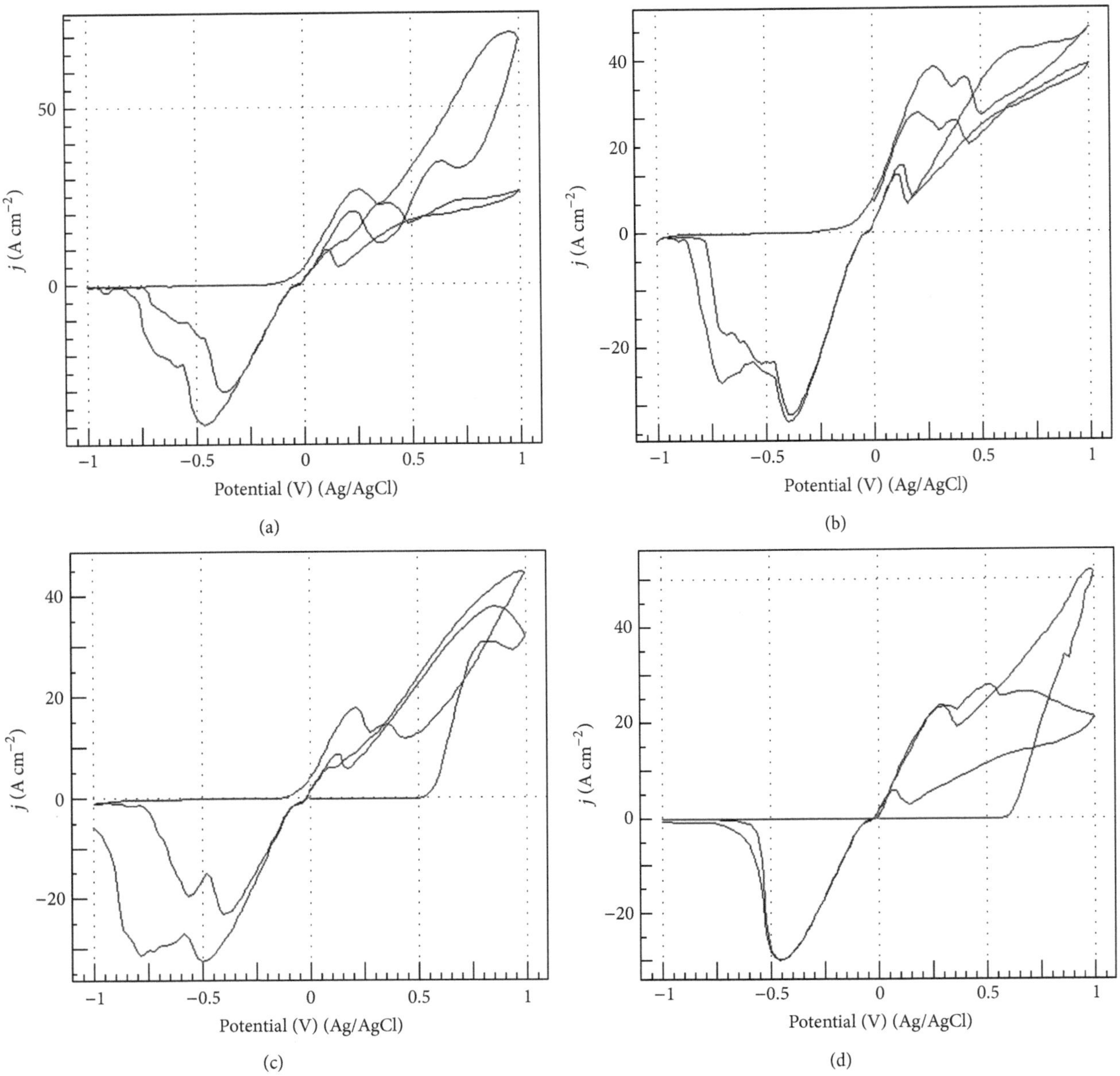

FIGURE 12: Cyclic voltammograms of Cu-Ni (90/10) alloy in synthetic seawater in the absence ((a): 1 h, (b): 48 h) and presence ((c): 1 h, (d): 48 h) of 3.33 mM BTAH. (Temperature: 30°C).

TABLE 13: Corrosion parameters of Cu-Ni (90/10) obtained by potentiodynamic polarization studies in seawater containing 10 ppm of inorganic sulphide in the absence of BTAH at different immersion periods. (Temperature: 30°C).

Immersion period (h)	E_{corr} (mV)	i_{corr} (μA cm^{-2})	β_c (mV dec^{-1})	β_a (mV dec^{-1})
1	−239.0	4.27	−138.0	78.0
12	−290.4	3.84	−130.4	81.6
24	−287.6	3.55	−122.0	65.9
48	−303.8	2.77	−191.0	93.4

anodic potentials in the presence of inhibitor at all immersion periods in comparison to the control solution consisting of seawater and 10 ppm sulphide. However, there is a shift in both the anodic and cathodic Tafel slopes, which infer that BTAH acts as a mixed inhibitor. The potentiodynamic polarization studies infer that the corrosion inhibition by BTAH in seawater polluted with 10 ppm sulphide also is more under anodic control.

3.3. Cyclic Voltammetric Studies. The cyclic voltammograms of Cu-Ni (90/10) alloy in seawater in the absence of BTAH in the potential range from −1.0 to +1.0 V versus Ag/AgCl after 1- and 48-hour immersion periods are given in Figures 12(a) and 12(b), respectively. At both immersion periods, in the first cycle, in the forward sweep, they show two anodic peaks. The first peak at +250 mV is due to Cu-Cu^+ couple and

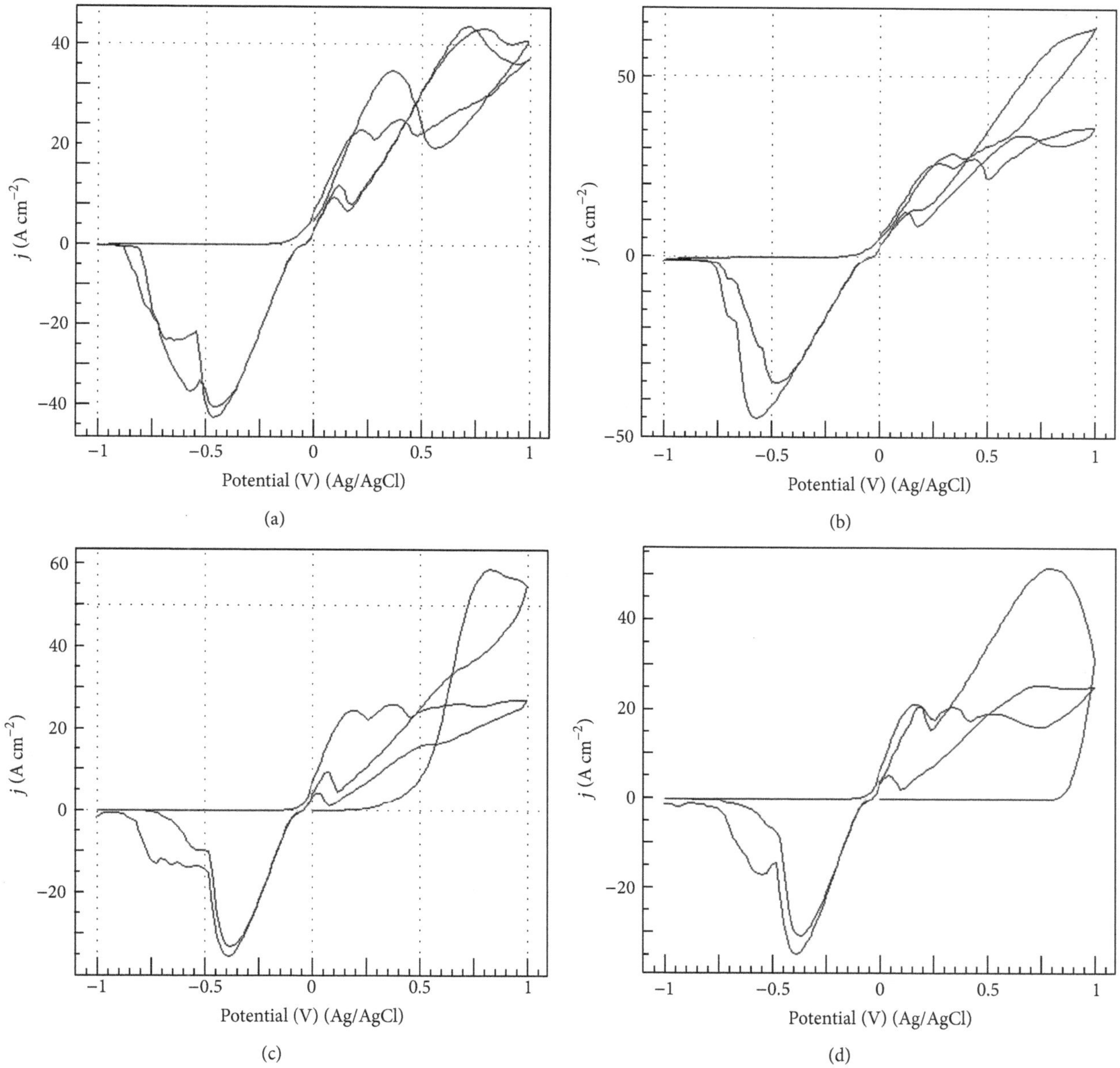

FIGURE 13: Cyclic voltammograms of Cu-Ni (90/10) alloy in synthetic seawater polluted with 10 ppm sulphide in the absence ((a): 1 h, (b): 48 h) and presence ((c): 1 h, (d): 48 h) of 3.33 mM BTAH. (Temperature: 30°C).

TABLE 14: Corrosion parameters of Cu-Ni (90/10) alloy obtained by potentiodynamic polarization studies in seawater containing 10 ppm of inorganic sulphide in the presence of 3.33 mM BTAH at different immersion periods. (Temperature: 30°C).

Immersion period (h)	E_{corr} (mV)	i_{corr} (μA cm^{-2})	β_c (mV dec^{-1})	β_a (mV dec^{-1})	IE (%)
1	−182.6	0.308	−119.0	135.0	92.79
12	−238.5	0.189	−88.5	81.8	95.07
24	−224.5	0.047	−76.7	113.0	98.68
48	−157.2	0.032	−45.2	62.6	98.84

the second one at +450 mV is due to Cu^{+}-Cu^{2+} couple. Two corresponding reduction peaks are observed in the reverse sweep. In the second cycle also the same behaviour is seen. Both the anodic and cathodic peak current densities are increased to some extent in the second cycle.

In the presence of 3.33 mM BTAH after 1-hour immersion period (Figure 12(c)), in the first cycle there is no anodic peak up to +500 mV. With the increase in immersion period up to 48 h (Figure 12(d)), the film became more protective even up to +550 mV. In the second cycle, similar

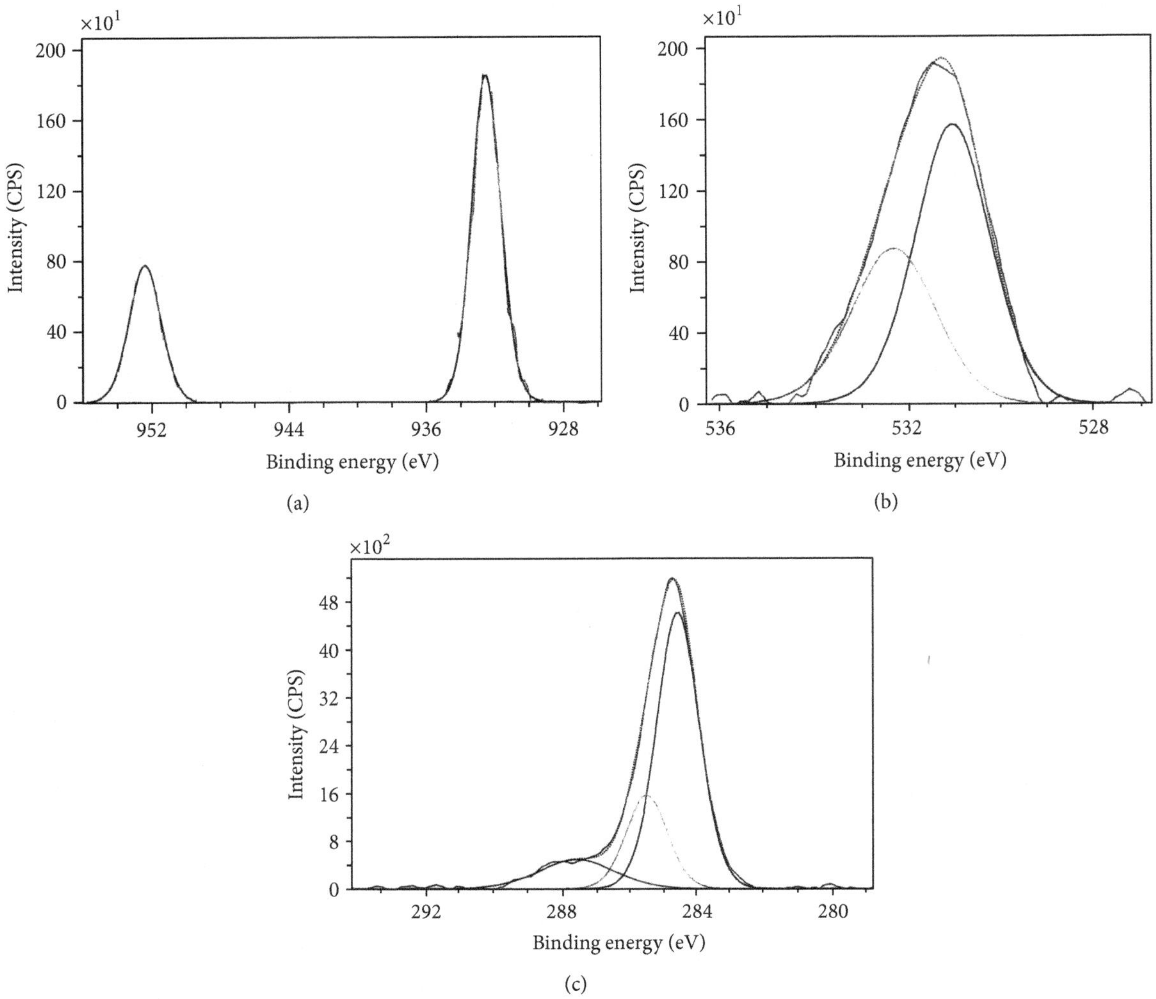

FIGURE 14: XPS deconvolution spectra of different elements present on the Cu-Ni (90/10) alloy in synthetic seawater in the absence of BTAH for a period of 3 days. ((a): Cu 2p, (b): O 1s, and (c): C 1s).

TABLE 15: Weight-loss data of Cu-Ni (90/10) alloy in seawater and 10 ppm of inorganic sulphide containing seawater in the absence and presence of different concentrations of BTAH after 30-day immersion period. (Temperature: 30°C).

Environment	Weight before immersion (W_1) (g)	Weight after immersion (W_2) (g)	Weight loss (ΔW) (g)	Corrosion rate (mmpy)	IE (%)
Seawater	8.10845	8.09889	0.00956	0.01295	—
Seawater + 10 ppm sulphide	8.26840	8.25173	0.01667	0.02258	—
Seawater + 2.5 mM BTAH	8.15888	8.15825	0.00063	0.00085	93.41
Seawater + 3.33 mM BTAH	8.56288	8.56283	0.00005	0.00007	99.48
Seawater + 10 ppm sulphide + 2.5 mM BTAH	7.71100	7.70865	0.00235	0.00318	85.90
Seawater + 10 ppm sulphide + 3.33 mM BTAH	8.28198	8.28191	0.00007	0.00009	99.58

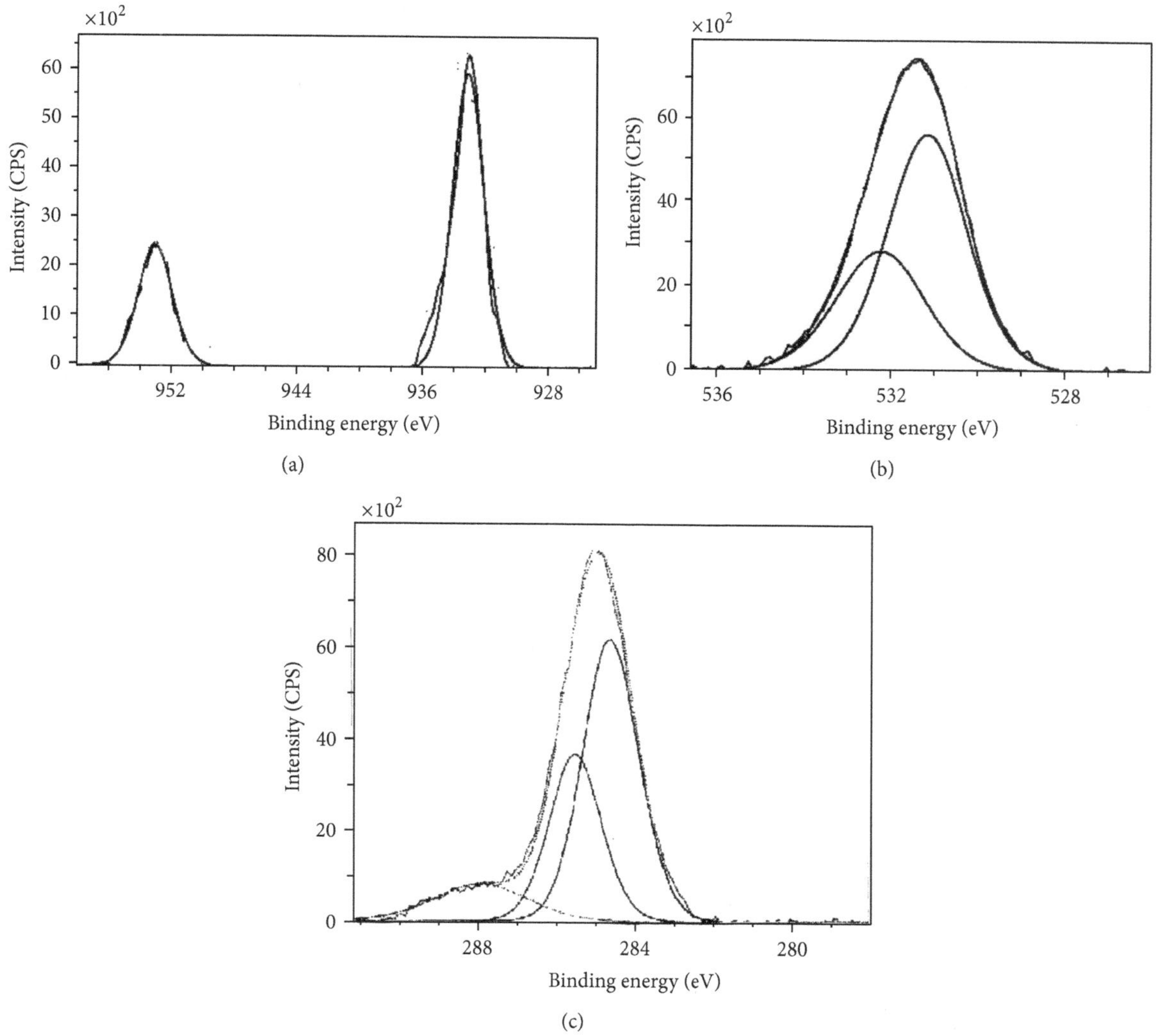

FIGURE 15: XPS deconvolution spectra of different elements present on the Cu-Ni (90/10) alloy in synthetic seawater containing 10 ppm sulphide in the absence of BTAH for a period of 3 days. ((a): Cu 2p, (b): O 1s, and (c): C 1s).

behaviour of bare alloy in the absence of BTAH is seen because of the breakdown of the film as a consequence of the application of higher potentials up to 1.0 V versus Ag/AgCl. The CV profiles confirm that BTAH inhibitor functions excellently and protects the Cu-Ni (90/10) alloy from corrosion in seawater environment. Similar behaviour is also exhibited in the presence of 10 ppm sulphide-polluted seawater environment, both in the absence (Figures 13(a) and 13(b)) and presence (Figures 13(c) and 13(d)) of BTAH.

3.4. Gravimetric Studies. The results of gravimetric studies of Cu-Ni (90/10) alloy in seawater and 10 ppm sulphide containing seawater in the absence and presence of two different concentrations of BTAH after an immersion period of 30 days are shown in Table 15. The concentrations of BTAH are chosen based on the inhibition efficiencies obtained from electrochemical studies. Essentially these studies are carried out to find out the efficiency of the inhibitor after a long immersion period, such as 30 days. When 2.5 mM BTAH is added to seawater and sulphide-polluted seawater, the corrosion rates are decreased. When the concentration of BTAH is increased to 3.33 mM, negligible weight loss is observed in both the environments. At 3.33 mM concentration of BTAH, an inhibition efficiency of more than 99% is obtained in both seawater and sulphide-polluted seawater environments. Thus, the results of gravimetric studies revealed that the inhibitor, BTAH, proved to be an excellent inhibitor even after an immersion period of 30 days.

3.5. X-Ray Photoelectron Spectroscopic Studies (XPSs). In the XPS survey spectrum of Cu-Ni (90/10) alloy in seawater in the absence of BTAH after 3-day immersion period, the peaks due to copper, oxygen, and carbon are detected. No appreciable nickel signal is seen. The computer deconvolution spectra for copper, oxygen, and carbon are shown in Figures 14(a), 14(b), and 14(c), respectively. The high resolution Cu $2p_{3/2}$ peak at 932.4 eV and Cu $2p_{1/2}$ peak at 952.4 eV are attributed to Cu_2O [43]. The O 1s peak at 531.0 eV is ascribed to the

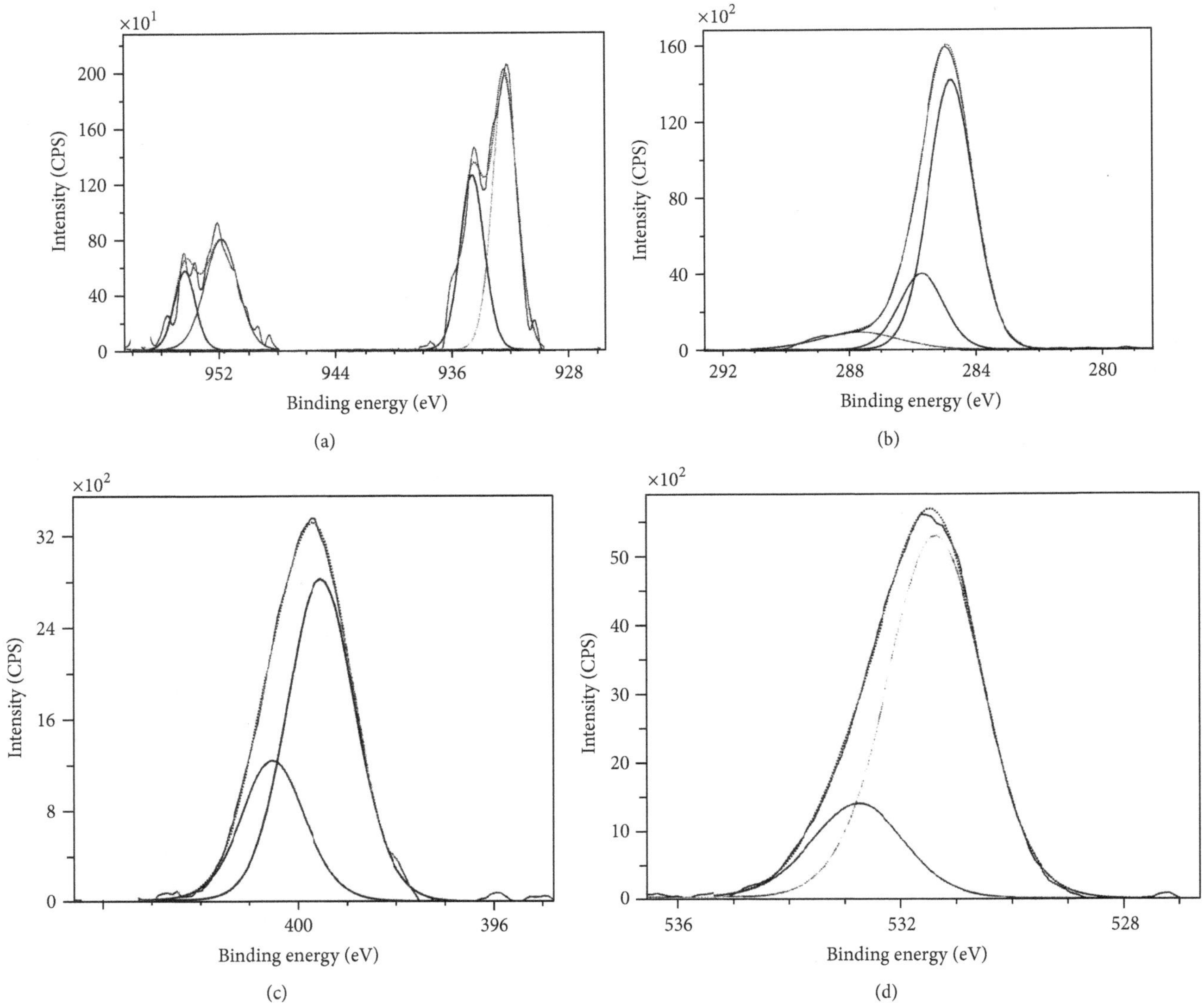

FIGURE 16: XPS deconvolution spectra of different elements present on the Cu-Ni (90/10) alloy in synthetic seawater in the presence of 3.33 mM BTAH for a period of 3 days. ((a): Cu 2p, (b): C 1s, (c): N 1s, and (d): O 1s).

presence of Cu_2O [44]. The O 1s peak at 532.3 eV is due to the presence of hydroxide [45]. From these observations, it is concluded that the outermost corrosion product is Cu_2O. The C 1s peaks at 284.6 eV, 285.5 eV, and 287.6 eV are due to contaminant carbon, which is likely due to the cracking of vacuum oil used during the operation of XPS instrument [46].

The survey scan of the film formed on Cu-Ni (90/10) alloy in 10 ppm sulphide containing seawater in the absence of BTAH also shows the peaks due to copper, oxygen, and carbon. The computer deconvolution spectra for copper, oxygen, and carbon are shown in Figures 15(a), 15(b), and 15(c), respectively. The sulphur signal is not observed because the sulphide is oxidized by dissolved oxygen in seawater. The half-life of sulphide studied by Bates and Popplewell [47] at a concentration of 3.8 ppm in the air-saturated seawater is of the order of 20 min. Ostland and Alexander [48] investigated that 10 ppm sulphide decayed to 0.1–1.0 ppm after 1 day in the air-saturated seawater. In our present study, the immersion period is 3 days. Hence, the sulphur signal is not detected. The presence of Cu $2p_{3/2}$ peak at 932.9 eV and Cu $2p_{1/2}$ peak at 952.9 eV confirms the presence of Cu (I) [43]. The O 1s peaks at 531.2 eV and 532.2 eV are due to the presence of oxide and hydroxide, respectively. From these observations, it is inferred that in sulphide containing seawater also, the major corrosion product is found to be Cu_2O. The C 1s peaks at 284.6 eV, 285.5 eV, and 287.9 eV are due to contaminant carbon.

The XPS survey spectrum of Cu-Ni (90/10) alloy in seawater in the presence of BTAH after 3-day immersion period shows peaks corresponding to copper, carbon, nitrogen, and oxygen. The corresponding computer deconvolution spectra are shown in Figures 16(a), 16(b), 16(c), and 16(d). The Cu 2p spectrum shows peaks due to Cu $2p_{3/2}$ at 932.2 eV with a satellite peak at 934.5 eV and Cu $2p_{1/2}$ at 952.3 eV with a satellite peak at 954.7 eV. The shake-up satellite structures

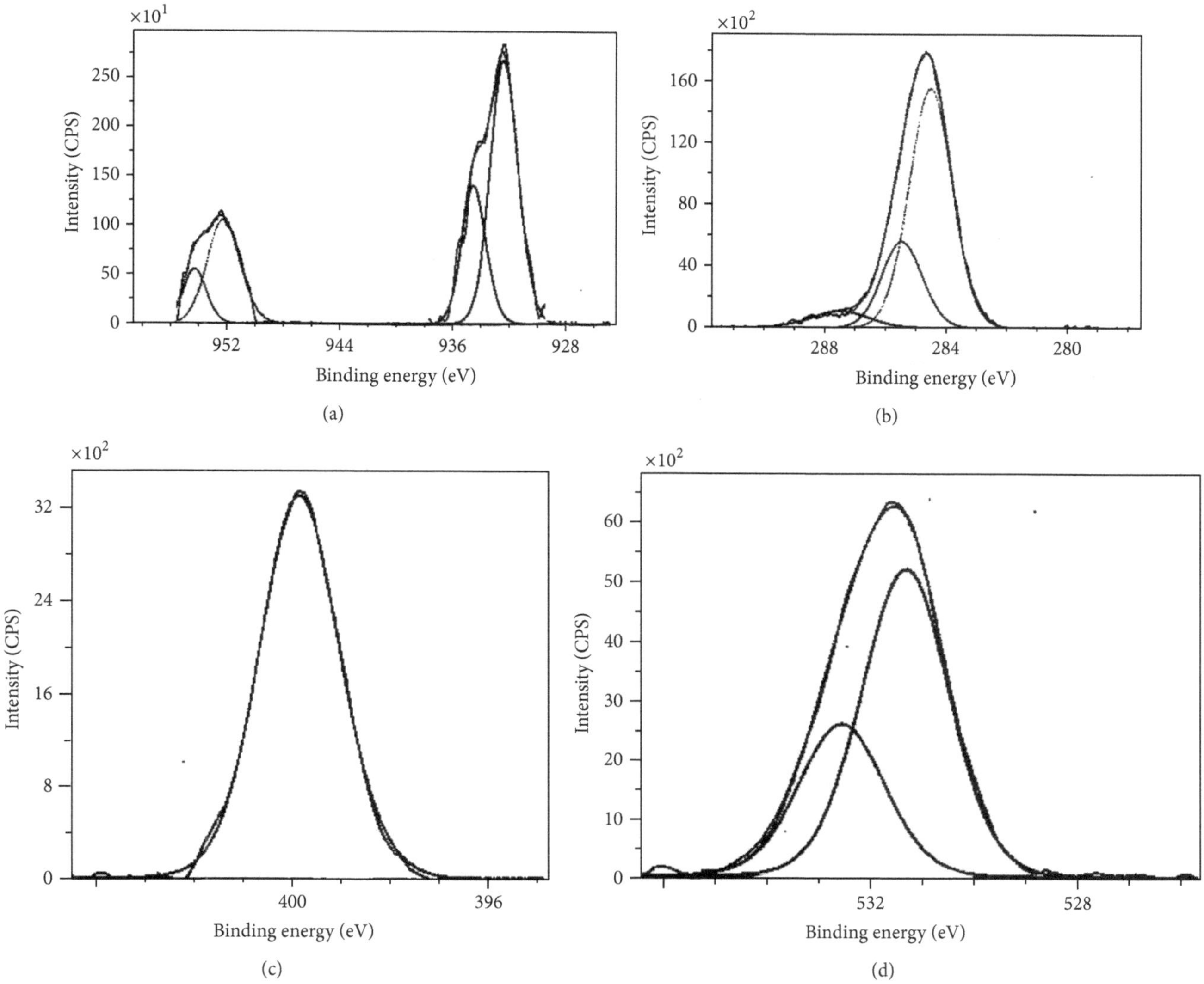

FIGURE 17: XPS deconvolution spectra of different elements present on the Cu-Ni (90/10) alloy in synthetic seawater containing 10 ppm sulphide in the presence of 3.33 mM BTAH for a period of 3 days. ((a): Cu 2p, (b): C 1s, (c): N 1s, and (d): O 1s).

are closely related to the oxidation state of copper [49]. All cupric compounds whose Cu 2p spectra have been studied show satellite peaks arise due to multielectron excitation process. From these observations, the presence of Cu (II)-BTA complex is inferred. In fact, there is a formation of Cu (I)-BTA complex, which is oxidized to Cu (II)-BTA complex in the course of time. The observed transition from a Cu (I) species to a Cu (II) species is in agreement with the results of studies reported in literature [50, 51]. C 1s shows three peaks, one at 285.6 eV and the other two at 284.6 eV and 287.6 eV. The C 1s peak at 285.6 eV is characteristic of carbon present in aromatic ring [52]. The C 1s peaks at 284.6 eV and 287.6 eV are due to contaminant carbon. N 1s spectrum shows two peaks, one at 399.6 eV and the other at 400.6 eV, which are due to the presence of nitrogen atoms in BTAH in two different chemical environments. The peak at 399.6 eV is assigned to "N=N" group and the one at 400.6 eV is assigned to "HN–N" group. Two O 1s peaks are also observed in the presence of BTAH. This result infers the formation of Cu_2O film over which Cu (I)-BTA complex is formed.

The XPS deconvolution spectra (Figures 17(a), 17(b), 17(c), and 17(d)) of Cu-Ni (90/10) alloy in seawater containing 10 ppm sulphide in the presence of BTAH after 3-day immersion period show peaks corresponding to Cu $2p_{3/2}$, Cu $2p_{1/2}$, C 1s, N 1s, and O 1s. The Cu $2p_{3/2}$ at 932.3 eV associated with a satellite peak at 934.5 eV and the Cu $2p_{1/2}$ at 952.5 eV associated with a satellite peak at 954.6 eV infer the presence of Cu (II)-BTA complex. The C 1s peak at 285.6 eV is characteristic of carbon present in aromatic ring [52]. The C 1s peaks at 284.6 eV and 287.7 eV are due to contaminant carbon. The N 1s peak at 399.8 eV is due to the delocalization of the three nitrogen atoms present in the ring, providing negative charge to the ring. Al Kharafi et al. studied the effect of sulphide pollution on the stability of the protective film of benzotriazole on copper surface [53]. They obtained a single peak at 400.0 eV and interpreted

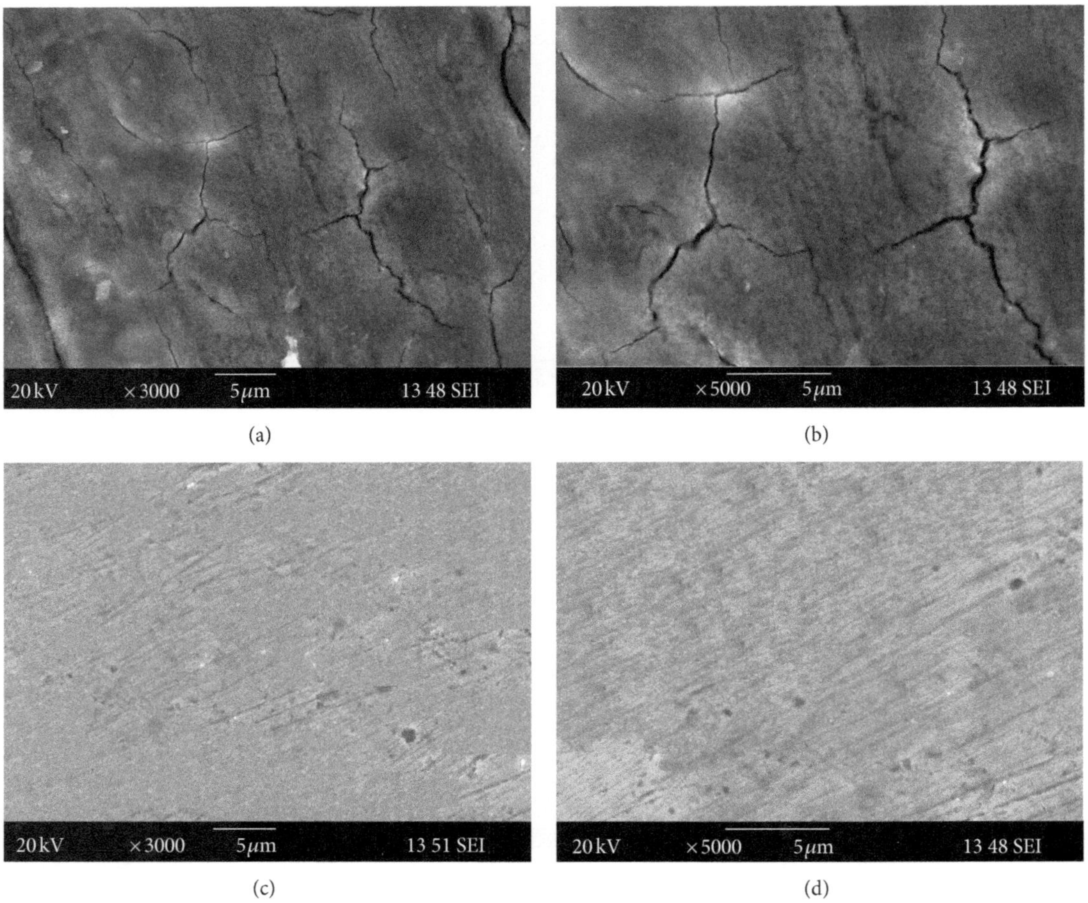

FIGURE 18: SEM images of Cu-Ni (90/10) alloy in synthetic seawater in the absence ((a): ×3000, (b): ×5000) and presence ((c): ×3000, (d): ×5000) of BTAH for a period of 30 days.

the same to the three nitrogen atoms of the benzotriazole ring. Souto et al. studied the corrosion inhibition of copper by benzotriazole and potassium ethyl xanthate in aqueous solutions containing 0.1 M NaCl [54]. They reported a single N 1s peak due to the presence of benzotriazole on the copper surface.

3.6. Scanning Electron Microscopy (SEM)/Energy Dispersive X-Ray Analysis (EDX). Scanning electron micrographs of Cu-Ni (90/10) alloy in seawater in the absence of BTAH after 30-day immersion at different magnifications are shown in Figure 18(a). The SEM image of Cu-Ni (90/10) alloy in seawater in the absence of BTAH clearly shows that the surface is severely corroded and there is a formation of different forms of corrosion products on the alloy surface. Breakdown of oxide film is also seen. To increase the aggressiveness of the environment, 10 ppm of inorganic sulphide is added to seawater (Figure 19(a)). In this environment, the corrosion damage on the alloy surface is relatively more severe. The entire surface is covered by a scale-like black corrosion product. However, in the presence of BTAH in both seawater (Figure 18(b)) and sulphide containing seawater (Figure 19(b)), the surface is smooth with small notches. These micrographs reveal the absence of any corrosion product.

The EDX spectra in the absence of BTAH in seawater (Figure 20(a)) confirm the presence of a large amount of Cu_2O as indicated by the very high intensity of O signal. In sulphide-polluted seawater (Figure 21(a)), in addition to O signal, S signal is also detected. Hence, it confirms that the corrosion products are both Cu_2O and Cu_2S in sulphide-polluted seawater. However, in the presence of BTAH in both seawater (Figure 20(b)) and sulphide-polluted seawater (Figure 21(b)), the spectra show signals, which are characteristic of the existence of C and N present in the BTAH molecule. In addition, the intensity of O signal is significantly reduced. The S signal is completely absent in sulphide-polluted seawater in the presence of BTAH. The EDX spectra indicate the presence of protective BTAH film on alloy surface. The elemental composition of Cu-Ni (90/10) alloy obtained during all these studies is given in Table 16.

4. Conclusions

(1) All the three electrochemical methods reveal that 1,2,3-benzotriazole is proved to be an excellent inhibitor for Cu-Ni (90/10) alloy in synthetic seawater and seawater containing sulphide ions.

(c) (d)

FIGURE 19: SEM images of Cu-Ni (90/10) alloy in synthetic seawater containing 10 ppm sulphide in the absence ((a): ×3000, (b): ×5000) and presence ((c): ×3000, (d): ×5000) of BTAH for a period of 30 days.

TABLE 16: Elemental composition of Cu-Ni (90/10) alloy obtained by SEM-EDX in seawater and sulphide-polluted seawater in the absence and presence of BTAH after 30-day immersion period.

Seawater		Seawater + 10 ppm sulphide		Seawater + 3.33 mM BTAH		Seawater + 10 ppm sulphide + 3.33 mM BTAH	
Element	Atom (%)	Element	Atom (%)	Element	Atom (%)	Element	Atom (%)
O	54.5	O	39.37	C K	30.47	C K	28.58
S	0.6	S	17.34	N K	4.14	N K	3.25
Cl	1.29	Cl	2.02	O K	2.64	O K	6.51
Fe	0.97	Fe	0.56	Fe K	0.85	Fe K	0.87
Ni	6.17	Ni	4.59	Ni K	7.1	Ni K	7.81
Cu	36.47	Cu	36.12	Cu K	54.8	Cu K	52.98

(2) BTAH is proved to be an excellent inhibitor even after an immersion period of 30 days, as evidenced by weight-loss studies.

(3) Electrochemical impedance studies showed that BTAH molecules increase both the charge transfer resistance and the film resistance of the Cu-Ni (90/10) alloy. The protective BTAH film withstands even up to 60°C.

(4) Potentiodynamic polarization studies indicated that BTAH considerably shifts the corrosion potential to less negative values and greatly decreases the corrosion current density. These studies give valuable information regarding the anodic behavior of Cu-Ni (90/10) alloy in the absence and presence of BTAH, which is proved to be a mixed inhibitor, predominantly anodic in nature.

(5) Cyclic voltammeric studies show the stability of the protective film even at anodic potentials of 550 mV versus Ag/AgCl.

(6) Surface examination studies by XPS and SEM-EDX confirm the presence of protective BTAH film on the alloy surface.

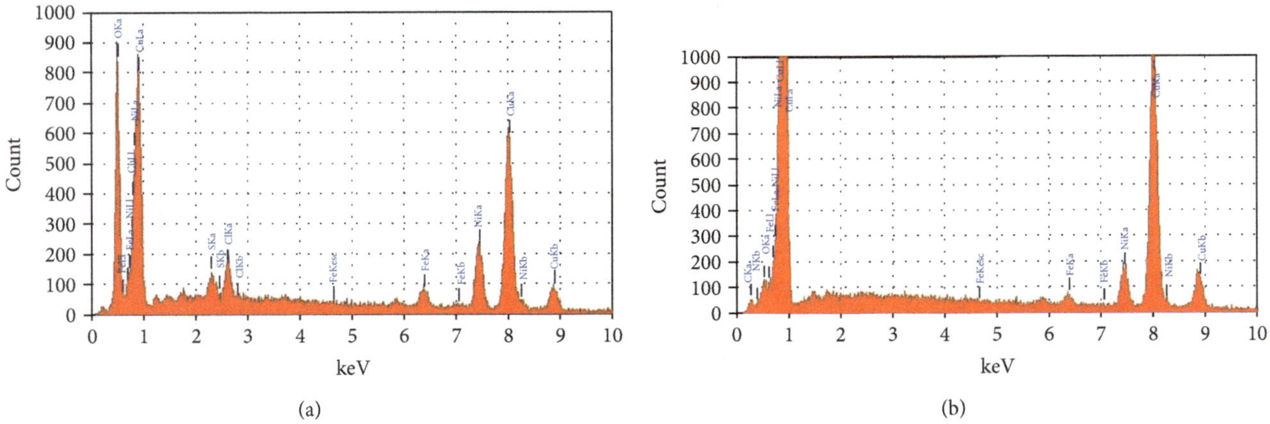

Figure 20: EDX spectra of Cu-Ni (90/10) alloy in synthetic seawater in the (a) absence and (b) presence of BTAH for a period of 30 days.

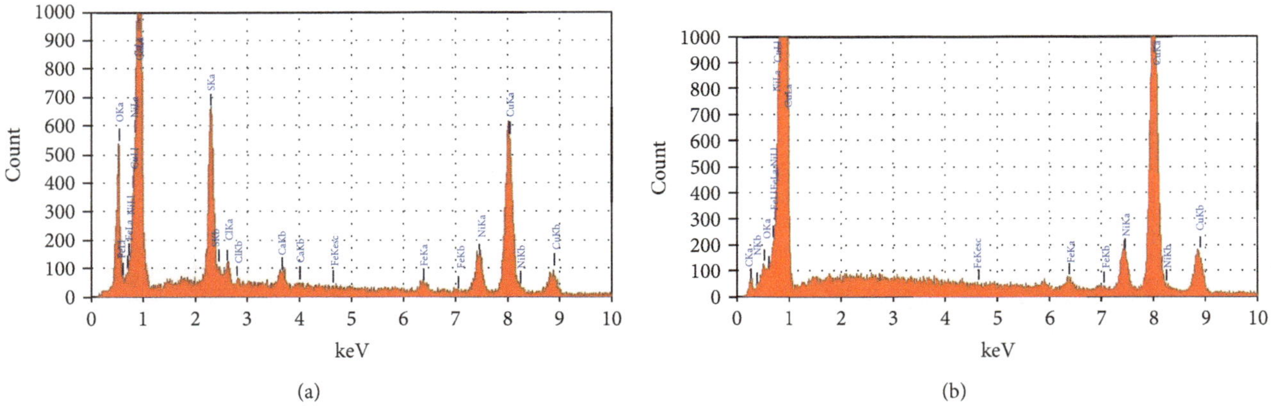

Figure 21: EDX spectra of Cu-Ni (90/10) alloy in synthetic seawater containing 10 ppm sulphide in the (a) absence and (b) presence of BTAH for a period of 30 days.

Acknowledgments

The authors are grateful to Naval Research Board (NRB), Government of India and Rajiv Gandhi National Fellowship (RGNF), UGC, Government of India for giving the financial assistance.

References

[1] R. W. Cahn, P. Hassen, and E. J. Kramer, *Materials Science and Technology, A Comprehensive Treatment, Structure and Properties of Non-Ferrous Alloys*, vol. 8, VCH, New York, NY, USA, 1996.

[2] E. G. West, *Copper and Its Alloys*, Ellis Horwood Ltd., Halsted Press, New York, NY, USA, 1982.

[3] B. Todd and P. A. Lovett, "Marine engineering practice: selecting materials for seawater systems," Tech. Rep., Institute of Marine Engineers, London, UK, 1976.

[4] C. A. Powell, "Marine applications of copper-nickel alloys, section 1: copper-nickel alloys-resistance to corrosion and biofouling," Tech. Rep., Copper Development Association, Potters Bar, UK, 1998.

[5] F. Mansfeld and B. J. Little, "Microbiologically influenced corrosion of copperbased materials exposed to natural seawater," *Electrochimica Acta*, vol. 37, no. 12, pp. 2291–2297, 1992.

[6] A. Hall and A. J. M. Baker, "Settlement and growth of coppertolerant *Ectocarpus siliculosus* (Dillw.) Lyngbye on different copper-based antifouling surfaces under laboratory conditions," *Journal of Material Science*, vol. 20, no. 3, pp. 1111–1118, 1985.

[7] C. Kato, B. G. Ateya, J. E. Castle, and H. W. Pickering, "On the mechanism of corrosion of Cu-9.4Ni-1.7Fe alloy in air saturated aqueous NaCl solution—I. Kinetic investigations," *Journal of the Electrochemical Society*, vol. 127, no. 9, pp. 1890–1896, 1980.

[8] S. Cere and M. Vazquez, "Properties of the passive films present on copper and copper-nickel alloys in slightly alkaline solutions," *Journal of Materials Science Letters*, vol. 21, no. 6, pp. 493–495, 2002.

[9] J. O. M. Bockris, B. T. Rubin, A. Despic, and B. Lovrecek, "The electrodissolution of coppernickel alloys," *Electrochimica Acta*, vol. 17, no. 5, pp. 973–999, 1972.

[10] R. F. North and M. J. Pryor, "The influence of corrosion product structure on the corrosion rate of Cu-Ni alloys," *Corrosion Science*, vol. 10, no. 5, pp. 297–311, 1970.

[11] T. D. Burleigh and D. H. Waldeck, "Effect of alloying on the resistance of Cu-10% Ni alloys to seawater impingement," *Corrosion*, vol. 55, no. 8, pp. 800–804, 1999.

[12] S. M. Sayed, E. A. Ashour, and G. I. Youssef, "Effect of sulfide ions on the corrosion behaviour of Al-brass and Cu10Ni alloys

in salt water," *Materials Chemistry and Physics*, vol. 78, no. 3, pp. 825–834, 2003.

[13] H. Hack, "Susceptibility of 17 machinery alloys to sulphide induced corrosion in seawater," Tech. Rep. AWFAL-TR-B1-4019, Bethesda, Md, USA, 1980.

[14] S. R. De Sanchez and D. J. Schiffrin, "The flow corrosion mechanism of copper base alloys in sea water in the presence of sulphide contamination," *Corrosion Science*, vol. 22, no. 6, pp. 585–607, 1982.

[15] M. Benmessaoud, K. Es-salah, N. Hajjaji, H. Takenouti, A. Srhiri, and M. Ebentouhami, "Inhibiting effect of 2-mercaptobenzimidazole on the corrosion of Cu-30Ni alloy in aerated 3% NaCl in presence of ammonia," *Corrosion Science*, vol. 49, no. 10, pp. 3880–3888, 2007.

[16] W. A. Badawy, K. M. Ismail, and A. M. Fathi, "Environmentally safe corrosion inhibition of the Cu-Ni alloys in acidic sulfate solutions," *Journal of Applied Electrochemistry*, vol. 35, no. 9, pp. 879–888, 2005.

[17] E. M. M. Sutter, F. Ammeloot, M. J. Pouet, C. Fiaud, and R. Couffignal, "Heterocyclic compounds used as corrosion inhibitors: correlation between ^{13}C and 1H NMR spectroscopy and inhibition efficiency," *Corrosion Science*, vol. 41, no. 1, pp. 105–115, 1999.

[18] D. Tromans and R. H. Sun, "Anodic polarization behavior of copper in aqueous chloride/benzotriazole solutions," *Journal of the Electrochemical Society*, vol. 138, no. 11, pp. 3235–3244, 1991.

[19] I. H. Omar, F. Zucchi, and G. Trabanelli, "Schiff bases as corrosion inhibitors of copper and its alloys in acid media," *Surface and Coatings Technology*, vol. 29, no. 2, pp. 141–151, 1986.

[20] W. A. Badawy, K. M. Ismail, and A. M. Fathi, "Corrosion control of Cu-Ni alloys in neutral chloride solutions by amino acids," *Electrochimica Acta*, vol. 51, no. 20, pp. 4182–4189, 2006.

[21] R. Babic, M. Metikos-Hukovic, and M. Loncar, "Impedance and photoelectrochemical study of surface layers on Cu and Cu-10Ni in acetate solution containing benzotriazole," *Electrochimica Acta*, vol. 44, no. 14, pp. 2413–2421, 1999.

[22] N. K. Allam, E. A. Ashour, H. S. Hegazy, B. E. El-Anadouli, and B. G. Ateya, "Effects of benzotriazole on the corrosion of Cu10Ni alloy in sulfide-polluted salt water," *Corrosion Science*, vol. 47, no. 9, pp. 2280–2292, 2005.

[23] J. M. Maciel and S. M. L. Agostinho, "Use of a rotating cylinder electrode in corrosion studies of a 90/10 Cu-Ni alloy in 0.5 mol L^{-1} H_2SO_4 media," *Journal of Applied Electrochemistry*, vol. 30, no. 8, pp. 981–985, 2000.

[24] S. J. Yuan and S. O. Pehkonen, "Surface characterization and corrosion behavior of 70/30 CuNi alloy in pristine and sulfide-containing simulated seawater," *Corrosion Science*, vol. 49, no. 3, pp. 1276–1304, 2007.

[25] ASTM Standard G 31-72, *Standard Practice for Laboratory Immersion Corrosion Testing of Metals*, Annual Book of ASTM Standards, ASTM, Philadelphia, Pa, USA, 1990.

[26] R. A. Freeman and D. C. Silverman, "Error propagation in coupon immersion tests," *Corrosion*, vol. 48, no. 6, pp. 463–466, 1992.

[27] A. L. Bacarella and J. C. Griess, "The anodic dissolution of copper in flowing sodium chloride solutions between 25° and 175°C," *Journal of the Electrochemical Society*, vol. 120, no. 4, pp. 459–465, 1973.

[28] G. Kear, B. D. Barker, and F. C. Walsh, "Electrochemical corrosion of unalloyed copper in chloride media—a critical review," *Corrosion Science*, vol. 46, no. 1, pp. 109–135, 2004.

[29] F. Bentiss, M. Lagrenee, M. Traisnel, and J. C. Hornez, "The corrosion inhibition of mild steel in acidic media by a new triazole derivative," *Corrosion Science*, vol. 41, no. 4, pp. 789–803, 1999.

[30] Y. S. Tan, M. P. Srinivasan, S. O. Pehkonen, and S. Y. M. Chooi, "Effects of ring substituents on the protective properties of self-assembled benzenethiols on copper," *Corrosion Science*, vol. 48, no. 4, pp. 840–862, 2006.

[31] X. Wu, H. Ma, S. Chen, Z. Xu, and A. Sui, "General equivalent circuits for faradaic electrode processes under electrochemical reaction control," *Journal of the Electrochemical Society*, vol. 146, no. 5, pp. 1847–1853, 1999.

[32] B. C. Syrett, "The mechanism of accelerated corrosion of coppernickel alloys in sulphidepolluted seawater," *Corrosion Science*, vol. 21, no. 3, pp. 187–209, 1981.

[33] L. E. Eiselstein, B. C. Syrett, S. S. Wing, and R. D. Caligiuri, "The accelerated corrosion of Cu-Ni alloys in sulphide-polluted seawater: mechanism no. 2," *Corrosion Science*, vol. 23, no. 3, pp. 223–239, 1983.

[34] E. M. Sherif, R. M. Erasmus, and J. D. Collins, "Inhibition of corrosion processes on copper in aerated sodium chloride solutions by 5-(3-aminophenyl)-tetrazole," *Journal of Applied Electrochemistry*, vol. 39, no. 1, pp. 83–91, 2009.

[35] A. M. Alfantazi, T. M. Ahmed, and D. Tromans, "Corrosion behavior of copper alloys in chloride media," *Materials and Design*, vol. 30, no. 7, pp. 2425–2430, 2009.

[36] Y. Z. Wang, A. M. Beccaria, and G. Poggi, "The effect of temperature on the corrosion behaviour of a 70/30 Cu-Ni commercial alloy in seawater," *Corrosion Science*, vol. 36, no. 8, pp. 1277–1288, 1994.

[37] P. Druska, H. H. Strehblow, and S. Golledge, "A surface analytical examination of passive layers on Cu-Ni alloys—part I. Alkaline solution," *Corrosion Science*, vol. 38, no. 6, pp. 835–851, 1996.

[38] J. M. Popplewell, R. J. Hart, and J. A. Ford, "The effect of iron on the corrosion characteristics of 90-10 cupro nickel in quiescent 3.4% NaCl solution," *Corrosion Science*, vol. 13, no. 4, pp. 295–309, 1973.

[39] W. S. Tait, *An Introduction to Electrochemical Corrosion Testing for Practicing Engineers and Scientists*, University of Wisconsin, Madison, Wis, USA, 1994.

[40] N. K. Allam and E. A. Ashour, "Promoting effect of low concentration of benzotriazole on the corrosion of Cu10Ni alloy in sulfide polluted salt water," *Applied Surface Science*, vol. 254, no. 16, pp. 5007–5011, 2008.

[41] J. N. Al-Hajji and M. R. Reda, "On the effects of common pollutants on the corrosion of copper-nickel alloys in sulfide polluted seawater," *Journal of Electrochemical Society*, vol. 142, no. 9, pp. 2944–2953, 1995.

[42] C. Kato, H W. Pickering, and J. E. Castle, "Effect of sulfide on the corrosion of Cu-9.4Ni-1.7Fe Alloy in aqueous NaCl solution," *Journal of Electrochemical Society*, vol. 131, no. 6, pp. 1225–1229, 1984.

[43] N. S. Mclntyre and M. G. Cook, "X-ray photoelectron studies on some oxides and hydroxides of cobalt, nickel and copper," *Analytical Chemistry*, vol. 47, no. 13, pp. 2208–2213, 1975.

[44] N. S. Mclntyre, T. E. Rummery, M. G. Cook, and D. Owen, "X-ray photoelectron spectroscopic study of the aqueous oxidation of monel-400," *Journal of Electrochemical Society*, vol. 123, pp. 1164–1170, 1976.

[45] S. Satish Laxman and K. Kar, "Facile synthesis of large area porous Cu_2O as super hydrophobic yellow-red phosphors," *RSC Advances*, vol. 2, no. 9, pp. 3647–3650, 2012.

[46] G. P. Cicileo, B. M. Rosales, F. E. Varela, and J. R. Vilche, "Comparative study of organic inhibitors of copper corrosion," *Corrosion Science*, vol. 41, no. 7, pp. 1359–1375, 1999.

[47] J. F. Bates and J. M. Popplewell, "Corrosion of condenser tube alloys in sulfide contaminated brine," *Corrosion*, vol. 31, no. 8, pp. 269–275, 1975.

[48] H. E. Ostland and J. Alexander, "Oxidation rate of sulfide in sea water, a preliminary study," *Journal of Geophysical Research*, vol. 68, no. 13, pp. 3995–3997, 1963.

[49] P. E. Larson, "X-ray induced photoelectron and auger spectra of Cu, CuO, Cu_2O, and Cu_2S thin films," *Journal of Electron Spectroscopy and Related Phenomena*, vol. 4, no. 3, pp. 213–218, 1974.

[50] D. Chadwick and T. Hashemi, "Benzotriazole adsorption on copper studied by X-ray photoelectron spectroscopy," *Journal of Electron Spectroscopy and Related Phenomena*, vol. 10, no. 1, pp. 79–83, 1977.

[51] D. Chadwick and T. Hashemi, "Adsorbed corrosion inhibitors studied by electron spectroscopy: benzotriazole on copper and copper alloys," *Corrosion Science*, vol. 18, no. 1, pp. 39–51, 1978.

[52] J. F. Moulder, W. F. Stickle, P. E. Sobol, and K. D. Bomben, in *Handbook of X-ray Photoelectron Spectroscopy*, J. Chastain and R. C. King Jr., Eds., Physical Electronics, Inc., Eden Prairie, Minn, USA, 1995.

[53] F. M. Al Kharafi, A. M. Abdullah, I. M. Ghayad, and B. G. Ateya, "Effect of sulfide pollution on the stability of the protective film of benzotriazole on copper," *Applied Surface Science*, vol. 253, no. 22, pp. 8986–8991, 2007.

[54] R. M. Souto, V. Fox, M. M. Laz, M. Pérez, and S. González, "Some experiments regarding the corrosion inhibition of copper by benzotriazole and potassium ethyl xanthate," *Journal of Electroanalytical Chemistry*, vol. 411, no. 1-2, pp. 161–165, 1996.

The Development of a Mathematical Model for the Prediction of Corrosion Rate Behaviour for Mild Steel in 0.5 M Sulphuric Acid

I. Y. Suleiman,[1] O. B. Oloche,[2] and S. A. Yaro[3]

[1] *Department of Metallurgical Engineering, Waziri Umaru Federal Polytechnic, Birnin Kebbi, Nigeria*
[2] *Department of Mechanical Engineering, University of Abuja, Abuja, Nigeria*
[3] *Department of Metallurgical and Materials Engineering, Ahmadu Bello University, Zaria, Nigeria*

Correspondence should be addressed to I. Y. Suleiman; onoruoizadanjumas@yahoo.co.uk

Academic Editors: Q. Qu and E. Stupnisek-Lisac

The effect of varying temperature, concentration, and time on the corrosion rate of mild steel in 0.5 M H_2SO_4 acid with and without (wild jute tree) *grewia venusta* plant extract has been investigated by weight loss. The temperature, concentration of inhibitor and time were varied in the range of 0–10% v/v at 2% v/v interval, 30–70°C at 20°C interval, and 45–270 minutes at 45 minutes interval respectively. Scanning electron microscope was used to analyze the morphology of the sample surface. Linear regression equation and analysis of variance (ANOVA) were employed to investigate the influence of process parameters on the corrosion rate of the samples. The predicted corrosion rate of the samples was found to lie close to those experimentally observed ones. The confirmation of the experiment conducted using ANOVA to verify the optimal testing parameters shows that the increase in inhibitor concentration above 2% v/v and time would reduce the corrosion rate. The results also showed that the increase in temperature would also increase the corrosion rate greatly and that the plant extract was very effective for the corrosion inhibition of mild steel in acidic medium.

1. Introduction

Steels are the most extensively used structural materials in industry. Mild steel is the most versatile general purpose material due to its good mechanical strength, easy fabricability, formability and weldability, abundance and low cost [1].

In corrosive environments, mild steel structures can be saved by coating and/or cathodic protection. The use of inhibitors is one of the most practical methods for protection against corrosion and prevention of unexpected metal dissolution and acid consumption, especially in acid solutions. Different organic and inorganic compounds have been studied as inhibitors to protect metals from corrosive attack [2].

Such compounds can adsorb onto the metal surface and block the active surface sites, thus reducing the corrosion rate. Although many synthetic compound show good anticorrosive activity, most of them are highly toxic to both human beings and the environment [3], and they are often expensive and non-biodegradable. Thus, the use of natural products as corrosion inhibitors has become a key area of research because plant extracts are viewed as an incredibly rich source of naturally synthesized chemical compounds that are biodegradable in nature and can be extracted by simple procedures with low cost.

Corrosion of mild steel and its alloys in different acid media have been extensively studied [4–6]. Recently considerable interest has been generated in the use of nitrogen, oxygen and sulphur containing organic compounds as corrosion inhibitor for mild steel in different acids [7–9].

In this work, green wild jute tree (*Grewa venusta*) extract was used as inhibitor. Wild jute tree from the bark of *Grewa venusta*, is a shrub or small tree to 10.5 m tall called wild jute in English, Dargaza in Hausa and belong to Tiliaceae family. It is used as fibre and the phytochemical analysis showed that both the Leaves and bark contain many compounds, such as polysaccharides, volatile oils, vitamins, tannins, minerals, alkaloids (e.g., caffeine) and polyphenols (catechins and flavonoids). It is found in the northern part of the country [10].

Thermodynamics and kinetics are useful parameters for analyzing systems undergoing chemical reactivity. Corroding systems are not in equilibrium and therefore thermodynamic calculation cannot be applied. Hence, from the engineering point of view, the major interest is the kinetics or rate of corrosion [11].

In this work, wild jute tree (*Grewa venusta*) extract was used as inhibitor. Wild jute tree from the bark of *Grewa venusta* is a shrub or small tree to 10.5 m tall called wild jute and belongs to Tiliaceae family. It is used as fibre and the phytochemical analysis showed that both the leaves and bark contain many compounds, such as polysaccharides, volatile oils, vitamins, tannins, minerals, alkaloids (e.g., caffeine), and polyphenols (catechins and flavonoids). It is found in the northern part of the country [10].

2. Experimental Procedure

2.1. Preparation of Mild Steel Specimen. Mild steel rods were mechanically cut into cylindrical shape of 20 mm by 10 mm with the following chemical compositions: 0.16% C, 0.38% Mn, 0.18% Si, 0.035% S, 0.034% P and the remainder Fe. The specimen were polished mechanically with emery papers of 80 to 800 grades and subsequently degreased and stored in the desiccators to avoid re-oxidation. Weight of the samples was taken before and after the test.

2.2. Preparation of the Plant Extract. The leaves of the plant *Grewa venusta* was taken and cut into small pieces and they were dried in an air for three days and ground well into powder. The refluxed solution using ethanol was then filtered and the concentration of the stock solution is expressed in terms of (% v/v). From the stock solution, 2–10% v/v concentration of the extract was prepared using 0.5 M sulphuric acid. Similar kind of preparation has been reported in studies using aqueous plant extracts in the recent years [12, 13].

2.3. Weight Loss Method. The pretreated specimens' initial weights were noted and were immersed in the experimental solution (in triplicate) with the help of glass hooks at 30°C for a period of 270 minutes. The experimental solution used was 0.5 M H_2SO_4 in the absence and presence of various concentrations of the plant extract. After 270 minutes, the specimens were taken out, washed thoroughly with distilled water, dried completely and their final weights were noted. From the initial and final weights of the specimen, the loss in weight was calculated and tabulated. From the weight loss, the corrosion rate (mpy), inhibition efficiency (%) and surface coverage (θ) of plant extract were calculated using the formula,

$$\mathrm{CR} = \frac{534W}{DAT}\,(\mathrm{mpy}), \quad (1)$$

where W—Weight loss in milligrams (mg), D—Density in grams per cubic centimeter (g/cm^3), A—Area of the specimen exposed in square inches (in^2) and T—Time of immersion in hours (h)

$$(\mathrm{IE\%}) = \left[1 - \frac{\mathrm{CR}i}{\mathrm{CR}I}\right] \times 100, \quad (2)$$

$$\text{surface coverage } (\theta) = \left[1 - \frac{\mathrm{CR}i}{\mathrm{CR}I}\right], \quad (3)$$

where CR*i* and CR*I* are corrosion rates in the absence and presence of the inhibitors.

2.4. Characterization of the Coupons. A Philips model XL30SFEG scanning electron microscope with an energy dispersive X-ray analyzer attached was used in this study. It is a highly-resolution field emission scanning electron microscope with analytical capability. The surface analyses of the coupons before and after corrosion were analyzed for the morphology and the inhomogeneity in the chemical composition. The scanning electron microscope (SEM) was equipped with Energy Dispersive X-ray Spectrometry (EDS) [14].

2.5. Development of Mathematical Model. The experimentations were conducted as per standard L8 orthogonal array, so as to investigate which corrosion control parameters significantly affects the corrosion rate and the independently controllable predominant process parameters considered for the investigation are temperature, concentration and time. Two levels of each of the three factors were used for the statistical analysis. The levels for the three factors are entered in Table 1 and the treatment combinations for the two levels and three factors are tabulated in Table 2.

The model equation was obtained by representing the corrosion rate value by CR, the response function can be expressed by equation below:

$$\mathrm{CR} = f(A, B, C), \quad (4)$$

where A is the temperature, B is the inhibitor and C is the time. The model selected includes the effects of main variables first order and second-order interactions of all variables. Hence the general model is written as [11, 15]

$$\mathrm{CR} = \beta_0 + \beta_1 A + \beta_2 B + \beta_3 C + \beta_4 AB + \beta_5 AC + \beta_6 BC + \beta_7 ABC, \quad (5)$$

where β_0 is average response of CR and β_1, β_2, β_3, β_4, β_5, β_6, and β_7 are coefficients associated with each variable A, B, and C and interactions. The test results were recorded against the standard order of sequence as shown in Table 3. The sum of squares for main and interaction effects was calculated using Yates algorithm. The significant factors (main and interaction) were identified by analysis of variance (ANOVA) technique [16].

3. Results and Discussion

3.1. Results. In order to compare the factors that influence corrosion rate of mild steel in 0.5 M sulphuric acid, the corrosion rate of the experimental specimens immersed in

TABLE 1: Percentage of crude chemical constituents in the plants investigated.

S/No.	Plants	Alkaloids	Tannins	Saponins	Flavonoids	Steroids	Volatile
1	A.S	1.43 ± 0.32	15.25 ± 0.11	3.67 ± 0.33	0.00	0.65 ± 0.12	0.65 ± 0.24
2	A.T	1.12 ± 0.39	13.45 ± 0.34	3.33 ± 0.21	0.98 ± 0.33	0.45 ± 0.11	0.86 ± 0.56
3	G.V	1.21 ± 0.45	14.65 ± 0.52	2.99 ± 0.87	0.96 ± 0.59	0.43 ± 10	0.77 ± 0.21

TABLE 2: Result of corrosion rate without and with inhibitor at different temperature and time.

Inhibitor concentration (% v/v)	Mild steel corrosion rate (mpy)			
	30°C	50°C	70°C	Time (mins)
0	51.71	62.41	70.43	
2	23.26	28.72	35.20	
4	17.06	21.84	27.16	135
6	14.47	19.35	25.35	
8	11.94	17.47	21.14	
10	11.02	14.09	18.61	
0	49.81	60.23	69.11	
2	16.28	23.08	28.81	
4	9.93	12.69	17.23	180
6	8.61	11.88	14.71	
8	6.92	10.82	13.15	
10	7.18	9.98	12.74	
0	43.63	56.41	63.60	
2	10.57	14.54	17.95	
4	8.29	10.72	13.35	270
6	6.83	10.35	12.50	
8	6.61	8.95	10.27	
10	6.73	8.79	10.23	

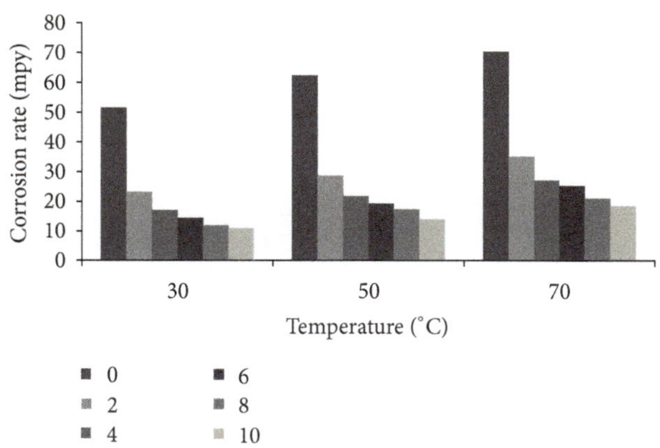

FIGURE 1: Variation of corrosion rate with temperature after 135 minutes of exposure time.

the corrosive reagent with and without inhibitor at varied temperature and time were determined using (1) above were shown in Figures 1, 2, 3, 4, 5, 6, and 7.

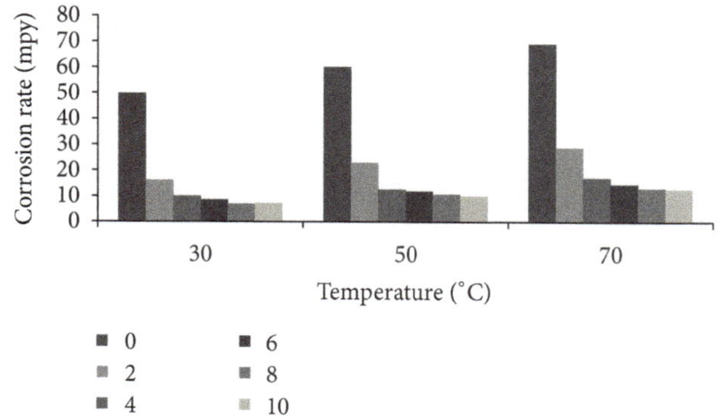

FIGURE 2: Variation of corrosion rate with temperature after 180 minutes of exposure time.

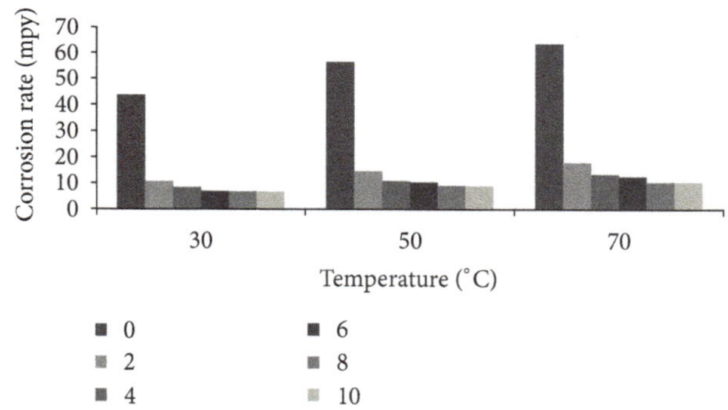

FIGURE 3: Variation of corrosion rate with temperature after 270 minutes of exposure time.

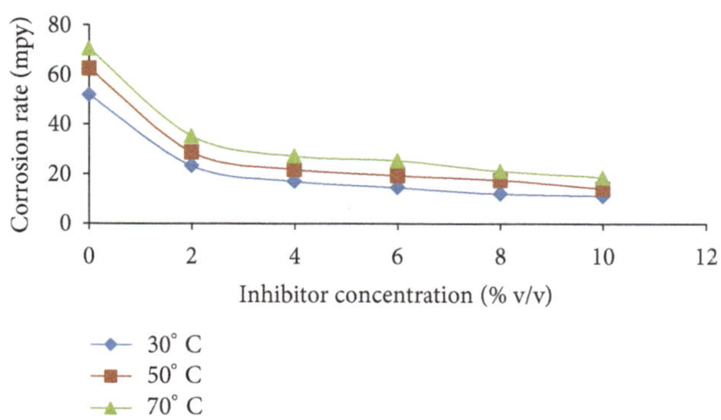

FIGURE 4: Variation of corrosion rate with inhibitor concentration after 135 minutes exposure time.

TABLE 3: Result of apparent activation energy for the absence and presence of inhibitor at different exposure time.

Inhibitor concentration (% v/v)	Mild steel (apparent activation energy, KJ/mol)	Time (mins)
0	2.55	
2	4.77	
4	5.73	135
6	10.19	
8	10.21	
10	10.95	
0	2.45	
2	4.62	
4	5.65	180
6	10.02	
8	10.15	
10	10.75	
0	2.35	
2	4.45	
4	5.55	270
6	9.96	
8	10.02	
10	10.44	

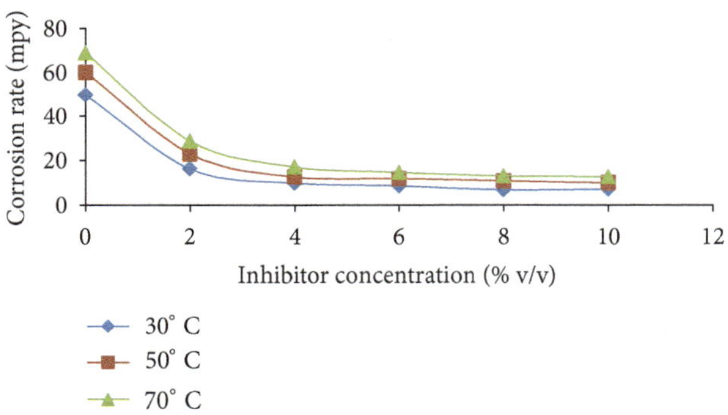

FIGURE 5: Variation of corrosion rate with inhibitor concentration after 180 minutes exposure time.

3.1.1. Kinetics Studies. Although, kinetics models are useful tool to discuss the mechanism of corrosion inhibition of grewa venusta. Arrhenius equation was used to determine the corrosion rate using the expression in (4) and also presented in Figures 8-9

$$A = A_0 e^{-Ea/RT}. \quad (6)$$

The logarithm of A could be represented as a linear equation given below in (5)

$$\ln A = -\frac{Ea}{RT} + \ln A_o. \quad (7)$$

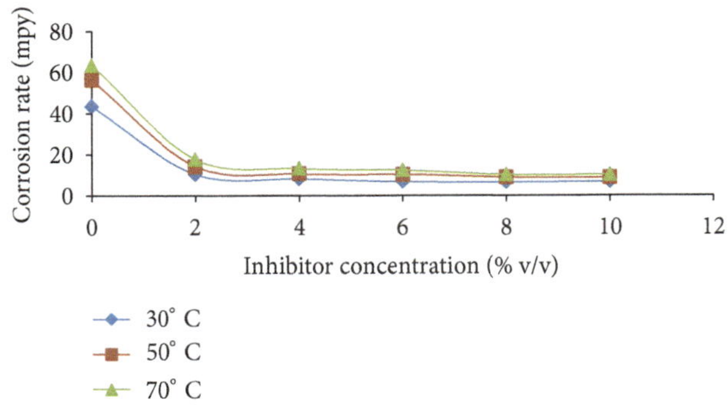

FIGURE 6: Variation of corrosion rate with inhibitor concentration after 270 minutes exposure time.

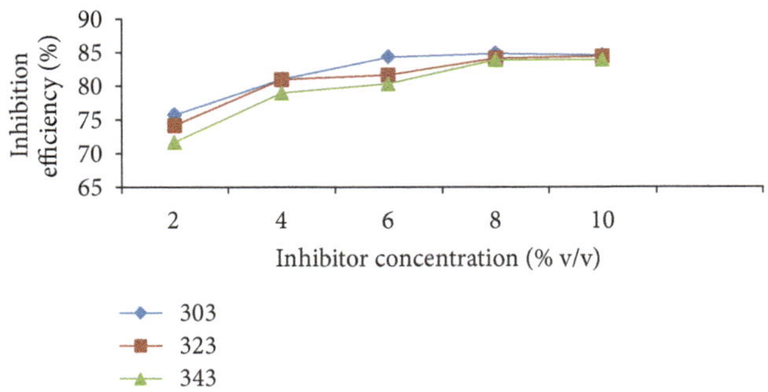

FIGURE 7: Variations of IE % against concentration of G.V after 270 minutes of exposure at 30, 50, and 70°C.

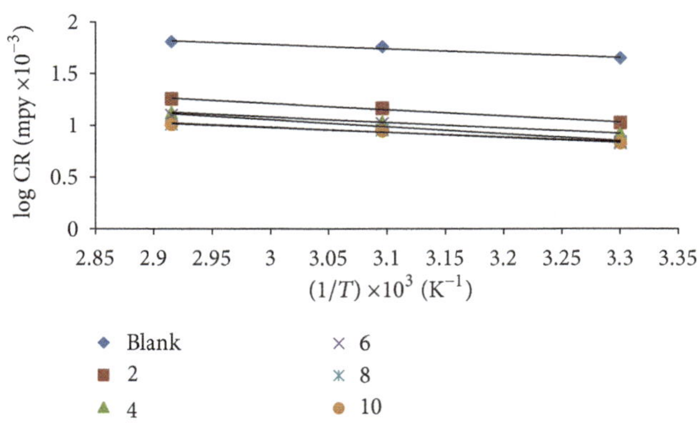

FIGURE 8: logCR against 1/Temperature for mild steel dissolution process in 0.5 M H_2SO_4 containing different concentrations of Grewa venusta at 30, 50, and 70°C.

3.1.2. SEM/EDS. The morphology of the polished, with and without inhibitor of grewa venusta were examined and presented in Figures 10–12.

3.1.3. Development of Model. The results of the statistical model were shown in Tables 7–9 and Figure 13 showed the graph of actual values and predicted values.

3.2. Discussion. Figures 1–3 showed the variation of corrosion rate with temperature after 135, 180 and 270 minutes and

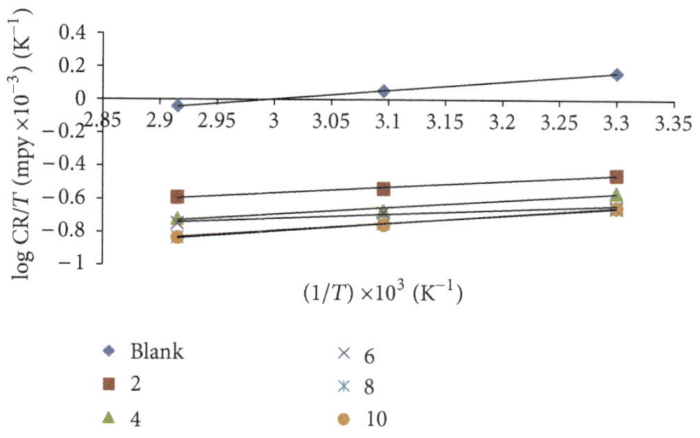

FIGURE 9: logCR/T against 1/Temperature for mild steel dissolution process in 0.5 M H_2SO_4 containin different concentrations of grewa venusta at 30, 50, and 70°C.

TABLE 4: Factorial design of the corrosion rate.

Factors	Low level	High level
Temperature (A)	30°C	70°C
Inhibitor (B)	0	10% v/v
Time (C)	45 minutes	270 minutes

also in Table 4, it can be seen from these figures and table that the corrosion rate increases with increasing temperature while increase in the plant extract leads to decrease in corrosion rate. From the figures; corrosion rate of 70.43, 69.11 and 63.60 mpy on the specimen immersed in the absence of inhibitor when the temperature was raised from 30 to 70°C at exposure time of 135, 180 and 270 minutes respectively while corrosion rate of 11.02, 7.18 and 6.61 mpy at an experimental temperature of 30°C when the inhibitor concentration were raised from 2 to 10% v/v at an exposure time of 135, 180 and 270 minutes. The figures indicated that temperature, inhibitor and time are significant parameters for the corrosion rate control. Variation of corrosion rate with inhibitor concentration after 135, 180 and 270 minutes of exposure time were shown in Figures 4–6 also confirm that corrosion rates decrease with increase in inhibitor concentration and increase with increase in temperature. This also supported the findings of [8, 9, 17]. Figure 7 shows the variation of inhibition efficiency (IE %) against Concentration of grewa venusta after 270 mins of exposure at 30, 50 and 70°C. From the Table 5, it can be concluded that the presence of phytochemical constituents (tannins, alkaloids, saponins, flavonoids) were responsible for the reduction of corrosion rate thereby increasing the efficiency (IE %) of the extract. This is also in support of the findings of [5, 12, 18].

Figures 8–9 show the linear regression of logCR and $1/T$ and logCR/T and $1/T$ for the specimen immersed in sulphuric acid with and without inhibitor [5, 6]. From the Table 6, the results showed that the apparent activation energy in the absence of inhibitor is lower than that in the presence of the inhibitor which implies physiorption adsorption isotherm [7, 8] and this suggests that the anticorrosion inhibitor is very active for mild steel in acidic medium.

TABLE 5: Factorial design of the corrosion process showing treatment combination.

Experimental no.	Temperature level	Concentration level	Time level
1	−1	−1	−1
A	+1	−1	−1
B	−1	+1	−1
AB	+1	+1	−1
C	−1	−1	+1
AC	+1	−1	+1
BC	−1	+1	+1
ABC	+1	+1	+1

Coded = −1 (low level), +1 (upper level or high).

Figures 10–12 show different morphologies structures of the coupons of polished, without and with inhibitor. The morphology of the uninhibited surface was altered during corrosion and as expected rough, uneven surface covered and pits and cracks were seen (see Figure 11). However, no pits and cracks were observed in the morphologg of sample with inhibitor (see Figure 12). The protective film formed on the surface of the mild steel was confirmed by SEM studies. Where as in the presence of the optimum extract, mild steel immersed in acidic medium plant extract show the presence of a protective film and smooth surfaces over the surface of the mild steel in the presence of the inhibitor as shown in Figure 12. This shows that the plant extracts inhibit corrosion of mild steel in acidic solution. This is in line with earlier work of (3, 19).

The results of ANOVA were presented in Table 7, the analysis was evaluated for a confidence level of 95%, that is for significance level of $\alpha = 0.05$. It can be observed from the results obtained that inhibitor was the most significant parameter having the highest statistical influence (73.28%) on the corrosion control followed by time 16.39% and temperature 7.48% respectively.

When the F-value for the models was less than 0.05, then the parameter or interaction can be considered as statistically significant [19]. From Table 8, it is observed that the temperature (A), inhibitor (B) and time (C) are significant model terms influencing corrosion rate of mild steel. Although the interaction effect of temperature with inhibitor (AB) and inhibitor with time (BC) were considered statistically insignificant since their F-values are greater than 0.05, and hence they are neglected. The coefficient of determination (R^2) is defined as the ratio of the explained variation to the total variation. It is a measure of the degree of fitness. When coefficient of determination R^2 approaches unity, a better response model results and it fits the actual data. The value of R^2 calculated for this model was 0.9165 which means that the developed model has high correlation with the experimental value. It demonstrates that 91.65% of the variability in the data can be explained by this model. Thus, it confirmed that the model provides reasonably good explanation of the relationship between the independent factors and their responses [20]. A multiple linear regression model developed

TABLE 6: Experimental condition and corrosion rate for each shown condition using A.S. extract

Experimental condition	Temperature level (°C)	Concentration level (% v/v)	Time level (mins)	Corrosion rate (mpy × 10^{-2})
1	30	0	45	43.63
A	70	0	45	75.32
B	30	10	45	38.52
AB	70	10	45	63.60
C	30	0	270	26.48
AC	70	0	270	59.65
BC	30	10	270	6.73
ABC	70	10	270	10.23

TABLE 7: Analysis of variance (ANOVA) for corrosion rate in the presence of *Grewia venusta* extract.

Source of variation	Sum of squares	Degree of freedom (DF)	Mean square	F_{cal} = Ms/Error Ms	*F*-value	*F* (%)
Main effect						
A	327.42	1	327.42	28.65	0.0332	7.48
B	3209.61	1	3209.61	280.86	0.0035	73.28
C	717.83	1	717.83	62.81	0.0155	16.39
Interaction						
AB	50.50	1	50.50	4.42	0.1703	1.15
BC	51.51	1	51.51	4.51	0.1677	1.18
Residual	22.86	2	11.43			0.52
Cor. Total	4379.73	7				100

TABLE 8: Effect of the variables at 95% confidence level for *Grewia venusta* extract.

Factor	Coefficient estimate	Degree of freedom	Standard error	95% CI low	95% CI high
Intercept	40.52	1	1.20	35.58	45.66
Temperature (*A*) 1.00	6.40	1	1.20	1.26	11.54
Inhibitor (*B*) 1.00	−20.03	1	1.20	−25.17	−14.89
Time (*C*) 1.00	−9.47	1	1.20	−14.61	−4.33
AB 1.00	−2.51	1	1.20	−7.65	2.63
BC 1.00	−2.54	1	1.20	−7.68	2.60

TABLE 9: Comparison of the actual and the predicted result for mild steel using *Grewia venusta*.

Exp. no.	Temperature (°C)	Inhibitor (% v/v)	Time (mins)	Corrosion rate (mpy × 10^{-3})		
				Actual	Predicted	Residual
S11	−1	−1	−1	59.65	58.58	1.07
S12	+1	−1	−1	75.32	76.40	−1.08
S13	−1	+1	−1	26.48	28.62	−2.14
S14	+1	+1	−1	38.52	36.38	2.14
S15	−1	−1	+1	43.63	44.71	−1.08
S16	+1	−1	+1	63.60	62.53	1.07
S17	−1	+1	+1	6.73	4.60	2.13
S18	+1	+1	+1	10.23	12.37	−2.14

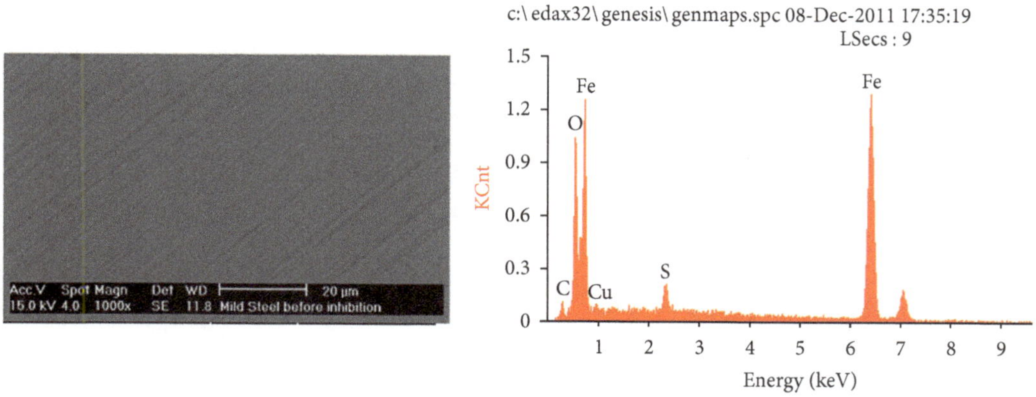

FIGURE 10: SEM/EDS microstructure of mild steel before corrosion test.

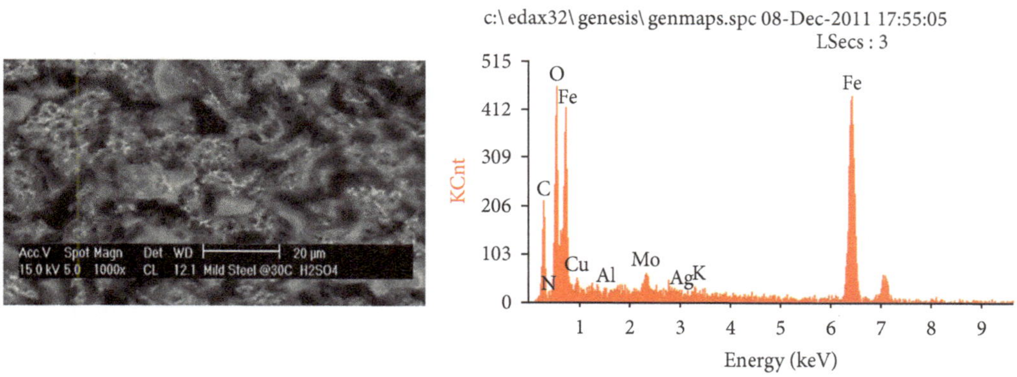

FIGURE 11: SEM/EDX microstructure of mild steel immersed in 0.5 M H_2SO_4.

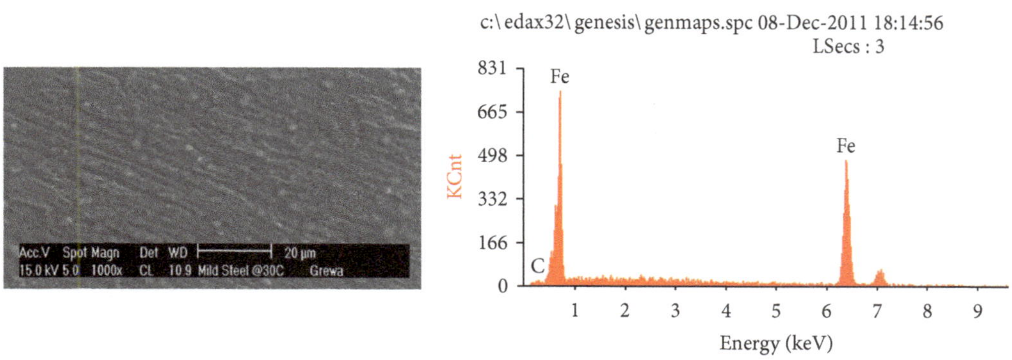

FIGURE 12: SEM/EDX micrograph of mild steel immersed in 0.5 M H_2SO_4 + 10% v/v G.V.

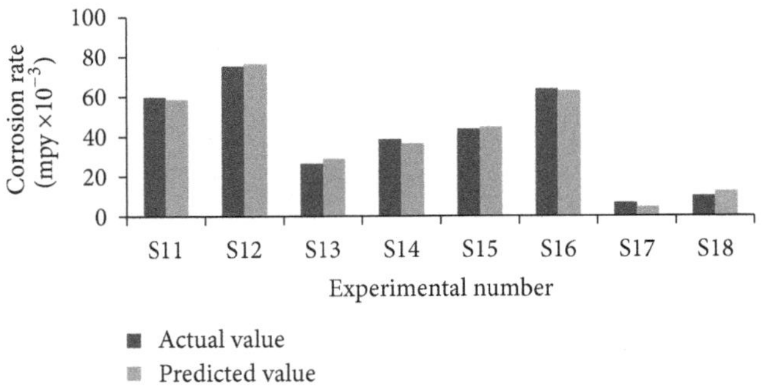

Figure 13: Variation of experimental number with corrosion rate for mild steel with G.V.

and the effect of 95% confidence levels for the extract was presented in Table 8. A regression equation thus generated establishes correlation between the significant terms obtained from ANOVA, namely, temperature, inhibitor and time. Therefore, it was concluded that the influence of temperature, inhibitor concentration and time on the corrosion rate were statistically significant. The model equation was obtained after calculating each of the coefficients of (7). The developed models equations for the corrosion behaviour of the mild steel in the acidic environment in the presence of the extract can be expressed as:

$$\mathrm{CR\,(G.V)} = 40.52 + 6.40A - 20.03B - 9.47C - 2.51AB - 2.54BC. \tag{8}$$

The results of linear regression model was presented in Table 8 for the extract of G.V which showed that the inhibitor appears to be the most important variable with main effect of −20.03 mpy followed by time (C) with −9.47 mpy and temperature (A) with 6.40 mpy. Similar results have been observed by [11, 21, 22].

The results of linear regression model Table 8 for the extract of G.V showed that the inhibitor appears to be the most important variable with main effect of −20.03 mpy followed by time (C) with −9.47 mpy and temperature (A) with 6.40 mpy. The regression revealed that raising the temperature from 30 to 70°C would result in an increase in the corrosion rates by 6.40 mpy while raising the time from 45 to 270 minutes would result to decrease in corrosion rates by 9.47 mpy and increasing the inhibitor concentrations from 0 to 10% v/v would also result to decrease in the corrosion rates by 20.03 mpy respectively.

The interaction effect of the variables temperature, inhibitor concentration and time are also quite significant and one must take into account these factors for predicting the combined effect of temperature, inhibitor and time on the corrosion rate of the material. The interactive effects are between temperature and inhibitor concentration (AB) and inhibitor concentration and time (BC) are −2.51 and −2.54 mpy. Similar results have been observed by (17, 22, 23).

In order to validate the regression model, confirmation test was conducted with parameter levels that were used for analysis. The different parameter levels chosen for the confirmation test are shown in Table 9. Residual variation estimated in (8) for the corrosion rate is in the range of −2.24 to 2.13. The results of the confirmation tests were obtained and comparisons were made between the actual corrosion rate values and the predicted values obtained from the regression models as shown in Table 9. The residual (error) associated with the relationship between the experimental values and the computed values from the regression models for mild steel were very less (less than 4% error). This is in line with findings of [23]. Hence, the regression models developed demonstrated feasible and effective way to predict the corrosion rate of the mild steel. Thus the developed equations can be used to predict corrosion for any combination of factor levels in the specified range. The actual and predicted corrosion rates values are presented in the form of histogram in Figure 13.

4. Conclusion

(1) Experimental data showed that in the presence of different concentration (2–10% v/v), plant extract grewa venusta inhibited the corrosion of mild steel in acidic medium. The inhibition efficiency increased with increase in the extracts concentration and with decrease in temperature leading to a physical adsorption.

(2) The highest efficiency of 86.47% was observed at the optimum of 8% v/v for grewa venusta extract in the acid solution and the effect of immersion time of the plant extracts at these optimums was attained at 180 minutes immersion time at 30°C which was sufficient for pickling process.

(3) The value of activation energy Ea calculated from Arrhenius equations revealed that Ea increases in presence of the plant extract in the acid solution and <80 KJ/molK, suggesting that the corrosion inhibition occurred through physical adsorption.

(4) The SEM micrographs revealed the presence of a protective layers over the metal surface in the presence of the extracts through an adsorption process, hence confirmed the high performance of inhibitive effect of the plant extract.

(5) EDS results also showed an enhancement of iron in the presence of the extract of grewa venusta.

(6) ANOVA results revealed that the parameters (A, B, and C) are statistically significant with F-values less than 0.05 and F (%) for the inhibitor grewa venusta are greater than 70 followed by time which is greater than 16 and temperature greater than 7. The interactions exhibited only minor influence and not statistically significant.

(7) The results obtained by regression equations closely correlate each other which validate the regression equations developed. A good agreement between the predicted and actual corrosion rate was observed.

(8) The results obtained from the statistical analysis are in good agreement with the experimental findings for

the temperature, inhibitor and time. It was found that corrosion rate increases with increasing temperature and decrease with increase in both the inhibitor and time.

(9) The developed mathematical models can be used to predict the corrosion values in terms of corrosion control process parameters obtained from any combinations within the ranges studied and also employed for optimization of the process parameters of mild steel with respect to corrosion control values.

(10) The gravimetric method is in a good agreement with the statistical analysis and this improves the validity of the overall results obtained.

References

[1] R. Tripathi, A. Chaturvedi, and R. K. Upadhayay, "Corrosion inhibitory effects of some substituted thiourea on mild steel in acid media," *Research Journal of Chemical Sciences*, vol. 2, no. 2, pp. 18–27, 2012.

[2] S. Rekkab, H. Zarrok, R. Salghi et al., "Green corrosion inhibitor from essential oil of *Eucalyptus globulus* (Myrtaceae) for C38 steel in sulfuric acid solution," *Journal of Materials and Environmental Science*, vol. 3, no. 4, pp. 613–627, 2012.

[3] A. Ostovari, S. M. Hoseinieh, M. Peikari, S. R. Shadizadeh, and S. J. Hashemi, "Corrosion inhibition of mild steel in 1 M HCl solution by henna extract: a comparative study of the inhibition by henna and its constituents (Lawsone, Gallic acid, α-d-Glucose and Tannic acid)," *Corrosion Science*, vol. 51, no. 9, pp. 1935–1949, 2009.

[4] S. A. Abd El-Maksoud, "The effect of organic compounds on the electrochemical behaviour of steel in acidic media. A review," *International Journal of Electrochemical Science*, vol. 3, no. 5, pp. 528–555, 2008.

[5] A. Begum, S. Harikrishna, I. Khan, and K. Veena, "Enhancement of the inhibitor efficiency of atropine methochloride in corrosion control of mild steel in sulphuric acid," *E-Journal of Chemistry*, vol. 5, no. 4, pp. 774–781, 2008.

[6] A. A. Rahim and J. Kassim, "Recent development of vegetal tannins in corrosion protection of iron and steel," *Recent Patents on Materials Science*, vol. 1, no. 3, pp. 223–231, 2008.

[7] A. M. Al-Turkustani, S. T. Arab, and R. H. Al-Dahiri, "Aloe plant extract as environmentally friendly inhibitor on the corrosion of aluminum in hydrochloric acid in absence and presence of iodide ions," *Modern Applied Science*, vol. 4, no. 5, pp. 105–124, 2010.

[8] A. Singh, V. K. Singh, and M. A. Quraishi, "Aqueous extract of kalmegh (*Andrographis paniculata*) leaves as green inhibitor for mild steel in hydrochloric acid solution," *International Journal of Corrosion*, vol. 2010, Article ID 275983, 10 pages, 2010.

[9] C. A. Loto, R. T. Loto, and A. Popoola, "Synergistic effect of tobacco and kola tree extracts on the corrosion inhibition of mild steel in acid chloride," *International Journal of Electrochemcal Science*, vol. 6, no. 9, pp. 3830–3843, 2011.

[10] B. O. Obadoni and P. O. Ochuko, "Phytochemical studies and comparative efficacy of the crude extracts of some haemostatic plants in edo and delta states of Nigeria," *Global Journal of Pure and Applied Sciences*, vol. 8, no. 2, pp. 203–208, 2001.

[11] O. B. Oloche, S. A. Yaro, and E. G. Okafor, "Analytical correlation between varying corrosion parameters and corrosion rate of Al-4.5Cu/10%$ZrSiO_4$ composite in hydrochloric acid by rare earth chloride," *Journal of Alloys and Compounds*, vol. 472, no. 1-2, pp. 178–185, 2009.

[12] A. M. Abdel-Gaber, B. A. Abd-El-Nabey, and M. Saadawy, "The role of acid anion on the inhibition of the acidic corrosion of steel by lupine extract," *Corrosion Science*, vol. 51, no. 5, pp. 1038–1042, 2009.

[13] G. Ilayaraja, A. R. Sasieekhumar, and P. Dhanakodi, "Inhibition of mild steel corrosion in acidic medium by aqueous extract of tridax procumbens L," *E-Journal of Chemistry*, vol. 8, no. 2, pp. 685–688, 2011.

[14] J. B. Wachtman and R. A. Haber, *Ceramic Films and Coatings*, Noyes Publications, Park Ridge, NJ, USA, 1993.

[15] V. S. Aigbodion, S. B. Hassan, E. T. Dauda, and R. A. Mohammed, "The development of mathematical model for the prediction of ageing behaviour for Al-Cu-Mg/bagasse ash particulate composites," *Journal of Minerals and Materials Characterization & Engineering*, vol. 9, no. 10, pp. 907–917, 2010.

[16] I. Miller and J. E. Freund, *Probability and Statistics for Engineers*, Prentice Hall India, New Delhi, India, 2001.

[17] L. E. Umoru, I. A. Fawehinmi, and A. Y. Fasasi, "Investigation of the inhibitive influence of theobroma cacao and cola acuminata leaves extracts on the corrosion of a mild steel in sea water," *Journal of Applied Sciences Research*, vol. 2, no. 4, pp. 200–204, 2006.

[18] N. O. Obi-Egbedi, K. E. Essien, and I. B. Obot, "Computational simulation and corrosion inhibitive potential of alloxazine for mild steel in 1M HCl," *Journal of Computational Methods in Molecular Design*, vol. 1, no. 1, pp. 26–43, 2011.

[19] S. M. Ross, *Introduction to Probability and Statistics for Engineers and Scientists*, Elsevier Academic Press, 3rd edition, 2004.

[20] S. Venkat Prasat, R. Subramanian, N. Radhika, B. Anandavel, L. Arun, and N. Praveen, "Influence of parameters on the dry sliding wear behaviour of aluminium/fly ash/graphite hybrid metal matrix composites," *European Journal of Scientific Research*, vol. 53, no. 2, pp. 280–290, 2011.

[21] E. E. Oguzie, "Corrosion inhibitive effect and adsorption behaviour of Hibiscus sabdariffa extract on mild steel in acidic media," *Portugaliae Electrochimica Acta*, vol. 26, no. 3, pp. 303–314, 2008.

[22] M. A. Amin, S. S. Abd El-Rehim, E. E. F. El-Sherbini, and R. S. Bayoumi, "Chemical and electrochemical (AC and DC) studies on the corrosion inhibition of low carbon steel in 1.0 M HCl solution by succinic acid-temperature effect, activation energies and thermodynamics of adsorption," *International Journal of Electrochemcal Science*, vol. 3, no. 2, pp. 199–215, 2008.

[23] P. G. Kochure and K. N. Nandurkar, "Mathematical modeling for selection of process parameters in induction hardening of EN8 D steelJournal of Mechanical and Civil Engineering," vol. 1, no. 2, pp. 28–32, 2012.

Permissions

The contributors of this book come from diverse backgrounds, making this book a truly international effort. This book will bring forth new frontiers with its revolutionizing research information and detailed analysis of the nascent developments around the world.

We would like to thank all the contributing authors for lending their expertise to make the book truly unique. They have played a crucial role in the development of this book. Without their invaluable contributions this book wouldn't have been possible. They have made vital efforts to compile up to date information on the varied aspects of this subject to make this book a valuable addition to the collection of many professionals and students.

This book was conceptualized with the vision of imparting up-to-date information and advanced data in this field. To ensure the same, a matchless editorial board was set up. Every individual on the board went through rigorous rounds of assessment to prove their worth. After which they invested a large part of their time researching and compiling the most relevant data for our readers. Conferences and sessions were held from time to time between the editorial board and the contributing authors to present the data in the most comprehensible form. The editorial team has worked tirelessly to provide valuable and valid information to help people across the globe.

Every chapter published in this book has been scrutinized by our experts. Their significance has been extensively debated. The topics covered herein carry significant findings which will fuel the growth of the discipline. They may even be implemented as practical applications or may be referred to as a beginning point for another development. Chapters in this book were first published by Hindawi Publishing Corporation; hereby published with permission under the Creative Commons Attribution License or equivalent.

The editorial board has been involved in producing this book since its inception. They have spent rigorous hours researching and exploring the diverse topics which have resulted in the successful publishing of this book. They have passed on their knowledge of decades through this book. To expedite this challenging task, the publisher supported the team at every step. A small team of assistant editors was also appointed to further simplify the editing procedure and attain best results for the readers.

Our editorial team has been hand-picked from every corner of the world. Their multi-ethnicity adds dynamic inputs to the discussions which result in innovative outcomes. These outcomes are then further discussed with the researchers and contributors who give their valuable feedback and opinion regarding the same. The feedback is then collaborated with the researches and they are edited in a comprehensive manner to aid the understanding of the subject.

Apart from the editorial board, the designing team has also invested a significant amount of their time in understanding the subject and creating the most relevant covers. They scrutinized every image to scout for the most suitable representation of the subject and create an appropriate cover for the book.

The publishing team has been involved in this book since its early stages. They were actively engaged in every process, be it collecting the data, connecting with the contributors or procuring relevant information. The team has been an ardent support to the editorial, designing and production team. Their endless efforts to recruit the best for this project, has resulted in the accomplishment of this book. They are a veteran in the field of academics and their pool of knowledge is as vast as their experience in printing. Their expertise and guidance has proved useful at every step. Their uncompromising quality standards have made this book an exceptional effort. Their encouragement from time to time has been an inspiration for everyone.

The publisher and the editorial board hope that this book will prove to be a valuable piece of knowledge for researchers, students, practitioners and scholars across the globe.

List of Contributors

M. A. Afifi
Modern University for Technology and Information (MTI University), Cairo, Egypt

C. Mary Anbarasi
PG Department of Chemistry, Jayaraj Annapackiam College for Women, Periyakulam 625601, India

Susai Rajendran
Corrosion Research Centre, PG and Research Department of Chemistry, GTN Arts College, Dindigul 624005, India
Department of Chemistry, RVS School of Engineering and Technology, Dindigul 624005, India

M. P. Chakravarthy and K. N. Mohana
Department of Studies in Chemistry, University of Mysore, Manasagangotri, Mysore 570 00, India

Rajasekar Aruliah
Department of Chemical & Bio molecular Engineering, National University of Singapore, Singapore 117576
Environmental Molecular Microbiology Lab, Department of Biotechnology, Thiruvalluvar University, Serkkadu, Vellore 632 115, India

Yen-Peng Ting
Department of Chemical & Bio molecular Engineering, National University of Singapore, Singapore 117576

J. C. Valle-Quitana and J. G. Gonzalez-Rodriguez
Universidad Autonoma del Estado de Morelos, CIICAP, Avenida Universidad 1001, 62209 Cuernavaca, MOR, Mexico

G. F. Dominguez-Patiño
Facultad de Ciencias Biologicas, Universidad Autonoma del Estado de Morelos, Avenida Universidad 1001, 62209 Cuernavaca, MOR, Mexico

Akshya Kumar Guin, Suryakanta Nayak, Manish Kumar Bhadu, Veena Singh and Tapan Kumar Rout
Research and Development, Tata Steel Ltd., Jamshedpur 831001, India

Shyamala R. Krishnamurthy and Meenakshi H. N. Parameswaran
Department of Chemistry, Avinashilingam Institute for Home Science and Higher Education forWomen, Coimbatore, Tamil Nadu 641043, India

S. T. Aruna, G. Savitha, Jyothi Shedthi and V. K. William Grips
National Aerospace Laboratories, Council of Scientific and Industrial Research, Post Bag No. 1779, Bangalore 560 017, India

Vikas Chawla
D.A.V. College of Engineering and Technology, Kanina 123027, India

I. M. Pohrelyuk, O. V. Tkachuk and R. V. Proskurnyak
Karpenko Physico-Mechanical Institute, National Academy of Sciences of Ukraine, 5 Naukova Street, Lviv 79060, Ukraine

Chikkur B. Pradeep Kumar and Kikkeri N. Mohana
Department of Studies in Chemistry, University of Mysore, Manasagangotri, Mysore 570006, India

V. Sribharathy
PG and Research Department, GTN Arts College, Dindigul 624005, Tamil Nadu, India

Susai Rajendran
Department of Chemistry, RVS School of Engineering and Technology, Dindigul 624005, Tamil Nadu, India

Ramana M. Pidaparti, Brian Hinderliter and Darshan Maskey
Department of Mechanical and Nuclear Engineering, Virginia Commonwealth University, 401West Main Street, Richmond, VA 23284, USA

Vinod P. Raphael, Joby Thomas Kakkassery, Shaju K. Shanmughan and Aby Paul
Research Division, Department of Chemistry, St. Thomas' College (University of Calicut), Thrissur, Kerala 680005, India

N. Kumar, A. Jyothirmayi, K. R. C. Soma Raju, V. Uma and R. Subasri
International Advanced Research Centre for Powder Metallurgy and New Materials (ARCI), Balapur, Hyderabad, Andhra Pradesh 500005, India

Manish Kumar Bhadu, Akshya Kumar Guin, Veena Singh and Shyam K. Choudhary
R&D and Scientific Services, TATA Steel Ltd., Jamshedpur 831001, India

Sudesh Kumar and Suraj Prakash Mathur
Department of Pure and Applied Chemistry, Maharshi Dayanand Saraswati University, C-698, Panchsheel Nagar, Ajmer, Rajasthan 305009, India

C. N. Machio
Council for Scientific and Industrial Research CSIR, Materials Science and Manufacturing, PO Box 35, Pretoria 0001, South Africa

D. S. Konadu
DST/NRF Centre of Excellence in Strong Materials, and School of Chemical and Metallurgical Engineering, University of Witwatersrand, Johannesburg, Wits 2050, South Africa
Materials Science and Engineering Department, University of Ghana, Legon, Accra, Ghana

J. Van der Merwe
DST/NRF Centre of Excellence in Strong Materials, and School of Chemical and Metallurgical Engineering, University of Witwatersrand, Johannesburg, Wits 2050, South Africa

J. H. Potgieter
School of Research, Enterprise and Innovation, Manchester Metropolitan University, Oxford Road, Manchester M1 5GD, UK

S. Potgieter-Vermaak
School of Science and the Environment, Manchester Metropolitan University, Oxford Road, Manchester M1 5GD, UK
School of Chemistry, University of the Witwatersrand, Johannesburg, Private Bag X3, Wits 2050, South Africa

Venkata Appa Rao Boyapati and Chaitanya Kumar Kanukula
Department of Chemistry, National Institute of Technology Warangal (NIT Warangal), Warangal, Andhra Pradesh 506 004, India

I. Y. Suleiman
Department of Metallurgical Engineering,Waziri Umaru Federal Polytechnic, Birnin Kebbi, Nigeria

O. B. Oloche
Department of Mechanical Engineering, University of Abuja, Abuja, Nigeria

S. A. Yaro
Department of Metallurgical and Materials Engineering, Ahmadu Bello University, Zaria, Nigeria